Methods of
Experimental Physics

VOLUME 13

SPECTROSCOPY

PART B

METHODS OF EXPERIMENTAL PHYSICS:

L. Marton, *Editor-in-Chief*

Claire Marton, *Assistant Editor*

1. Classical Methods
   Edited by Immanuel Estermann
2. Electronic Methods, Second Edition (in two parts)
   Edited by E. Bleuler and R. O. Haxby
3. Molecular Physics, Second Edition (in two parts)
   Edited by Dudley Williams
4. Atomic and Electron Physics—Part A: Atomic Sources and Detectors, Part B: Free Atoms
   Edited by Vernon W. Hughes and Howard L. Schultz
5. Nuclear Physics (in two parts)
   Edited by Luke C. L. Yuan and Chien-Shiung Wu
6. Solid State Physics (in two parts)
   Edited by K. Lark-Horovitz and Vivian A. Johnson
7. Atomic and Electron Physics—Atomic Interactions (in two parts)
   Edited by Benjamin Bederson and Wade L. Fite
8. Problems and Solutions for Students
   Edited by L. Marton and W. F. Hornyak
9. Plasma Physics (in two parts)
   Edited by Hans R. Griem and Ralph H. Lovberg
10. Physical Principles of Far-Infrared Radiation
    L. C. Robinson
11. Solid State Physics
    Edited by R. V. Coleman
12. Astrophysics—Part A: Optical and Infrared
    Edited by N. Carleton
    Part B: Radio Telescopes, Part C: Radio Observations
    Edited by M. L. Meeks
13. Spectroscopy (in two parts)
    Edited by Dudley Williams

Volume 13

Spectroscopy

PART B

Edited by

DUDLEY WILLIAMS

Department of Physics
Kansas State University
Manhattan, Kansas

1976

ACADEMIC PRESS · New York   San Francisco   London
A Subsidiary of Harcourt Brace Jovanovich, Publishers

ACADEMIC PRESS, INC.
111 Fifth Avenue, New York, New York 10003

United Kingdom Edition published by
ACADEMIC PRESS, INC. (LONDON) LTD.
24/28 Oval Road, London NW1

Library of Congress Cataloging in Publication Data

Main Entry under title:

Spectroscopy.

    (Methods of experimental physics ; v. 13)
    Includes bibliographical references and index.
    1.   Spectrum analysis.   I.   Williams, Dudley,
(date)   II.   Series.
QC451.S63   535'.84   76-6854
ISBN 0−12−475954−8 (pt. B)

2782   9-12-75   MZ4   Academic 5372 . . . . . .

J. L. − 9-12-75

CONTENTS

CONTRIBUTORS vii

FOREWORD ix

PREFACE xi

CONTENTS OF VOLUME 13, PART A xiii

CONTRIBUTORS TO VOLUME 13, PART A xv

4. Molecular Spectroscopy
    4.1. Infrared Region 1
    by DUDLEY WILLIAMS

        4.1.1. Infrared Sources 2
        4.1.2. Detection and Measurement of Infrared
                Radiation 6
        4.1.3. Optical Components for the Infrared 13
        4.1.4. Resolving Instruments 15
        4.1.5. The Infrared Spectra of Gases 28
        4.1.6. Studies of Molecular Interactions 37
        4.1.7. Vibrational Spectra of Larger Polyatomic
                Molecules 43
        4.1.8. Molecules in Condensed Phases 44
        4.1.9. Applications to Astronomy 48

    4.2. Far-Infrared and Submillimeter-Wave Regions 50
    by D. OEPTS

        4.2.1. Introduction 50
        4.2.2. Microwave and Laser Methods 54
        4.2.3. Far-Infrared Grating Spectroscopy 57
        4.2.4. Fourier Transform Spectroscopy 60
        4.2.5. Other Methods 83
        4.2.6. Applications 87

4.3. Microwave Region 102
   by DONALD R. JOHNSON AND RICHARD PEARSON, JR.

      4.3.1.   Introduction 102
      4.3.2.   Sources 104
      4.3.3.   Detectors 109
      4.3.4.   Modulation 114
      4.3.5.   Practical Spectrometers 121
      4.3.6.   Applications 131

4.4. Radio-Frequency Region 134
   by J. B. HASTED

      4.4.1.   Introduction 134
      4.4.2.   Experimental Techniques of
               Radio-Frequency Spectroscopy 136
      4.4.3.   The Physical Basis of Dielectric
               Relaxation Spectra 168
      4.4.4.   Molecular Structure and Dielectric
               Relaxation 194

5. Recent Developments
   5.1. Beam-Foil Spectroscopy 213
      by C. LEWIS COCKE

      5.1.1.   Introduction and History 213
      5.1.2.   General Characteristics of
               Radiation Source 216
      5.1.3.   Beam-Foil Spectra 218
      5.1.4.   Transition Probabilities 234
      5.1.5.   Quantum Beats 244
      5.1.6.   High-Z Few-Electron Systems 256

   5.2. Tunable Laser Spectroscopy 273
      by MARVIN R. QUERRY

      5.2.1.   Introduction 273
      5.2.2.   Tunable Lasers 275
      5.2.3.   Spectroscopic Applications 323

AUTHOR INDEX 343

SUBJECT INDEX FOR PART B 357

SUBJECT INDEX FOR PART A 361

CONTRIBUTORS

Numbers in parentheses indicate the pages on which the authors' contributions begin.

C. LEWIS COCKE, *Department of Physics, Kansas State University, Manhattan, Kansas* (213)

J. B. HASTED, *Department of Physics, Birkbeck College, University of London, London, England* (134)

DONALD R. JOHNSON, *National Bureau of Standards, Molecular Spectroscopy Section, Optical Physics Division, Washington, D.C.* (102)

D. OEPTS, *Association Euratom-FOM, FOM-Instituut voor Plasmafysica, Rijnhuizen, Nieuwegein, The Netherlands* (50)

RICHARD PEARSON, JR., *National Bureau of Standards, Molecular Spectroscopy Section, Optical Physics Division, Washington, D.C.* (102)

MARVIN R. QUERRY, *Department of Physics, University of Missouri, Kansas City, Missouri* (273)

DUDLEY WILLIAMS, *Department of Physics, Kansas State University, Manhattan, Kansas* (1)

FOREWORD

Several aspects of spectroscopy have been treated in some of our earlier volumes (see Volumes 3A and 3B, Molecular Physics, second edition; Volume 10, Far Infrared; Volume 12A, Astrophysics). The rapid expansion of physics made it desirable to issue a separate treatise devoted to spectroscopy only, emphasizing such aspects which may not have been treated adequately in the volumes dealing essentially with other facets of physics. The present volumes contain a much more thoroughgoing treatment of the spectroscopy of photons of all energies. It is our intention to follow this with a volume devoted to particle spectroscopy.

Professor Dudley Williams, who is already well known to readers of "Methods of Experimental Physics" as editor of our Molecular Physics volumes, was kind enough to accept the editorship of the Spectroscopy volumes. His knowledge of the field and his excellent judgment will, no doubt, be appreciated by the users of "Spectroscopy" methods. We wish to express our profound gratitude to him and to all contributors to these volumes for their untiring efforts.

L. MARTON
C. MARTON

PREFACE

Spectroscopy has been a method of prime importance in adding to our knowledge of the structure of matter and in providing a basis for quantum physics, relativistic physics, and quantum electrodynamics. However, spectroscopy has evolved into a group of specialties; practitioners of spectroscopic arts in one region of the electromagnetic spectrum feel little in common with practitioners studying other regions; in fact, some practitioners do not even realize that they are engaged in spectroscopy at all!

In the present volumes we attempt to cover the entire subject of spectroscopy from pair production in the gamma-ray region to dielectric loss in the low radio-frequency region. Defining spectroscopy as the study of the emission and absorption of electromagnetic radiation by matter, we present a general theory that is applicable throughout the entire range of the electromagnetic spectrum and show how the theory can be applied in gaining knowledge of the structure of matter from experimental measurements in all spectral regions.

The books are intended for graduate students interested in acquiring a general knowledge of spectroscopy, for spectroscopists interested in acquiring knowledge of spectroscopy outside the range of their own specialties, and for other physicists and chemists who may be curious as to "what those spectroscopists have been up to" and as to what spectroscopists find so interesting about their own work! The general methods of spectroscopy as practiced in various spectral regions are remarkably similar; the details of the techniques employed in various regions are remarkably different.

Volume A begins with a brief history of spectroscopy and a discussion of the general experimental methods of spectroscopy. This is followed by a general theory of radiative transitions that provides a basis for an understanding of and an interpretation of much that follows. The major portion of the volumes is devoted to chapters dealing with the spectroscopic methods as applied in various spectral regions and with typical results. Each chapter includes extensive references not only to the original literature but also to earlier books dealing with spectroscopy in various regions; the references to earlier books provide a guide to readers who may wish to go more deeply into various branches of spectroscopy. The final chapters of Volume 13 are devoted to new branches of spectroscopy involving beam foils and lasers.

The list of contributors covers a broad selection of competent active research workers. Some exhibit the fire and enthusiasm of youth; others are at the peak of the productive activity of their middle years; and still others are battle-scarred veterans of spectroscopy who hopefully draw effectively on long experience! All contributors join me in the hope that the present volumes will serve a useful purpose and will provide valuable insights into the general subject of spectroscopy.

DUDLEY WILLIAMS

CONTENTS OF VOLUME 13, PART A

1. Introduction
   by DUDLEY WILLIAMS

   1.1. History of Spectroscopy
   1.2. General Methods of Spectroscopy

2. Theory of Radiation and Radiative Transitions
   by BASIL CURNUTTE, JOHN SPANGLER, AND
   LARRY WEAVER
   2.1. Introduction
   2.2. Light
   2.3. Interaction of Light and Matter
   2.4. Applications
   2.5. Conclusion

3. Nuclear and Atomic Spectroscopy
   3.1. Gamma-Ray Region
        by JAMES C. LEGG AND GREGORY G. SEAMAN
   3.2. X-Ray Region
        by ROBERT L. KAUFFMAN AND PATRICK RICHARD
   3.3. Far Ultraviolet Region
        by JAMES A. R. SAMSON
   3.4. Optical Region
        by P. F. A. KLINKENBERG

AUTHOR INDEX—SUBJECT INDEXES FOR PARTS A AND B

CONTRIBUTORS TO VOLUME 13, PART A

BASIL CURNUTTE, *Department of Physics, Kansas State University, Manhattan, Kansas*

ROBERT L. KAUFFMAN, *Department of Physics, Kansas State University, Manhattan, Kansas*

P. F. A. KLINKENBERG, *Zeeman-Laboratorium, University of Amsterdam, The Netherlands*

JAMES C. LEGG, *Department of Physics, Kansas State University, Manhattan, Kansas*

PATRICK RICHARD, *Department of Physics, Kansas State University, Manhattan, Kansas*

JAMES A. R. SAMSON, *Behlen Laboratory of Physics, University of Nebraska, Lincoln, Nebraska*

GREGORY G. SEAMAN, *Department of Physics, Kansas State University, Manhattan, Kansas*

JOHN SPANGLER, *Department of Physics, Kansas State University, Manhattan, Kansas*

LARRY WEAVER, *Department of Physics, Kansas State University, Manhattan, Kansas*

DUDLEY WILLIAMS, *Department of Physics, Kansas State University, Manhattan, Kansas*

4. MOLECULAR SPECTROSCOPY

4.1. Infrared Region*

During the closing decades of the 19th century and the early decades of the present century infrared spectroscopy began to take form; developments during this period have been summarized in several books.[1-4] The volume by Schaefer and Matossi is especially valuable in giving an excellent history of the subject along with a wealth of detail regarding the experimental methods employed in the era prior to 1930. In the period immediately following 1940, infrared spectroscopy advanced from a series of beautiful experiments conducted by painstaking individuals having an abundant supply of patience to an established branch of spectroscopy practiced in industrial laboratories as well as in laboratories devoted to pure research. Much of this rapid expansion came about as a result of a wider appreciation of the importance of infrared methods to molecular physics and chemistry.[5,6] This was also the period when recording spectrographs or spectrophotometers were developed to replace earlier instruments which had relied on galvanometers to provide data on a point-by-point basis; commercial manufacture of recording spectrographs made infrared spectroscopy a tool of industry. Williams[7] and Sutherland and Lee[8] have provided excellent surveys of the rapid developments of experimental techniques during this period and have included extensive references to the pertinent literature.

Great impetus was given to the development of infrared techniques by the recognition during World War II of the military importance of infrared

[1] J. Lecomte, "Le Spectre Infrarouge." Presses Univ. de France, Paris, 1928.

[2] C. Schaefer and F. Matossi, "Das Ultrarote Spektrum." Springer-Verlag, Berlin and New York, 1930.

[3] F. I. G. Rawlins and A. M. Taylor, "Infrared Analysis of Molecular Structure." Cambridge Univ. Press, London and New York, 1929.

[4] G. B. B. M. Sutherland, "Infrared and Raman Spectra." Methuen, London, 1935.

[5] G. Herzberg, "Spectra of Diatomic Molecules," 2nd ed. Van Nostrand-Reinhold, Princeton, New Jersey, 1950.

[6] G. Herzberg, "Infrared and Raman Spectroscopy." Van Nostrand-Reinhold, Princeton, New Jersey, 1945.

[7] V. Z. Williams, *Rev. Sci. Instrum.* **19**, 135 (1948).

[8] G. B. B. M. Sutherland and E. Lee, *Rep. Progr. Phys.* **11**, 144 (1948).

*Chapter 4.1 is by Dudley Williams.

radiation as a means of detection and signal transmission. The new methods of detecting and measuring infrared radiation that were developed during this period have been described in considerable detail in the excellent book by Smith and his colleagues.[9] Since 1960 numerous books on infrared spectroscopy have appeared and provide extensive references to the more recent literature.[10-15] There are other recent useful books dealing primarily with surveys of experimental results and theoretical interpretations of the results.[16-20]

In this chapter we shall discuss the various types of sources, detectors, resolving instruments, and optical materials used in the infrared and shall attempt to present typical experimental results along with their interpretation in terms of the theory presented in Chapter 2, Part A. We make no attempt to be encyclopedic; the range of topics treated, which reflects to some extent the present author's personal interests, is also limited to subjects not previously covered in detail in other recent books.

4.1.1. Infrared Sources

The range of quantum energies encompassed by the infrared region represents energies well below those separating the lowest energy levels of most atoms. Therefore, the atomic lines appearing in this region are associated with transitions between highly excited states of atoms and can usually be studied only in emission. For example, in the hydrogen spectrum the Paschen, Brackett, Pfund, and Humphreys[21] series with lower quantum

[9] R. A. Smith, F. E. Jones, and R. P. Chasmar, "The Detection and Measurement of Infrared Radiation." Oxford Univ. Press, London and New York, 1957. (Revision, 1968.)

[10] P. W. Kruse, E. D. McGlauchlin, and R. B. McQuistan, "Elements of Infrared Technology." Wiley, New York, 1962.

[11] J. A. Jamieson, R. H. McFee, G. N. Plass, and R. G. Richards, "Infrared Physics and Engineering." McGraw-Hill, New York, 1963.

[12] H. A. Szymanski, "Theory and Practice of Infrared Spectroscopy." Plenum Press, New York, 1964.

[13] J. E. Stewart, "Infrared Spectroscopy." Dekker, New York, 1970.

[14] K. D. Möller and W. D. Rothschild, "Far-Infrared Spectroscopy." Wiley (Interscience), New York, 1971.

[15] L. C. Robinson, "Physical Principles of Far-Infrared Radiation." Academic Press, New York, 1973.

[16] H. H. Nielsen, "Handbuch der Physik" (S. Flugge, ed.), Vol. 37/1, p. 153. Springer-Verlag, Berlin and New York, 1959.

[17] H. C. Allen and P. C. Cross, "Molecular Vib-Rotors." Wiley, New York, 1966.

[18] J. E. Wollrab, "Rotational Spectra and Molecular Structure." Academic Press, New York, 1967.

[19] R. T. Conley, "Infrared Spectroscopy," Allyn & Bacon, Boston, Massachusetts, 1966.

[20] G. Amat, H. H. Nielsen, and G. Tarrago, "Higher-Order Rotation–Vibration Energies of Polyatomic Molecules." Dekker, New York, 1971.

[21] C. J. Humphreys, *J. Opt. Soc. Amer.* **42**, 432 (1952).

numbers $n = 3, 4, 5,$ and 6, respectively, appear in the infrared. Since more interesting phenomena are usually encountered in the transitions between lower atomic states, the infrared portions of atomic emission spectra have been somewhat neglected as compared with the visible and ultraviolet portions. In general, however, the sources employed for studies of atomic spectra in the infrared represent only modifications of those used in other regions.

The quantum energy range in the infrared does encompass the energies associated with transitions between vibrational and rotational energy states of molecules in their electronic ground states. Many of these transitions can be studied in emission from flames. In the emission spectrum of nearly every hydrocarbon–oxygen flame, H_2O vapor and CO_2 emission spectra can be observed; these spectra are most useful in providing information regarding these two molecules in highly excited vibrational and rotational states. By proper adjustment of the flames, emission from CO and the free radicals OH and CH can be observed.[22, 23] By injection of other molecules into the flame from a burner it is possible to observe the emission spectra of these molecules. Various types of nonhydrocarbon flames can also be employed for other molecules. Since many of the lower vibrational states of molecules are excited at ambient temperatures, transitions between these lowest states and the ground states can be studied by comparing the emitted radiation with that emitted by a blackbody at low temperature.[24]

Infrared spectroscopy as generally practiced, however, has been concerned chiefly with absorption. In the remainder of this section we shall discuss the sources commonly employed in laboratory investigations. In later sections, we shall return briefly to some of the problems encountered in studies of the emission spectra of astronomical and atmospheric sources.

In investigations of absorption spectra it is usually desirable to have a source that provides intense continuous spectral emission $I(v)$ throughout the region to be investigated.[†] In view of Kirchhoff's law, it would thus appear that a blackbody would be not only the perfect radiator but also the ideal source for use in infrared spectroscopy. In many respects, this is true, and blackbody radiators furnish excellent references with which other radiators can be compared. It is not necessary, however, to employ a blackbody cavity radiator to obtain a satisfactorily continuous spectral source. Any sufficiently hot opaque solid will serve satisfactorily; the radiation from a narrow V-shaped slot cut into the wall of such a solid gives an extremely

[22] R. C. Herman and G. Hornbeck, *Astrophys. J.* **118**, 214 (1953).

[23] G. Herzberg, "The Spectra and Structures of Simple Free Radicals," p. 3. Cornell Univ. Press, Ithaca, New York, 1971.

[24] R. Sloan, J. H. Shaw, and D. Williams, *J. Opt. Soc. Amer.* **45**, 455 (1955).

[†] Noncontinuous sources are discussed in Chapter 5.2.

close approximation of a blackbody. Similarly, if a metal ribbon is folded to form a narrow V, the emission from the V-shaped opening closely resembles the emission from a blackbody at the temperature of the ribbon. No thermal source can have a greater emissive power than that of a blackbody at the same temperature; therefore knowledge of Planck's radiation law can serve as a guide to the design of sources.

The experimental methods of studying blackbody radiation have been described by Schaefer and Matossi[2] and the derivation of the radiation laws have been briefly summarized by Smith *et al.*[9] It will be recalled that the total radiant power from a blackbody is proportional to the fourth power of its absolute temperature. The peak of the radiation curve[†] giving $I(\lambda)$ versus λ shifts toward shorter wavelengths as the temperature increases; the product $\lambda_{max} T = $ const, where λ_{max} is the wavelength at which the peak occurs. It is useful to remember that $\lambda_{max} = 1$ μm for $T = 2897$ K $\simeq$ 3000 K; the location of the peak at any other temperature can thus be conveniently computed. Blackbody radiation curves never cross; the blackbody radiation curve for a given temperature is at every wavelength above that for a lower temperature. Thus it is usually desirable to use a source with as high a temperature as possible. The spectral flux intensity $I(\lambda)$ at wavelengths long as compared with the wavelength at the peak is directly proportional to the source temperature in accordance with the Rayleigh–Jeans law, which applies in good approximation at long wavelengths. In the vicinity of the peak, $I(\lambda)$ increases rapidly with increasing temperature.

Of the commercially available sources, the one most widely used is probably the *globar*, which consists of a rod of silicon carbide several centimeters in length and approximately 6 mm in diameter. Heated electrically by a current of about 5 A, the globar operates satisfactorily at a maximum temperature of 1400 K; above this temperature oxidation becomes a serious problem and the binding material boils out. The globar mount must usually be water cooled in order to maintain good electrical contact between the rod and the external circuit. Since the diameter of the globar is larger than necessary for use with most spectrometers, it is wasteful of electrical power; however, its large thermal capacity minimizes the effect of short-term variations in line voltage; for long-term stability it is desirable to operate the globar from a voltage-regulated power supply or constant-voltage transformer. Silverman[25] has shown that the emissivity of a globar is approximately 80% of that of a blackbody over most of the range between 2

[25] S. Silverman, *J. Opt. Soc. Amer.* **38**, 989 (1948).

[†]In keeping with modern spectroscopic practice it would be preferable to present a plot of $I(v)$ versus v for the blackbody, but the $I(\lambda)$ versus λ plot is in such wide use that there is no move toward change. In fact, the $I(v)$ versus v plot would not be immediately recognized by the uninitiated.

and 16 μm (5000–600 cm^{-1}). The globar is useful to 100 cm^{-1} in the far infrared.

Another commonly used infrared source is the *Nernst glower*, which was initially developed by Nernst and Bose[26]; it consists of a rod approximately 3 cm long and 1 mm in diameter and is composed of a mixture of zirconium and yttrium oxides. Platinum wires are attached to the ends of the rod,[27] which is heated electrically. With properly attached platinum wires, the Nernst glower can operate satisfactorily at 2000 K for prolonged periods without water cooling. At room temperature the Nernst glower is non-conducting and must be preheated by a flame or by a nearby electrical heater in order to make it a conductor. In view of its large negative temperature coefficient of resistance, it is usually necessary to operate it in series with a current-limiting device; in commercial devices operated from ac lines or regulated power supplies, this is usually accomplished by means of barretters or by more elaborate control devices. We find, however, that it gives completely satisfactory performance when operated from the laboratory power line in series with a tungsten filament lamp. In view of its small thermal capacity, the Nernst glower should be shielded from air currents in the laboratory, which change its temperature and thus its resistance. Although Nernst glowers allegedly give off a fine powder when operated in vacuum, we have used these devices for prolonged periods in vacuum without encountering serious difficulties. The Nernst glower has an emission spectrum closely approximating that of a blackbody in the spectral range between the visible and 15 μm (665 cm^{-1}); in the far infrared it is inferior to the globar. It is a convenient and electrically efficient source with a size and shape admirably suited for use with most spectrometers.

Another commercially available source is the *carbon arc*, which can be operated with the positive crater of the arc at 4100 K. This source has been extensively investigated by Strong and his associates.[28] Although the carbon arc provides high spectral emission $I(v)$ throughout the near infrared (4000–600 cm^{-1}), it is rather bulky for many applications in spectroscopy and tends to be somewhat unstable as compared with other sources. Automatic feeding of the carbon rods employed in some commercial models provides fairly satisfactory short-term stability.

In the near infrared (10,000–3000 cm^{-1}) *tungsten filament* lamps are available commercially; in one form a tungsten ribbon filament requiring a current of 20 to 30 A is operated in a glass envelope equipped with a quartz window. Quartz-enclosed *iodine lamps with a tungsten filament* sold commercially for photographic purposes are excellent sources for the near infrared; selected lamps have stable radiant output and long life.

[26] W. Nernst and E. Bose, *Phys. Z.* **1**, 289 (1900).
[27] E. S. Ebers and H. H. Nielsen, *Rev. Sci Instrum.* **11**, 429 (1940).
[28] C. S. Rupert and J. Strong, *J. Opt. Soc. Amer.* **39**, 1061 (1949).

For use in the far infrared (250–10 cm^{-1}) the quartz-enclosed *mercury arc* is the best available source.[29] Most of the emission originates in the discharge plasma but emission from the quartz walls apparently makes some contribution. Although the quartz walls darken as the arc is used, the output of the arc in the far infrared remains nearly constant. Water cooling of the housing must be provided.

The so-called *carbon-rod furnace* is now widely used in high-resolution spectroscopy in the intermediate infrared. The use of carbon-rod sources was first discussed by Smith,[30] but considerable improvements have been made more recently by Rao and his associates,[31] who have used resistively heated rods at 3000 K; water-cooled mountings are required. By cutting a V-shaped slot in the wall of the carbon rod, blackbody emission at 3000 K can be closely approximated.

Another source, the Welsbach gas mantle, widely used in the early days of infrared spectroscopy, offers advantages in certain spectral regions,[32, 33] but has fallen into disuse with the advent of more convenient electrically powered sources.

4.1.2. Detection and Measurement of Infrared Radiation

The devices employed for the detection of infrared radiation can be classified as (a) *thermal detectors*, which depend for their operation on the increase in the temperature of the sensing element, and (b) *quantum detectors*, which usually depend for their operation on internal photoelectric processes producing changes in the resistance of the sensing element or producing photovoltaic effects in the sensing element. Thermal detectors were developed first and have survived with various refinements to the present. The quantum detectors are essentially products of modern solid-state physics and have come into wide use since World War II.

Detectors usually supply a dc voltage change that is proportional to the radiant flux reaching the sensing element; in the early work this voltage change was usually measured by the deflection of a sensitive galvanometer. Various cleverly designed dc amplification systems were developed and used to advantage; all of these were subject to the chronic drifts that usually plague dc amplifiers. Later it was recognized that periodically varying voltages could be produced by periodic interruption of the radiant flux reaching the detector and that these varying voltages could be amplified by conventional electronic amplifiers. The process of periodic interruption

[29] E. K. Plyler, D. J. C. Yates, and H. A. Gebbie, *J. Opt. Soc. Amer.* **52**, 859 (1962).

[30] L. G. Smith, *Rev. Sci. Instrum.* **13**, 63 (1942).

[31] R. Spanbauer, P. E. Fraley, and K. N. Rao, *Appl. Opt.* **2**, 340 (1963).

[32] R. B. Barnes, *Rev. Sci. Instrum.* **5**, 237 (1934).

[33] T. K. McCubbin and W. M. Sinton, *J. Opt. Soc. Amer.* **42**, 113 (1952).

is called *radiation chopping*; at low frequencies chopping is usually accomplished by a rotating toothed wheel and at higher frequencies by an electrically driven tuning fork with a small mirror attached to one prong of the fork. The choice of an appropriate chopping frequency is usually dictated by the *time constant* of the detector, which is defined as the time required by the detector to come to within $1/e$ of its maximum voltage response after being exposed to incident radiant flux. A reference voltage supplied from the rotating wheel or the tuning fork, can be used in "lock-in amplifiers" or "phase-sensitive detectors" with extremely narrow passbands. Suitably rectified amplifier output voltages can be used with conventional chart recorders or can be used to provide an output record in digital form.

4.1.2.1. Thermal Detectors. 4.1.2.1.1. THERMOCOUPLES. The earliest of the thermal detectors was the *thermocouple* devised in 1830 shortly after the discovery of the Seebeck effect; a number of thermocouples in series, called a *thermopile*, was effectively used as a radiation detector by Melloni. Although the thermopile was more satisfactory in practice because of the inadequacy of the 19th century devices for voltage measurement, there is no fundamental advantage in using more than a single thermocouple. The pairs of metals employed in the early thermocouples were antimony bismuth, bismuth–silver, or copper–constantan; various alloys were also used to advantage. The sensitivity of a thermocouple can be greatly increased by mounting it in vacuum; modern commercially available radiation thermocouples are usually supplied in permanently evacuated mounts equipped with windows that are transparent in the infrared.

The early thermocouples were usually fabricated from fine wires and had time constants of several seconds. With the advent of radiation-chopping techniques, shorter time constants have become highly desirable, and various evaporative techniques have been devised for fabrication of thermocouples with low thermal capacity. The metal junction in a radiation thermocouple must be equipped with a "black" radiation receiver of low thermal capacity; various metallic blacks deposited by evaporative procedures have been developed. Gold blacks with a surface density of 80 μgm/cm^2 can absorb 90% of the incident radiation in much of the infrared; metallic blacks are employed in most of the thermal detectors in current use. Certain modern thermocouples, notably the Reeder thermocouple, take advantage of the high thermoelectric powers of semiconductors such as Bi_2Te_3 and its alloys; in our own work we have achieved excellent results with selected thermocouples of this type.

4.1.2.1.2. BOLOMETERS. The chief early competitor of the thermocouple was the *bolometer*, which has had a very interesting history. Langley's original bolometer, which consisted of two nearly identical platinum strips in the arms of a Wheatstone bridge that was unbalanced when one of the strips was exposed to radiation, was superior to any thermocouple available in 1880. As further developed by Abbot, the bolometer became a standard detector in astrophysical work and was adapted for absolute radiation measurements by Callendar. Like the thermocouple, a bolometer works best in a vacuum. Although it can be proved theoretically that the metallic bolometer can achieve the same sensitivity as the thermocouple, the metallic bolometer's performance has in practice been generally disappointing. The resistive properties of thin evaporated layers of metals have different properties from those of the bulk metals; in general, bolometers with sufficiently small mass for use in radiation chopping are subject to unexpectedly large electrical noise. Metallic bolometers are rarely used in spectroscopy laboratories. Bolometer *thermistors* composed of semiconductors have been developed for various laboratory applications but are not widely used in spectroscopy.

Remarkable improvements in bolometer performance can be achieved by the use of low temperatures. Many metals and alloys become superconducting at very low temperatures; and the transition from normal conduction to superconduction takes place over a small but finite temperature range. D. H. Andrews and his colleagues devised bolometers composed of tantalum and of niobium nitride for use in the transition range; a further advantage of low-temperature operation is that the specific heat of a metal becomes very small and hence reduces the bolometer time constants so that radiation chopping techniques can be employed. A more recently developed bolometer is made of tin evaporated on a mica substrate and is operated at liquid helium temperatures; an ac bridge can be employed and, with radiation chopping, the out-of-balance voltage can be stepped up by a transformer operating at liquid helium temperature. Although the necessary bulky cryogenic apparatus is undesirable in most laboratories, the low-temperature bolometer has important special applications. E. P. Ney has made effective use of such a detector in his recent studies of planetary and stellar spectra.

Other recently developed low-temperature bolometers of the carbon-flake and single-crystal germanium types have been discussed by Smith[34] but have not been widely applied in spectroscopic studies.

4.1.2.1.3. PNEUMATIC DETECTORS. A modern thermal detector, which is an adaptation of the differential gas thermometer used in early infrared studies, was devised by Golay.[35, 36] The Golay cell, which has a sensitivity comparable with that of good thermocouples, has played an important role as a broad-band detector and can be employed from the ultraviolet to the microwave region provided it is equipped with suitable windows. Its receiving

[34] R. A. Smith, *Appl. Opt.* **4**, 631 (1965).
[35] H. Zahl and M. Golay, *Rev. Sci. Instrum.* **17**, 511 (1946).
[36] M. Golay, *Rev. Sci. Instrum.* **18**, 357 (1947).

element consists of a very thin layer of aluminum deposited on a 0.01-μm-thick collodion substrate; the thickness of the aluminum layer is adjusted to give a surface resistance designed for maximum absorption of electromagnetic waves and its absorption is nearly independent of spectral frequency.

The irradiated receiver heats the gas in a small chamber. A flexible membrane wall of the chamber is distorted by the increased pressure of the heated gas; this distortion provides a measure of absorbed radiant power. An optical arrangement involving the deflection of a light beam by the distorted wall is incorporated; distortion of the wall changes the amount of light reaching a photoelectric cell which provides a voltage change that can be amplified by conventional methods. The rather long time constant of the Golay cell necessitates the use of chopping frequencies of 13 Hz or lower. The spectral limitations of the cell are imposed only by the nature of the window transmitting incident radiant flux to the receiver.

The Golay cell, conveniently operated at room temperature, is a very useful one for many spectroscopic applications in which its slow response is not objectionable; it is excellent for use in the far infrared. Golay cells are commercially available under patent restrictions from the Eppley Laboratory in the U.S.A. and from Unicam in England. Some cells have objectionable microphonic properties and must be mounted with care on "vibrationless supports." The Golay cell can be permanently damaged when exposed to excessively large flux, which ruptures the flexible wall of the chamber; this feature can become a serious problem in Fourier transform spectroscopy.[†]

Whereas the Golay detector is a broad-band detector sensitive in principle to *all* electromagnetic waves, other types of pneumatic detectors can be highly selective in response. The general principles involved were discussed by Fastie and Pfund[37] in connection with their nondispersive gas analyzer. When an infrared quantum is absorbed in a radiative transition between vibration–rotation levels in a gas at pressures for which molecular collisions are frequent, the absorbed energy becomes rapidly "thermalized" with some considerable portion going into translational degrees of freedom; this increased translational energy produces an increase in gas pressure. By chopping the radiation reaching an enclosed absorbing gas sample, one can produce periodic pressure variations that can be detected by a sensitive microphone; proper selection of the sample tube length can produce acoustical resonance in the so-called spectrophone. Such a device will produce a signal only for infrared frequencies corresponding to the absorption lines of

[37] W. G. Fastie and A. H. Pfund, *J. Opt. Soc. Amer.* **37**, 762 (1947).

[†] See Chapter 4.2.

the enclosed gas; the spectrophone is thus a highly selective detector for infrared radiation.

Although commercially available spectrophones normally do not have sufficient sensitivity for spectroscopic purposes, the general method can be applied by removing the black receiver from a Golay cell and then filling the gas chamber of the cell with an absorbing gas; the modified Golay cell will then give response only at the characteristic frequencies of the enclosed absorbing gas.

4.1.2.2. Quantum Detectors. Although some of the quantum or photo-detectors used in the visible region involve the *ejection* of electrons from an illuminated surface, most photoconductive detectors used in the infrared depend on an *internal photoelectric effect* involving the transfer of photon energy to an electron within the conducting solid. The phenomenon was originally discovered many years ago when it was found that many minerals have marked changes in conductivity when they are illuminated.[38, 39] There was little progress in understanding these phenomena until measurements were made on pure or intentionally doped materials; early work of this type has been reviewed in considerable detail by Smith.[40, 41] The phenomena involved are complex; although they are understood in general principle,[9] many of the details involved in the operation of photoconductive detectors are still not fully understood. The quantum detectors in present use in the infrared can be classified as *intrinsic semiconductors*, pure substances which conduct without the addition of impurities, and *extrinsic semiconductors*, the conduction of which depends on the controlled addition of impurities.

4.1.2.2.1. INTRINSIC SEMICONDUCTORS. Photodetectors involving pure substances depend for their operation on the absorption of a quantum of energy greater than the energy gap ΔE between the valence band and the conduction band. Therefore, photodetectors to be used in the infrared are made of intrinsic semiconductors characterized by a small energy gap. The detecting elements used in spectroscopy are usually made in the form of a rectangle on which the image of the illuminated exit slit of a spectrometer can be focused. In view of the low-energy gap ΔE, the conduction band is appreciably populated at room temperature and it is usually desirable to cool the detector. This is accomplished by having the detector element deposited on the outside surface of the inner wall of a Dewar vessel; the radiant flux to be measured is directed to the element through an infrared-

[38] T. W. Case, *Phys. Rev.* **9**, 305 (1917).

[39] W. W. Coblentz and H. Kahler, *Bull. Nat. Bur. Std.* **15**, 121 (1919).

[40] R. A. Smith, *Advan. Phys.* **2**, 321 (1953).

[41] R. A. Smith, "Semiconductors." Cambridge Univ. Press, London and New York, 1959.

transparent window in the outer wall of the Dewar. In an intrinsic semi-conductor the conduction current is carried by the electrons in the conduction band and by the positive holes in the conduction band. The time constants involved in the operation of intrinsic photodetectors are small as compared with those of thermal detectors; this makes conveniently high radiation chopping speeds possible.

The currently employed photodetectors employing intrinsic semiconductors include:

(1) the PbS cell, which can be used for frequencies as low as 3300 cm^{-1} when operated at ambient laboratory temperature and as low as 2800 cm^{-1} when cooled to dry ice temperatures.

(2) the PbTe cell, which has a low-frequency cutoff of 1800 cm^{-1} at liquid nitrogen temperatures.

(3) the PbSe cell, which has a low-frequency limit of 1400 cm^{-1} at liquid nitrogen temperatures

(4) the InSb cell operated at liquid nitrogen temperatures has a sharp cutoff at 1800 cm^{-1}.

These detectors have a varying spectral response in contrast to the relatively constant spectral response of thermal detectors equipped with blackened receivers.

4.1.2.2.2. EXTRINSIC SEMICONDUCTORS. In order to use quantum detectors at lower frequencies in the infrared it is necessary to make use of extrinsic semiconductors. When a semiconductor like germanium is properly doped with an impurity, new energy levels are created between the conduction and valence bands of the host material; these levels can serve as "electron traps." If such a trap is close to the conduction band and is normally occupied, absorption of an infrared quantum can raise the electron to the conduction band; the resistance is thus reduced by increased electron mobility; *donor levels* of this type can be produced in Ge by doping with column V elements such as Sb, As, and P. On the other hand, doping Ge with column III elements such as Bi, In, and Ga produces *acceptor levels* close to the valence band, absorption of an infrared quantum raises an electron to the trap, thereby leaving a hole in the valence band; the resistance is thereby reduced by mobility of the hole. Doping Ge by other elements such as Zn, Au, and Cu gives sets of impurity levels intermediate in position between those just described; more than one level can be produced by each impurity.

Because the energy-level spacings between the impurity levels and the band levels associated with the Ge host is much smaller than the energy gap ΔE between the conduction band and the valence band in pure Ge, the extrinsic type of detectors must usually be cooled for successful operations in spectroscopy. Doping Ge with Au, Hg, Cu, and Zn progressively lowers the lower cutoff in spectral response; a Zn-doped Ge detector cooled to 6 K has a cutoff frequency of 250 cm^{-1}.

4.1.2.3. Comparison of Detectors. Because of the wide variety of physical processes employed in detectors, any meaningful comparison of different types of detectors, presents difficulties. One criterion that has frequently been employed is called *responsivity* R_V and is defined as the ratio of the root mean square (rms) output voltage to the rms radiant power reaching the detector. Since, in addition to the desired output signal from a detector, there is always a random voltage output due to thermal noise in the detector, the responsivity by itself is not a good figure of merit. Jones[42-44] has given considerable attention to the problems involved in comparing detectors, and, despite certain shortcoming noted by Smith and his colleagues,[9] Jones's criteria have now become more or less standard. One of the simplest figures of merit is the *minimum detectable power* defined as the incident radiant power that will give a signal-to-noise ratio of unity at the output terminals of the detector. Since the noise power in the amplifier output is dependent on the amplifier passband and since the desired signal is produced by a chopped beam of radiant flux, statements of minimum detectable power are usually stated for an amplifier passband of 1 Hz centered at the chopping frequency; this reduced noise equivalent power (NEP) is usually used in statements of minimum detectable power.

Jones has defined a figure of merit called the *detectivity* D defined as the reciprocal of the minimum detectable power

$$D = 1/\text{NEP} = R_V/\text{rms noise voltage from the detector.}$$

This is probably the most meaningful criterion for use in comparing detectors used in spectroscopic studies. The quantity usually stated is, however, a normalized detectivity $D*$ which refers to a detector of unit area and to a passband of 1 Hz:

$$D* = A^{1/2}D = A^{1/2}/\text{NEP.}$$

The use of $D*$ implies that the minimum detectable power is directly proportional to the square root of the sensitive area of the detector. Although this implied relationship applies in good approximation to many quantum detectors, it does not apply to thermocouples for which the NEP depends only on the nature of the metallic junction and is largely independent of the area of the blackened receiver.

[42] R. C. Jones, *J. Opt. Soc. Amer.* **42**, 286 (1952).
[43] R. C. Jones, *Advan. Electron.* **5**, 1 (1953).
[44] R. C. Jones, *Proc. Inst. Radio Eng.* **47**, 1495 (1959).

In Table I we list a set of characteristics of typical infrared detectors of various kinds based on a comparison by Kneubühl.[45] In the table we list the operating temperatures, responsivity R_V, the minimum detectable power NEP, the D^* values, and the time constants in seconds. The time constant τ is an important parameter for the choice of radiation chopping frequencies, which are usually of the order of $1/10\tau$. We note that the indicated frequency cutoff sometimes stated for the thermal detectors is rather meaningless; the receivers used are relatively black at all infrared frequencies and suitable windows are available for all spectral regions.

TABLE I. Infrared Detectors[a]

Detector	Operating temperature (K)	NEP (W)	R_V (V/W)	$D^{*\,b}$ (cm·Hz$^{1/2}$/W)	τ (sec)
Golay	300	5×10^{-11}	2×10^5	1.6×10^9	1.5×10^{-2}
Thermocouple	300	2.5×10^{-10}	1	3×10^9	3×10^{-2}
PbS	193	5×10^{-9}	10^6	4×10^{11}	4×10^{-3}
PbSe	77	—	10^6	1×10^{10}	5×10^{-5}
InSb	77	1×10^{-11}	8×10^3	4×10^{10}	5×10^{-6}
Ge:Au	77	1×10^{-10}	7×10^3	4×10^9	1×10^{-6}
Ge:Cu	5	6.6×10^{-11}	5×10^3	2×10^{10}	1×10^{-6}

[a] These are typical values. The precise values depend on types and manufacturer

[b] The values listed are at the peaks of the spectral response curves for the quantum detectors.

4.1.3. Optical Components for the Infrared

In infrared spectroscopy lenses are rarely used. Throughout most of the infrared nearly all metals are good reflectors, and their reflectivity increases with decreasing frequency. Nearly all focusing devices employed are front-surface mirrors. The mirrors are usually fabricated from Pyrex glass and are coated with evaporated layers of metal. Although a newly deposited silver surface has somewhat better reflective qualities than aluminum in the near infrared, silver tends to tarnish rapidly in the usual laboratory atmosphere. Aluminum coatings are in general use in most infrared instruments; a good aluminum surface has a reflectance of approximately 0.98 throughout the infrared at frequencies below 2000 cm^{-1}. Gold-coated mirrors are sometimes employed and are especially desirable when corrosive gases present problems.

The use of mirrors avoids the optical problems of chromatic aberration normally encountered with lenses in the visible region. Since, however, mirrors must usually be used in an off-axis mode, other problems are introduced. In the portions of the optical systems external to the spectrograph

[45] F. Kneubühl, *Appl. Opt.* **8**, 505 (1969).

itself, spherical mirrors are normally employed to produce an image of the source on the entrance slit of the spectrograph. Within the spectrograph aberrations must be minimized. This can be accomplished in a straight-forward manner by the use of off-axis paraboloidal mirrors as collimating and condensing devices.

The slit system is of extreme importance in the design of a good spectro-graph. All slits in modern use provide bilateral jaw movement so that the center of the slit remains fixed in position as the jaws open and close. The jaws must remain parallel and must be sufficiently movable to provide an opening as large as 1 or 2 mm; the slit mechanism must provide for accurate resetting. Since rays from different points along the entrance slit are deviated by slightly different angles by either a prism or a grating, a curved image of a straight entrance slit would be formed at the exit slit position. In most spectrographs this effect is usually avoided by using a curved entrance slit and a straight exit slit.

The dispersed radiant flux passing through the exit slit must be directed to the detector, which generally provides a very small rectangular detecting element or receiver. This necessitates the formation of a highly reduced image of the exit slit. This reduction is best accomplished by means of an ellipsoidal mirror. Although the ellipsoidal mirror is usually an on-axis device and some of the beam is obscured by the detector mount, good off-axis ellipsoidal mirrors are commercially available.

Windows must be provided for absorption cells and for some types of detectors; most window materials have a low-frequency cutoff imposed by strong characteristic absorption bands in the infrared. The cutoff for glass is approximately 4000 cm^{-1} and that for quartz approximately 3000 cm^{-1}. The alkali halide crystals are satisfactory window materials in much of the infrared and are widely used in spite of their hygroscopic properties. Com-mercially available synthetic crystals of this type include: LiF with a cutoff at 1600 cm^{-1}; NaCl with a cutoff at 600 cm^{-1}; KBr with a cutoff of 400 cm^{-1}; CsBr with a cutoff of 250 cm^{-1}; and CsI with a cutoff of 200 cm^{-1}. These alkali-halide materials are mechanically fragile and are subject to thermal fracture but can be polished easily. More mechanically rugged, nonhygro-scopic window materials include sapphire with a cutoff of 1600 cm^{-1} and fluorite CaF$_2$ with a cutoff of 1000 cm^{-1}. Two other window materials are AgCl and KRS-5, a mixture of thalium bromide and thalium iodide; both of these are rather soft but malleable materials with cutoff frequencies of 450 cm^{-1} and 250 cm^{-1}, respectively.[†] Both AgCl and KRS-5 have large refractive indices and therefore large surface reflectances, as have CsBr and CsI.

[†] The cutoff frequencies listed are for materials at 300 K; at cryogenic temperatures the cut-off frequencies are much lower.

In the far infrared at frequencies below 200 cm^{-1} quartz becomes transparent and is an excellent window material. Polyethylene, polypropylene, silicon, and germanium are also useful as cell windows for the far infrared. Diamond windows are used for certain receivers and for special purposes in the far infrared.

4.1.4. Resolving Instruments

The resolving instruments in present use for covering wide spectral ranges in the infrared are primarily prism and grating spectrophotometers. With the development of modern digital computers, however, the interferometer has become highly competitive; the computer supplies the spectrum in the form of a Fourier transform of the interferogram obtained directly from the initial measurements. Since Fourier transform spectroscopy has special advantages in the far infrared, it will be treated in Chapter 4.2. In this section, we give a brief discussion of prism and grating instruments and provide references to more detailed treatments. We also mention briefly certain other devices recently developed for special applications.

The history of the development of prisms and gratings has rather interesting aspects. The prism gives no direct wavelength or frequency measurement until it has been calibrated but has the advantage of dispersing all incident radiant flux into a single spectrum. The early grating techniques provided direct measurements of wavelength in terms of measured angles together with known regular spacings between the wires in a transmission grating or between the grooves in a reflection grating; the early gratings, however, had the great disadvantage of dispersing the incident flux in many overlapping orders. In a beautiful set of measurements in the closing years of the 19th century,[2] Paschen, Rubens and E. F. Nichols, Carvollo, and Langley used gratings to provide a set of *dispersion curves* giving refractive index n as a function of wavelength λ for quartz, fluorite, rocksalt, and certain other naturally occurring prism materials. Once these dispersion curves had been established, prism spectrometers could be used to provide reliable spectra at modest resolution. In nearly all prism instruments employed prior to 1940, the Wadsworth mounting was employed; this type of mounting provided a spectrum based on a single passage of radiant flux through the prism at the angle of minimum deviation, which is readily expressed in terms of n for each wavelength. By 1940 such a wealth of high-resolution spectral data had been amassed by grating studies that it became possible to base prism-spectrometer calibrations on empirical measurements of known spectral lines and bands; R. Bowling Barnes first took advantage of this situation and introduced a new era of prism-instrument design by passing the radiation through the prism more than once and by forsaking the earlier strict dependence on minimum-deviation considerations. At present extensive tables of standard wavelengths[46] have been established and are useful in the calibration of grating instruments as well as those employing prisms.

An important advance in grating technology was made in 1910 by Wood when he invented the *echelette grating*.[47] In this form of reflection grating the grooves or facets are shaped in such a way as to direct most of the incident flux in a particular direction called the *blaze angle*. The diffracted radiation in the vicinity of the blaze angle consists of many overlapping orders, which must be separated by some auxiliary device in the spectrograph. Although effective separation can now be accomplished by filters, order separation in grating spectrographs designed prior to 1960 was usually accomplished by means of a foreprism placed between the source and the entrance slit of the grating spectrograph proper. Various improvements in grating engines by Wood himself as well as by later workers have served to improve the qualities of echelette gratings.

In the period prior to 1940 all prisms for the intermediate infrared were fabricated from naturally occurring minerals. Although the supply of specimens of suitable optical quality was definitely limited, the cost of a good prism was usually much smaller than the cost of an original grating. This economic advantage in favor of the prism was considerably enhanced when synthetic crystals became commercially available shortly after World War II. During this time, however, methods of producing high-quality *replicas* from carefully ruled master gratings were being developed. In 1954 Cole[48] showed that it is possible to cover the spectral range 2.5–15 μm (4000–650 cm^{-1}) with a single grating at higher resolution and at lower costs than with prisms. Cole employed a replica echelette grating blazed for 10 μm in the first order and obtained good resolution in the following regions:

7–15 μm	(1400–670 cm^{-1})	1st order,
4–7 μm	(2500–1400 cm^{-1})	2nd order,
3–4 μm	(3300–2500 cm^{-1})	3rd order,
2.5–3 μm	(4000–3300 cm^{-1})	4th order.

Although blazed for 10 μm in the first order for an ideal grating with narrow slots, the range of wavelengths actually observed in the first order (7–15 μm) is considerably larger because of the superimposed diffraction pattern of each actually nonnarrow facet in the grating.[49] With improvements of ruling and replication techniques and with the development of effective

[46] K. N. Rao, C. J. Humphreys, and D. H. Rank, "Wavelength Standards in the Infrared." Academic Press, New York, 1966.

[47] R. W. Wood, *Phil. Mag.* **20**, 770 (1910).

[48] A. R. H. Cole, *J. Opt. Soc. Amer.* **44**, 741 (1954).

[49] R. W. Ditchburn, "Light," pp. 179–186. Blackie, Glasgow and London, 1963.

filters for order separation, the economic advantages of grating instruments become progressively greater. Prism instruments will, however, probably continue to find applications for many purposes.

4.1.4.1. **Prism Spectrophotometers.** The optical materials used as prism materials are characterized by strong absorption bands in the ultraviolet region and in the far infrared. The dispersion curves giving n versus λ for these materials are characterized by a decrease in n with increasing wavelength λ as the far infrared absorption band is approached. The angular dispersion $d\theta/d\lambda$ achieved by a given prism depends on the magnitude of $dn/d\lambda$ and on the prism angle. The Rayleigh criterion for the resolving power R obtainable with a prism as derived for a single passage of radiation through the prism is given by the expression

$$R = \lambda/\Delta\lambda = \nu/\Delta\nu = T_p|dn/d\lambda| \qquad (4.1.1)$$

where T_p is the thickness of the prism at its base. The prism angle itself does not affect the resolving power. In order to achieve high resolving power it is necessary to use a large prism in a spectral region where the dispersive power $|dn/d\lambda|$ of the prism material is large.

The Rayleigh expression for resolving power in Eq. (4.1.1) is based on consideration of diffraction limitations alone; it would imply that the optimum spectral region for a given prism is the region in which $|dn/d\lambda|$ is greatest. The dispersive power $|dn/d\lambda|$ is, however, greatest at wavelengths well *within* the strong infrared absorption band for the prism material, a spectral region where the prism would pass little spectral flux. The spectral region in which maximum resolving power is actually achieved is *near* the infrared absorption band, a region in which $|dn/d\lambda|$ is large but in which the prism material is still largely transparent. Gore and his colleagues[50] have devoted considerable attention to the question of maximum attainable resolving power. In Table II we list the commercially available prism materials, their optimum spectral ranges, and the regions in which maximum resolving power is actually achieved for each type of prism. Flint glass and quartz prisms are also available for effective use at higher frequencies in the near infrared.

Although the ultimate performance limitations of a prism spectrophotometer are imposed by properties of the source, the prism, and the detector employed, the actual optical design of the spectrometer is also of great importance. Most modern prism instruments usually employ rotation of a plane Littrow mirror behind the prism to provide spectral scanning.[51]

[50] R. C. Gore, R. S. MacDonald, V. Z. Williams, and J. U. White, *J. Opt. Soc. Amer.* **37**, 23 (1947).

[51] R. B. Barnes, R. S. MacDonald, V. Z. Williams, and R. F. Kinnaird, *J. Appl. Phys.* **16**, 77 (1945).

TABLE II. Prism Materials

| Material | Optimum range | | Region of maximum resolving power |
	λ (μm)	ν (cm^{-1})	(cm^{-1})
LiF	2–5.3	5000–1900	2300
CaF$_2$	5.3–8.5	1900–1250	1200
NaCl	8.5–15.4	1250–650	900
KBr	15.4–26	650–385	500
CsBr	14–38	720–270	350
CsI	20–50	500–200	250

In the simplest form of such an instrument, radiant flux from the source is directed to the entrance slit of the spectrometer by a spherical mirror. The solid angle subtended by the flux beam emerging from the entrance slit should be of such a size as to ensure a collimated beam that will fill the face of the prism; proper matching of solid angles is maintained throughout the spectrograph. Flux from the entrance slit passes to an off-axis paraboloidal mirror, which collimates the flux beam and directs it to the prism (Fig. 1). After traversal of the prism, the refracted beam strikes the plane Littrow mirror, which returns the radiation through the prism to the paraboloidal mirror, which then serves as a condenser in bringing the refracted radiation to focus at an exit slit slightly displaced optically from the entrance slit. Dispersed radiation emerging from the exit slit is directed to the detector by an ellipsoidal mirror that produces an image of the exit slit on the receiver of the detector. Because of their finite slit widths, the exit slit passes spectral flux in a narrow band of spectral frequencies to the detector; this narrow band of frequencies $\Delta \nu$ reaching the detector is called the *spectral slit width* of the instrument and should be small as compared with the widths of any spectral features being studied.

One of the problems that plague infrared spectroscopists involves stray radiation of high frequency reaching the detector. With thermal sources approximating blackbodies the spectral intensity $I_0(\nu)$ of the source in the vicinity of the radiation peak is much larger than the spectral intensity at lower frequencies in the infrared. Small amounts of this high-frequency flux can seriously contaminate the observed spectrum and become a major problem in the far infrared. In order to minimize stray radiation in a simple infrared instrument, the surfaces of all components such as prisms and mirrors should be kept clean and free of dust; blackened radiation baffles are usually placed at strategic locations within the spectrometer housing. Another method of handling this problem involves the use of a glass shutter to establish the "shutter zero" level from which chart recorder deflections are measured; the glass shutter is transparent to undesired high-frequency radiation but is opaque to the desired low-frequency radiation. Another approach to the problem is the use of a double monochromator such as the Leiss instrument described in some detail by Roberts.[52] In such an instrument a second prism spectrometer is placed beyond the exit slit of the simple instrument just described; most of the stray radiation has hopefully been removed before the radiation reaches the second prism. In a double mono-chromater provision must be made for synchronous spectral "tracking" of the two prisms; although achievement of this provision presented diffi-culties in early instruments, these difficulties have been virtually eliminated in modern commercial instruments.

Many of the advantages of a double monochromator can be achieved at considerably less expense by employing certain multiple-traversal and chopping techniques first described by Walsh.[53] The rationale of the Walsh monochromator can be understood from the schematic diagram in Fig. 1, which shows the passage of a single selected ray through the instrument.

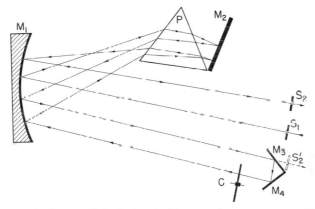

FIG. 1. Schematic diagram of the Walsh double-pass prism spectrometer. The instrument can be operated in the single-pass mode $S_1M_1PM_2PM_1S_2'$ by placing the exit slit S_2' near the line of intersection of the two plane mirrors constituting the "corner reflector."

Radiant flux through entrance slit S_1 passes to M_1, which is an off-axis paraboloidal mirror and directs a collimated beam of radiation to the prism. A dispersed beam of radiation from the prism strikes the Littrow mirror M_2, which sends the beam back through the prism to mirror M_1. Mirror M_1 then directs a converging beam of radiant flux to plane mirrors M_3 and M_4, which constitute a right-angled "corner reflector" that returns the flux to M_1 which sends a collimated beam for another passage through

[52] V. Roberts, *J. Sci. Instrum.* **29**, 134 (1952).
[53] A. Walsh, *J. Opt. Soc. Amer.* **42**, 94 (1952).

the prism to M_2. After a second reflection by Littrow mirror M_2 and return to M_1, the beam reflected by M_1 comes to focus at exit slit S_2, the illuminated image of which is focused on the detector by an ellipsoidal mirror not shown in the diagram. For a given orientation of prism and Littrow mirror, the selected ray shown in the diagram gives the path traversed by radiation of a given wavelength along the path $S_1 M_1 P\ M_2 P\ M_1 M_3 M_4 M_1 P\ M_2 P\ M_1 S_2$, which thus makes four traversals of the prism or two complete "passes" of the Littrow mounting before reaching the exit slit S_2.

Slits S_1 and S_2 and the line joining M_3 and M_4 are all in the focal plane of M_1; in order to minimize aberrations they should all be as close together as possible. The effects of aberrations are usually further reduced by giving appropriate curvature to the slits.

Mixed with the desired radiation emerging from S_2 will be stray radiation of other frequencies which has made a single pass through the prism as well as radiation scattered from the refracting and reflecting surfaces. In the Walsh instrument attainment of an effectively much purer spectrum can be achieved by placing the radiation chopper C in the beam reflected from M_4, which includes only radiation making a second pass or round trip through the prism. As the modulated voltage output of the detector serves as the input for a narrow-band amplifier tuned to the chopping frequency, the effects of the unchopped single-pass and stray radiation are removed. The Walsh scheme for modulating only desired radiation has been used in many later instruments involving gratings as well as prisms.

The extension of the double-pass features to systems using higher-multiple passes is obvious. Walsh and his colleagues have given details regarding the construction and performance of triple- and quadruple-pass mono-chromators.[54, 55]

The resolving power attained with most prism instruments is usually less than 1000 and is considerably lower than this in the far infrared $v < 500$ cm^{-1}. For investigations requiring higher resolving powers it is usually more convenient to employ a grating instrument.

4.1.4.2. Grating Spectrophotometers. Nearly all modern grating in-struments for the infrared employ echelette gratings. Schaefer and Matossi[2] give an interesting discussion of the preparation and use of early gratings of this type, and Ditchburn[49] gives a good introductory treatment of their properties. A detailed and rigorous treatment of echelettes and of the technical problems involved in their fabrication has been given by Stroke.[56] Whereas the early ruling engines[57] relied on calibrated precision screws to

[54] N. S. Ham, A. Walsh, and J. B. Willis, *J. Opt. Soc. Amer.* **42**, 496 (1952).

[55] A. Walsh and J. B. Willis, *J. Opt. Soc. Amer.* **43**, 989 (1953).

[56] G. W. Stroke, Diffraction gratings, *in* "Handbuch der Physik." Vol. 29. Springer-Verlag, Berlin and New York, 1967.

[57] J. Strong, *J. Opt. Soc. Amer.* **41**, 3 (1951); **50**, 1148 (1960).

provide uniform line spacings, modern ruling engines as developed by G. R. Harrison and his associates are controlled by servo mechanisms directly controlled by interferometric methods involving light waves in the visible region. The interesting details involved in the modern fabrication processes have been discussed by Stroke, who in collaboration with Harrison has made important contributions to the work.[58]

The fabrication of a diffraction grating by modern techniques is a long and enormously expensive process. Fortunately, methods have been developed for the preparation of high-quality replica gratings from original master gratings. The replica of an echelette consists of a thin layer of plastic resin "printed" from the master grating and then mounted on a glass optical flat; clean separation of the replica from the master grating without deformation is a key step in the successful replication. After the plastic layer has been attached to the optical flat, its surface is aluminized by evaporation. A good replica has optical properties comparable with and, in some respects, superior to the properties of the master grating. Modern replicas have remarkably long useful lives without apparent serious deterioration. A highly useful and readable introductory general treatment of the properties and uses of diffraction gratings has been prepared by the Bausch and Lomb Company.[59]

As with other difffraction gratings the angular dispersion of an echelette is directly proportional to the diffraction order number m and is inversely proportional to the grating space a. For a spectrograph employing the so-called Littrow configuration in which the angles of incidence and diffraction are each equal to the angle between the facets and the plane of the grating, the grating equation simplifies to

$$m\lambda = 2a \sin \theta \qquad (4.1.2)$$

and the angular dispersion becomes $d\theta/d\lambda = m/(a \cos \theta)$, where we use θ to denote the common value of the incidence, diffraction, and facet angles. The resolving power R of a grating as given by the Rayleigh criterion is given by the product of the order number m and the total number of lines N or

$$R = mN = mW/a, \qquad (4.1.3)$$

where W is the total width of the grating. Consideration of Eqs. (4.1.2) and (4.1.3) suggests two methods of achieving high resolving power with a grating of a given width W:

(1) A finely ruled grating with small grating space a can be used in a low order m; this requires a small blaze angle, which for a typical *echelette* is in the range 5–25°.

[58] G. W. Stroke, *Progr.* **2**, 1–72 (1963)
[59] "Diffraction Grating Handbook." Bausch and Lomb, Rochester, New York, 1970.

(2) A coarse grating with large grating space a can be used in a high-order m; this requires a large blaze angle in the neighborhood of 63°; a plane grating blazed for such a large angle is called an *echelle*.

For gratings of the same total width the angular dispersion $d\theta/d\lambda$ is much greater for the echelle than for the echelette. In actual use echelles have resolving powers closer to the theoretical values given in Eq. (4.1.3) than do more finely ruled echelettes. It is usually more difficult, however, to eliminate radiation from undesired overlapping orders in the case of echelles. These matters have been considered in detail by Rao.[60] Echelle replica gratings are now commercially available in sizes as large as 21 × 41 cm (8 × 16 in.) in height and width, respectively. There are prospects for even larger gratings of this type.[61] These gratings are ruled with 79.1 or 316 grooves/mm—spacings that are convenient multiples of wavelengths of light from the He–Ne laser used to provide servo control of the ruling engine.[†]

Spectrographs for use with modern gratings have been discussed in a book by Davis.[62] Even more than for prism spectrographs, the design of optical systems for grating instruments is of extreme importance if the diffraction limit of resolving power is to be approached. In attempting to design the requisite optical systems spectroscopists have attempted either (1) to eliminate spherical aberrations and related effects or (2) to live with spherical aberration and actually make use of it! One example of the first approach involved the use of an off-axis paraboloidal mirror as a collimator and preferably a second off-axis mirror of this type as a condensing mirror to

[60] K. N. Rao, Large plane gratings for use in high-resolution infrared spectrographs, *in* "Molecular Spectroscopy: Modern Research" (K. N. Rao and C. W. Mathews, eds.). Academic Press, New York, 1972.

[61] G. R. Harrison and S. W. Thompson, *J. Opt. Soc. Amer.* **60**, 591 (1970).

[62] S. P. Davis. "Diffraction Grating Spectrographs." Holt, New York, 1970.

[†] We should at this point mention the recently developed "holographic gratings" now available commercially. Such a grating is formed by exposing a "photosensitive layer" of material mounted on a glass blank to intersecting laser beams, which produce equally spaced interference fringes. After the photosensitive layer has been "developed and processed," it is covered by vacuum deposition of "reflectance and protective coatings." Shaping of the "holographic grooves" can be accomplished by unspecified methods. The gratings are reportedly free of ghosts and have very low levels of scattered light. Holographic gratings are available "in sizes much larger than the maximum size of ruled gratings." The fabrication of holographic gratings is "described" in a handbook recently published by Jobin Yvon Optical Systems; so many proprietary secrets are involved that the description is largely unenlightening!

The quality of gratings of this type is lauded by research workers who have used them in studies of the Raman effect. It is interesting that optical techniques can be applied directly in grating fabrication without the use of mechanical operations. It will be interesting to have comparisons of holographic gratings with high-quality replicas conducted by impartial investigators and to receive more enlightening information regarding the fabrication and limitations of holographic gratings!

focus the diffracted radiation on the exit slit[63, 64]; the chief difficulty with this straightforward approach is the technical and hence financial one involved in fabricating large off-axis paraboloids of high quality. Another approach to the problem is the Pfund[65, 66] system in which on-axis paraboloids were employed and plane mirrors with small central apertures were placed between the slits and the paraboloids in order to intercept collimated beams of radiation and to direct the beams in desired directions; a diverging beam from the entrance slit passed through the small central aperture in the first plane mirror to the collimator, and a converging beam from the condenser passed through the aperture in the second plane mirror to the exit slit. The collimated beam from the collimator was directed to the grating by the first plane mirror; the diffracted beam from the grating was directed to the condensing paraboloid by the second plane mirror.

In the more recent types of grating spectrographs, designers have followed the second course mentioned above and have employed spherical mirrors in rather clever ways. Spherical mirrors of the large diameters needed for use with large modern gratings can be fabricated with relative ease and do not contribute greatly to the cost of the spectrograph. One currently popular design is the one proposed by Czerny and Turner,[67] which employs spherical collimating and condensing mirrors in such a way that the effects of their spherical aberrations cancel. Another design of present interest is one that was proposed initially by Ebert[68] but never appreciated until Fastie[69] rediscovered the pertinent optical relationships and proceeded to build a spectrograph based on them.

The Ebert–Fastie system is somewhat similar to the Czerny Turner instrument but employs a single large spherical mirror as shown schematically in Fig. 2. Radiation from the entrance slit is incident on one half of the mirror, which reflects a collimated beam to the grating; dispersed radiation from the grating is incident on the other half of the mirror, which produces an image of the entrance slit at the exit slit position. Fastie showed that a long curved entrance slit would be imaged at a long curved exit slit; the curved exit and entrance slits are symmetrically spaced about the optical axis of the large mirror. The axis of rotation of the grating intersects the optical axis of the mirror and is perpendicular to this axis. It can be shown that, if the grating were a plane mirror, the image of any point in the focal plane of the large spherical mirror would be formed in the focal plane at the

[63] H. M. Randall, *Rev. Sci. Instrum.* **3**, 196 (1932).
[64] E. E. Bell, R. H. Noble, and H. H. Nielsen, *Rev. Sci. Instrum.* **18**, 48 (1947).
[65] A. H. Pfund, *J. Opt. Soc. Amer.* **14**, 337 (1927).
[66] J. D. Hardy, *Phys. Rev.* **38**, 2162 (1931).
[67] M. Czerny and A. F. Turner, *Z. Phys.* **61**, 792 (1930).
[68] H. Ebert, *Wied. Ann.* **38**, 489 (1889).
[69] W. G. Fastie, *J. Opt. Soc. Amer.* **42**, 641, 647 (1952).

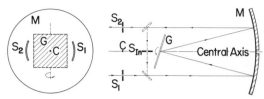

FIG. 2.   Schematic diagram of the Ebert–Fastie grating spectrograph, which can be operated in a double-pass mode as noted by the dotted rays by introducing the small plane mirrors. In the double-pass mode the entrance and exit slits as well as the plane mirrors are out of the plane of the diagram; use of the intermediate slit in the double-pass mode provides a "cleaner" spectrum. The cross-hatched arcs at the large Ebert mirror represent the smaller confocal spherical mirrors of the equivalent Czerny–Turner spectrograph.

opposite end of the diameter of a circle with its center located at the optical axis of the large mirror. Thus a curved slit along one arc of the circle would be imaged on a curved slit coinciding with the opposite arc of the circle. When the curved slits are arranged symmetrically with respect to the axis of rotation of the grating, this same relation applies in close approximation; there is some slight spherical aberration but coma and astigmatism are virtually eliminated.

The use of long slits is a great advantage in the far infrared, where spectral flux intensities $I(\nu)$ are low. Robinson[70] has designed an Ebert–Fastie instrument for the far infrared range 500–20 cm^{-1}. Rao and Nielsen[71] have designed a vacuum instrument employing an Ebert mirror with a diameter of 63.5 cm and a focal length of 6.4 m. This optical system used in conjunction with an echelle is capable of covering the spectral range between 10,000 and 70 cm^{-1} provided suitable arrangements are made for order separation. The curved slits in the instrument lie on a circle with a diameter of 25 cm. Special arrangements must be made for focusing the dispersed radiation from the long exit slit on the receiver of the detector.[72]

The resolving power attainable with grating spectrographs is of the order of several hundred thousand. In order to achieve resolving powers of this order it is, however, necessary to eliminate undesired radiation. Consideration of Eq. (4.1.2) indicates that if a grating is blazed for radiation of wavelength λ_1 in first order $m = 1$, it is also blazed for shorter wavelengths $\lambda_1/2$, $\lambda_1/3$, $\lambda_1/4$, . . . in orders $m = 2, 3, 4, . . .$, respectively. Effective provision must be made for the essential elimination of undesired short-wavelength radiation; this is a serious problem in view of the fact that $I(\lambda)$ is very large at short wavelengths in the vicinity of the peak of the radiation curves for commonly used sources. No serious problems are encountered as a result of

[70] D. W. Robinson, *J. Opt. Soc. Amer.* **49**, 966 (1959).
[71] K. N. Rao and H. H. Nielsen, *Appl. Opt.* **2**, 1123 (1963).
[72] W. Benesch and J. Strong, *J. Opt. Soc. Amer.* **41**, 252 (1951).

radiation with wavelengths much longer than λ_1, since they are not diffracted by the echelette; in fact, for wavelengths much longer than the grating space a, the grating acts as a mirror.

One method of ensuring the elimination of undesired radiation involves the use of a "fore prism" as a more or less crude prism spectrometer in series with the grating instrument; this method was employed in most of the early high-resolution instruments. Although the prism was usually placed between the source and the entrance slit of the grating instrument, it can equally well be placed between the exit slit and the detector. In normal practice, the prism was usually set by hand for a given spectral region; in more elaborate instruments the prism rotation can be tracked with the rotation of the grating so that both maintain optimum adjustment for each wavelength.

Another method involves the use of short-wavelength rejection filters. One such device is a finely ruled plane grating with grating space less than the desired wavelength and used as a plane mirror; the grating disperses short-wavelength radiation by diffraction but acts as a mirror for long wavelengths. A high-quality grating is not required. In fact, in the far infrared a metal plate with a matte surface can be substituted for the grating; provided the irregularities or "scratches" on the plate have mean spacings considerably smaller than the wavelength of interest, such a "scatter plate" is very effective.

The development of high-quality dielectric interference filters[9] in the years following 1949 has provided another means for the rejection of radiation of undesired wavelengths. Filters of this type are now commercially available in a wide variety of forms. Short-wave rejection filters of high quality are best suited for the purposes under present discussion. Bandpass interference filters are also available.

Other bandpass filters make use of the *reststrahlen* phenomena discovered by Rubens and Nichols,[73] who noted that crystals have nearly specular reflectivity at frequencies near their characteristic absorption bands and low reflectivity in other spectral regions. By arranging for several successive reflections, an effective reflection filter can be produced; such a filter has a well-defined pass band in the vicinity of the characteristic reflection peak of the crystal. Reststrahlen techniques are still used in grating instruments for the intermediate and far infrared.

Another type of bandpass filter is based on the Christiansen effect observed when small particles are suspended in an otherwise transparent medium. Radiation is effectively scattered by the suspension except at wavelengths for which the refractive indices of the particles and the medium are equal; for such wavelengths radiation is effectively transmitted by the suspension. As all materials exhibit "anomalous dispersion" in the vicinity

[73] H. Rubens and E. F. Nichols, *Phys. Rev.* **4**, 297, 314 (1897).

of absorption bands, Christiansen filters can be used effectively in such regions in the vicinity of major characteristic dispersion features.

Yoshinaga and his associates[74] have done excellent research on the fabrication of filters employing processes related to the reststrahlen and Christiansen effects. Yoshinaga filters are now commercially available and are widely applied; they prove especially useful in the far infrared.

The choice of a suitable resolving instrument is dictated to some extent by the phenomena being investigated and by any special restrictions that may be imposed by various experimental conditions. Although it is true to a certain extent that one can usually make a low-resolution instrument out of a high-resolution instrument by opening the slits or by some other adjustment, it is not only less expensive but sounder experimental policy to use as simple a resolving instrument as possible provided it is adequate for the problem at hand. It is not necessary to use a cannon to shoot humming birds!

The very highest resolving power is required in studies of the absorption spectra of small molecules. In early grating studies, line frequencies were determined in terms of angles and grating constants; the grating mount was equipped with a divided circle of the highest quality. In more recent high-resolution instruments infrared wavelengths observed in low order have been determined by comparisons with interferometrically determined atomic wavelengths observed in high order with the instrument. Instruments of the kind used in these ultrahigh-resolution studies are described in the book by Rao et al.[60] They have been developed by D. H. Rank, King McCubbin, and T. A. Wiggins at Pennsylvania State University, by Rao and his associates at Ohio State University, and by Plyler and his associates first at the National Bureau of Standards and later at Florida State University. Many of the listed standard infrared wavelength standards[60] are now recorded to eight significant figures. Rank[75] has measured two lines in the spectra of CO and HCN in terms of standard atomic wavelengths so precisely that his results can be used together with related microwave frequency measurements to determine the speed of light.

Since the establishment of suitable wavelength standards, the task of designing spectrometers with lower resolving powers has been simplified. Tubbs[76] of our laboratory has recently designed an instrument for use in the study of collisional line broadening that takes full advantage of the new standards. This instrument is one of the Czerny–Turner type but is operated in the Ebert mode by tilting the two spherical mirrors in such a way that their centers of curvature coincide. The optics are arranged for double-pass

[74] Y. Yamada, A. Mitsuishi, and H. Yoshinaga, *J. Opt. Soc. Amer.* **52**, 17 (1962).

[75] D. H. Rank, D. P. Eastman, B. S. Rao, and T. A. Wiggins, *J. Opt. Soc. Amer.* **51**, 929 (1961).

[76] L. D. Tubbs and D. Williams, *J. Opt. Soc. Amer.* **60**, 726 (1970).

at the grating and use an intermediate slit so as to incorporate certain advantages of a double monochromator. The grating drive mechanism employs a high-quality short precision screw, which can be kinematically mounted at various positions and seems to offer certain advantages over a single long screw. In view of the fact that the instrument is primarily intended for intensity studies as opposed to wavelength determinations, provision for slow scans has been made by turning the screw by means of a "stepping motor" of the type used in interferometer studies in the far infrared to improve signal-to-noise ratios. The instrument is calibrated by recording standard spectra[60] in terms of a revolution counter attached to the stepping motor; the instrument stably maintains its calibration curve giving the frequency in reciprocal centimeters as a function of step number. Since signal-to-noise ratios can be improved by lengthening the observation time, the master circuit controlling the stepping motor can be adjusted to provide extremely slow stepping rates in regions of particular interest and high stepping rates in other spectral regions. The diffraction limit of resolving power for the instrument is 80,000; with a cooled InSb detector the spectrograph can achieve a resolving power of 40,000–50,000 in routine operation fully adequate for its purpose.

Other instruments with comparable or somewhat lower resolving powers are available commercially. Such instruments are used along with prism instruments for studies of the contours of vibration bands of large polyatomic molecules, when complete resolution of the rotational fine structure cannot be achieved even with instruments of the highest available resolving power. Although modern high-resolution grating spectrographs are operated in evacuated enclosures, commercial instruments usually rely on dry nitrogen purging of nonevacuated housings to eliminate absorption by atmospheric gases such as CO_2 and H_2O vapor. Many of the commercial instruments are of the double-beam type and attempt to give a direct measure of the spectral transmittance of a sample by comparison of the spectral intensity of the beam transmitted through the sample with that transmitted through a reference beam.

4.1.4.3. Other Types of Spectrographs. Although the prism and grating instruments of the types described thus far are in general laboratory use, other devices have been developed for special applications. In certain studies of planetary and stellar spectra,[†] the incident radiant flux has such low intensity that the spectral slit widths even of prism instruments are too small for effective use. Ney[77] and his associates, however, have employed a set of

[77] E. P. Ney and W. A. Stein, *Astrophys. J. Lett.* **152**, L21 (1968); E. P. Ney, D. W. Strecker, and R. D. Gehrz, *Astrophys. J.* **180**, 809 (1973); D. W. Strecker and E. P. Ney, *Astron. J.* **79**, 797 (1974).

[†] See also Vol. 12A (Astrophysics), Parts 9–11.

bandpass interference filters in conjunction with low-temperature bolometers to map the gross features of the radiation curves and have made valuable contributions to our knowledge of stellar infrared sources. Bandpass filters have also been used on various space probes when weight requirements precluded the use of spectrographs or Michelson interferometers.

A crude form of spectrometer can be realized by the use of a tapered interference filter. If such a tapered filter is deposited as an annulus on a transparent wheel in such a way that the thickness does not vary with radius but varies in the tangential sense, a simple spectrograph can be achieved by allowing radiation to pass through the annular filter to a detector. As the wheel rotates the spectrum of the source is scanned by the detector. The spectral range covered is limited to a single octave with $v_{max} = 2v_{min}$, and the resolving power is usually less than 100. Interference spectrographs of this type have been used in flight instrumentation; their use in certain atmospheric pollution studies has been proposed.

Another technique being employed in atmospheric work is called gas-cell *correlation spectroscopy* and is an outgrowth of much earlier work on non-dispersive gas analyzers.[37] The widths of the absorption lines of gases at normal atmospheric or lower pressure are so narrow that at a constant pressure the spectral transmittances as measured with a prism spectrograph of two gases A and B are multiplicative[78, 79]; i.e., the measured fractional spectral transmittance T_{AB} of a mixture is given by the product $T_A \cdot T_B$ of the spectral transmittances of the two gases as measured separately, a relation called *Burch's law*.[80] In correlation spectroscopy the presence and concentration of an absorbing gas in a given sample can be measured without any resolving instrument other than the absorption spectrum of the gas itself in a correlation beam.

4.1.5. The Infrared Spectra of Gases

High-resolution studies of gases at low pressures have provided a wealth of knowledge regarding molecular moments of inertia—and thereby information regarding interatomic distance—and also regarding the force systems that hold atoms at equilibrium separations in molecules. In this section we show a few examples of some particularly simple spectra and indicate how information can be gathered in more complex cases. We omit consideration of many of the complexities that Herzberg[5, 6] has referred to as "finer details" although much modern research is being devoted to them.

---

[78] D. E. Burch and D. A. Gryvnak, *in* "Analytical Methods Applied to Pollution Measurements" (R. K. Stevens and W. F. Herget, eds.), pp. 193–231. Ann Arbor Sci., Ann Arbor, Michigan, 1974.

[79] D. E. Burch, J. N. Howard, and D. Williams, *J. Opt. Soc. Amer.* **46**, 452 (1956).

[80] G. M. Hoover, C. E. Hathaway, and D. Williams, *Appl. Opt.* **6**, 481 (1967).

4.1.5.1. **Pure Rotation Spectra.** The simplest infrared absorption spectrum is that associated with transitions between the quantized rotational energy levels of polar molecules in their ground vibrational and electronic energy states. The simplest rotational spectra are observed for polar diatomic and linear polyatomic molecules. Since the diatomic molecule CO will be treated in some detail in Chapter 4.2, we shall discuss here only the rotational spectrum of the linear polyatomic molecule nitrous oxide N_2O, which has the asymmetrical configuration NNO and is therefore polar; the pure rotation spectrum as observed in the far infrared is shown in Fig. 3. It consists of

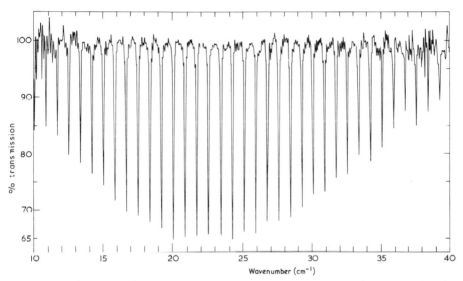

FIG. 3.   The pure rotational spectrum of N_2O (courtesy of Dr. John Fleming of the National Physical Laboratory, Teddington, Middlesex, England).

a set of individual absorption lines with nearly equal spacing and can be interpreted in terms of electric dipole transitions with $\Delta J = +1$ between rotational levels given by the expression

$$E_J = hc[J(J + 1)B - DJ^2(J + 1)^2] \qquad (4.1.4)$$

where B is related to the molecular *moment of inertia* and D a parameter giving a measure of the *centrifugal distortion* of the molecule with increasing angular momentum. The absorption lines shown in Fig. 3 represent transitions from lower states with J in the range 11–39. The intensities of these lines are related to the populations of the rotational states involved; the population of a given state is directly proportional to the statistical weight $(2J + 1)$ of the level and to the Boltzmann factor $\exp(-E_J/kT)$.

The pure rotational spectrum of a nonlinear polyatomic molecule depends on certain symmetry properties of the molecule. A symmetric top molecule is one in which two of the principal moments of inertia are equal to each other but are different from the third; typical examples of such molecules are the pyramidal ammonia molecule NH_3 and the methyl halide molecules CH_3X. The rotational spectrum of a typical symmetric top has a pure rotation spectrum with a deceptively simple appearance; under modest resolution the spectrum appears to consist of a set of fairly evenly spaced "lines." Each of these "lines" is, however, actually a manifold of more closely spaced lines that are unresolved. The absorption spectrum of a symmetric top can be interpreted in terms of electric dipole transitions between rotational energy states

$$E_{JK} = hc[J(J + 1)B + K^2(A - B) + \cdots], \qquad (4.1.5)$$

where B and A are related to the principal moments of inertia; the observed transitions involve selection rules $\Delta J = +1$ and $\Delta K = 0$. The second rotational quantum number K is a measure of the projection of the total angular momentum on the symmetry axis of the molecule. The spectrum of NH_3 is complicated to some extent as a result of the *inversion* effects arising from the fact that there are two stable configurations of the molecule involving the position of the N atom above or below the plane of the H atoms in the molecule.

Asymmetric top molecules are those with three different principal moments of inertia; the water vapor molecule H_2O is perhaps the most studied and the best understood asymmetric top. Its rotational spectrum[46] appears as a set of lines with apparently random spacings and random intensities throughout the far infrared. Although expressions for the energy levels are not given in closed form involving definite quantum numbers, the observed spectrum has been interpreted as a result of the brilliant theoretical work of Benedict.[81]

We note that nonpolar molecules such as H_2, N_2, and CH_4 have no readily observable pure rotational spectra since no electric dipole changes are associated with their rotational motion and electric dipole transitions are forbidden.

4.1.5.2. Vibration–Rotation Spectra. Vibration–rotation spectra of simple molecules appear characteristically in the near- and intermediate-infrared regions and involve simultaneous changes in vibrational and rotational quantum numbers. The spacings of the vibrational energy states are large as compared with the close spacings of rotational energy states.

In Fig. 4 we show a portion of the fundamental vibration–rotation band of carbon monoxide as given by the chart–recorder trace obtained with Tubbs' Czerny–Turner spectrometer[76] when operating in a rapid-scan mode.

---

[81] W. H. Benedict in the introduction to "The Solar Spectrum from 2.8 to 23.7 Microns," by M. Migeotte, L. Neven, and J. Swensson. Inst. Astrophys. Univ. de Liège, 1956.

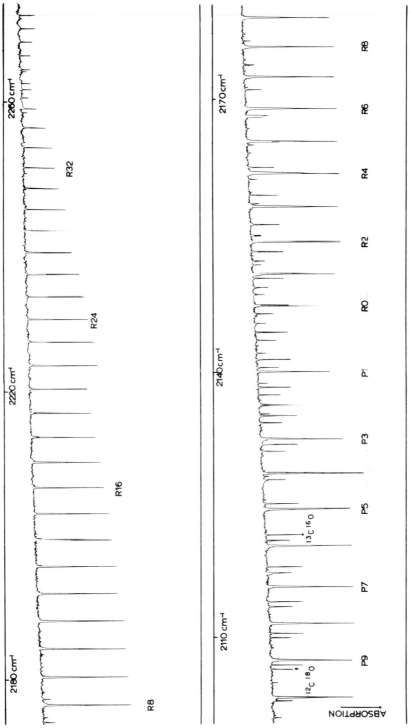

FIG. 4. The CO fundamental vibration–rotation band (courtesy of Dr. Lloyd D. Tubbs).

The major spectrum consists of a set of more or less equally spaced strong absorption lines constituting the R- and P-branches of the ^{12}CO band; the spacings of these lines is comparable with the line spacings in the pure rotation spectrum of CO. The R-branch is relatively "clean" but the P-branch is "contaminated" with weaker lines associated with the less abundant isotopic molecule ^{13}CO.

The major features of the vibration–rotation absorption bands of diatomic and linear molecules are readily interpreted in terms involving electric dipole transitions between levels given in gross approximation by the expression

$$E_{v_1 J} = (v + \tfrac{1}{2})hv_0 + hc[J(J + 1)B - J^2(J + 1)^2 D + \cdots] \qquad (4.1.6)$$

which involves a large vibrational term comparable with a harmonic oscillator of frequency v_0 and a much smaller rotational term. Transitions are subject to selection rules $\Delta v = 1, 2, 3, \ldots$ and to $\Delta J = +1$ for lines in the R-branch and $\Delta J = -1$ for lines in the P-branch. From the value of v_0 and the masses of the vibrating atoms the effective force constants involved in the vibration can be determined. The value of v_0 does not actually remain constant as suggested by the harmonic-oscillator term $(v + \tfrac{1}{2})hv_0$ in Eq. (4.1.6); i.e., the observed vibrational frequencies of the overtone bands with $\Delta v = 2, 3, 4, \ldots$ are not exact multiples of the frequency of the fundamental band with $\Delta v = 1$. Measurement of the frequencies of the overtones gives information regarding the potential energy of the atoms as a function of internuclear separation. From the spacings of the rotational lines in a vibration–rotation band, information regarding the moment of inertia of the molecule can be obtained; this information augments similar information derived from the pure rotation spectrum.

The vibration–rotation bands of linear polyatomic molecules have features that are similar in many ways to those of diatomic molecules but usually involve some complications in addition to those imposed by isotopic molecular species. Typical vibration–rotation spectra are shown in Figs. 5 and 6, which show the v_3 fundamental bands of N_2O and CO_2, respectively. Each of these molecules has three fundamental modes of oscillation; v_1 is a symmetric vibration in which the atoms move along the internuclear axis, v_2 a low-frequency bending mode in which the atoms move perpendicular

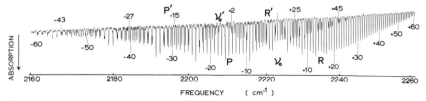

FIG. 5. The v_3 fundamental band of N_2O [from L. D. Tubbs and D. Williams, *J. Opt. Soc. Amer.* **63**, 859 (193)].

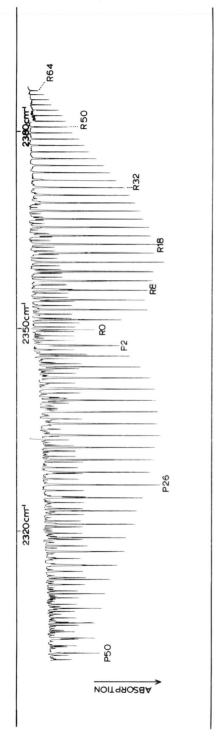

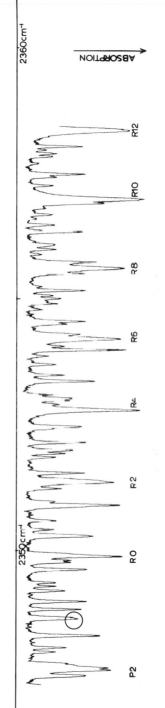

Fig. 6.   The ν_3 fundamental band of CO_2 (courtesy of Dr. Lloyd D. Tubbs).

to the internuclear axis, and v_3 an asymmetric vibration with atomic motion along the internuclear axis. At ambient laboratory temperatures appreciable numbers of molecules are in the first excited state of v_2 with quantum number $v_2 = 1$. Figure 5 shows the v_3 fundamental of N_2O, which has a major structure similar to that for the CO fundamental; the rotational lines in the N_2O spectrum are, however, more closely spaced because of the larger moment of inertia of the N_2O molecule. Superposed on the major band is a minor band; this minor band is called a "hot band" since it is associated with molecules initially in the first excited state $v_2 = 1$ of the v_2 mode and can be removed from the spectrum by cooling the sample.

The v_3 fundamental of CO_2 is shown in Fig. 6. Although the moments of inertia of the NNO and OCO molecules are nearly equal, the spacing of the CO_2 lines is nearly twice that of the N_2O lines; this results from the fact that half of the CO_2 lines are actually completely missing as a result of certain considerations[6] involving the statistical weights of the levels imposed by the symmetric positions of the ^{16}O nuclei, which have nuclear spin zero. Superposed on the v_3 fundamental are minor absorption bands, in some of which the rotational line spacing is smaller than their spacing in the $^{12}C^{16}O_2$ fundamental.

Although the major absorption bands shown in Figs. 4–6 are characterized by P- and R-branches, we note that in other vibration–rotation bands including the v_2 fundamentals of CO_2 and N_2O there are extremely closely spaced lines in the center of the band constituting a Q-branch characterized by a selection rule $\Delta J = 0$; in the absence of higher-order interactions, which we have not discussed, the entire Q-branch would appear as a single intense line.

The recorder tracing for the central portion of the central portion of the v_3 fundamental band of the methane molecule CH_4 as observed in fast scan is shown in the upper panel of Fig. 7. The band structure is obviously much more complex than the structure of the bands shown as typical of diatomic and linear polyatomic molecules. The individual lines in the P and R-branches obviously have a partially resolved structure; the Q-branch in the $3000–3020$ cm^{-1} consists of many closely spaced lines, some of which are resolved in the curve shown in the lower part of the figure, which represents a slower spectral scan of the Q-branch. There are numerous minor absorption lines in the band that have no obvious association with the major lines. The structure of the absorption bands of symmetric top molecules have many features similar to those in the spectrum of the spherical top molecule CH_4. Most of the complexities in such absorption bands can be interpreted in terms of higher-order interactions between rotation and vibration.[82]

[82] G. Amat, H. H. Nielsen, and G. Tarrago, "Higher-Order Rotation–Vibration Energies in Polyatomic Molecules." Dekker, New York, 1971.

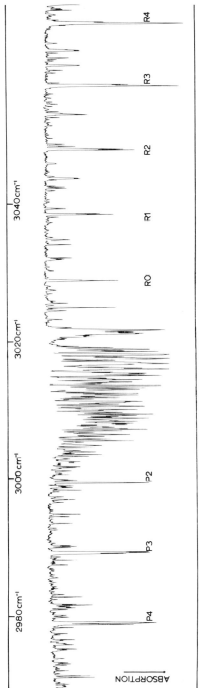

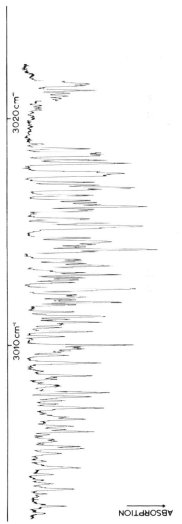

Fig. 7. The ν_3 fundamental band of CH_4 (courtesy of Dr. Lloyd D. Tubbs).

The modern theories of rotation–vibration interactions in polyatomic molecules have been recently reviewed and summarized by Blass and Nielsen,[83] who give extensive references to the original theoretical treatments. These authors point out the strong influence of the resolving power of the spectrograph on the appearance of the observed spectrum of polyatomic molecules. In the case of a symmetric top molecule Blass and Nielsen give computer-generated profiles for certain bands as based on the theoretical equations. We repeat one of their plots in Fig. 8, which shows in the lower

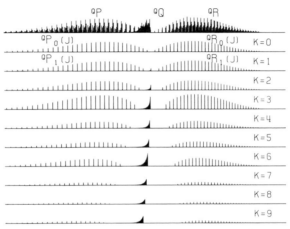

FIG. 8.   Computer-generated parallel band of a prolate symmetric top molecule such as CH_3F. The K subband structure is shown in the panels below the composite spectrum at the top (courtesy of Drs. W. E. Blass and A. H. Nielsen of the University of Tennessee).

panels the structure of the P-, Q-, and R-branches to be expected for various values of quantum number K; the top panel gives the composite spectrum to be expected by combining the spectra shown in the lower panels. In Fig. 9 we repeat Blass's computer-generated spectrum to be expected for the composite spectrum in the top panel of Fig. 8 when it is mapped by spectrometers characterized by the indicated spectral slit widths. With large spectral slit widths, the P-, Q-, and R-branches appear to exhibit relatively simple structures; their true complexities are revealed only when a spectrograph with sufficiently high resolving power is employed.

The vibration–rotation bands of asymmetric tops like the water vapor molecule have extremely complex structure when mapped at even modest resolution; their rotational structure has a seemingly random structure simi-

[83] W. E. Blass and A. H. Nielsen, in "Molecular Physics" (D. Williams, ed.), 2nd ed., pp. 126–202. Academic Press, New York, 1974.

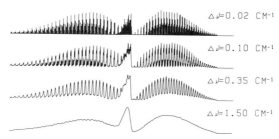

FIG. 9.   Computer-generated spectra showing the composite spectrum given in the top panel of Fig. 8 as it would be observed by spectrographs providing various spectral slit widths Δv.

lar to that of their pure rotational bands. The water vapor vibration–rotation bands have been successfully analyzed by Benedict.[81]

The use of modern spectrographs with extremely high resolving power has brought to light numerous previously unobserved phenomena. One such phenomenon was discovered by Rank[84] when he was able to observe the absorption of infrared radiation by nonpolar diatomic molecules such as H_2 as a result of electric quadrupole transitions. The absorption lines are so narrow that they were never observed until instruments with sufficiently narrow spectral slit widths were developed; selection rules $\Delta J = 0, \pm 2$ are involved in quadrupole transitions. With modern spectrographs it may soon be possible to study the absorption of molecular hydrogen H_2 in interstellar space and to identify homopolar molecules like N_2 in planetary atmospheres!

Although in any spectroscopic study it is always desirable to use instruments that directly provide sufficiently high resolving power to observe the phenomena of interest, it is possible to extend effective resolving power by the use of so-called *deconvolution techniques*, which employ computer techniques to remove the effects of instrumental slit function. The deconvolution process is essentially the reverse of the process employed by Blass and Nielsen in degrading the spectrum in Fig. 8 by imposing various instrumental slit widths to obtain the curves in Fig. 9. Considerable progress in developing computer programs for use in deconvolution has been made by various investigators including Jansson.[85]

4.1.6. Studies of Molecular Interactions

Thus far in our discussion of infrared spectra we have treated the molecules as essentially isolated closed systems, each of which is assumed to interact independently with the electromagnetic field. This is actually the situation only for molecular-beam studies in which the molecules are not

[84] U. Fink, T. A. Wiggins, and D. H. Rank, *J. Mol. Spectrosc.* **18**, 384 (1965).
[85] Peter A. Jansson, R. H. Hunt, and E. K. Plyler, *J. Opt. Soc. Amer.* **60**, 596 (1970).

subject to collisions. In this section we consider the effects of molecular collisions first in gases at atmospheric pressure and below; then we discuss more highly compressed gases. We note that the high-resolution studies discussed thus far provide precise information regarding the frequencies of absorption lines but usually provide only qualitative information regarding line intensities; studies of this type are sometimes referred to as *frequency spectroscopy* as opposed to *intensity spectroscopy*, some examples of which will be discussed in this and later sections.

4.1.6.1. Collisional Broadening of Spectral Lines. After appropriate corrections have been made for instrumental slit widths, the small finite widths of absorption lines in gas samples at extremely low pressure can, in the absence of wall collisions, be attributed to the limited radiative lifetimes of molecules in initial and final states and to the Doppler effect. At higher sample pressures, molecular collisions serve to broaden the lines. The true fractional spectral transmittance $T(v)$ of a gas sample is given by the relation: $T(v) = \exp[-\alpha(v)w]$, where $\alpha(v)$ is the Lambert absorption coefficient and w the absorber thickness or optical density of the sample which is proportional to pl, the product of the partial pressure of the absorbing gas and the path length in the sample. At pressures of 1 atm or less, the shape of a collisionally broadened line in the vicinity of its central frequency v_0 is described in good approximation by the Lorentz expression

$$\alpha(v) = (S/\pi)\{\gamma/[(v_0 - v)^2 + \gamma^2]\}, \qquad (4.1.7)$$

where the line strength $S = \int \alpha(v)\, dv$ depends on the quantum-mechanical probability for radiative transitions between the initial and final states and on the populations of these states; γ represents the half-width of the line between frequencies at which $\alpha(v) = \alpha(v_0)/2$. In terms of simple collision theory, $\gamma = 1/2\pi T_c = f_c/2\pi$, where T_c is the mean time between collisions and f_c the corresponding mean collision frequency. The mean collision frequency is given by the relation $f_c = N\bar{V}\sigma$, where N is the number of molecules per unit volume and thus proportional to the sample pressure, σ the collision cross section, and $\bar{V}$ the most probable relative speed of the molecules experiencing collision which can be expressed as $(2kT/\mu)^{1/2}$, where T is the sample temperature and μ the reduced mass of the collision pair.

In view of the small values of γ it is very difficult to map the contours of individual absorption lines even when deconvolution techniques are applied to data provided by high-resolution spectrographs; in general, indirect methods such as curve-of-growth measurements must be applied.[86−88] Much

[86] J. R. Nielsen, V. Thornton, and E. B. Dale, *Rev. Mod. Phys.* **16**, 307 (1944).

[87] S. S. Penner, "Quantitative Molecular Spectroscopy and Gas Emissivities." Addison-Wesley, Reading, Massachusetts, 1959.

[88] R. M. Goody, "Atmospheric Radiation." Oxford Univ. Press, London and New York, 1964.

early information was obtained by application of "band-model theory" to low-resolution measurements of the type reported by Howard and Burch and their colleagues.[89, 90] More recently studies of the broadening of individual rotational lines of CO have been made in numerous studies reviewed in a paper by Williams et al.[91]; the collision cross sections for the broadening of lines involving transitions between lower rotational states are larger than those between rotational states of higher energy; the corresponding cross sections are approximately the same for lines in different vibrational bands and in the pure rotation spectrum.† The observed values of the cross sections of CO for self-broadening and foreign gas broadening are plotted in Fig. 10, in which they are compared with values based on a simple semiempirical theory[91]; this theory involves two kinds of collisions:

(1)   adiabatic collisions in which the absorbing molecule remains in its initial state after collision and

(2)   diabatic collisions in which the rotational energy of the absorbing molecule changes during the collision.

The collisions of the second kind involve molecular energy and angular momentum transfer between translation and rotation. Certain predictions based on this simple theory have been verified in studies of collision broadening of CO at radically reduced temperatures.[92]

In the collisional self-broadening of lines in the spectrum of the highly polar molecule HCl[93] it is necessary to include a resonant dipole interaction involving the transfer of rotational energy from one molecule to another during collision; the observed broadening is compared in Fig. 11 with the simple empirical theory[91] with the inclusion of a resonant dipole term. A rigorously general detailed impact theory of line broadening has been developed by Anderson[94] and adapted to molecular absorption lines by Tsao and Curnutte[95]; unfortunately, this theory involves numerous molecular parameters, many of which have not been accurately determined.

[89] J. N. Howard, D. E. Burch, and D. Williams, *J. Opt. Soc. Amer.* **46**, 186, 237, 242, 334 (1956).

[90] D. E. Burch, E. B. Singleton, D. Williams, *Appl. Opt.* **1**, 359 (1962); D. E. Burch and D. Williams, *Appl. Opt.* **1**, 473 (1962); **1**, 587 (1962); D. E. Burch, D. Gryvnak, and D. Williams, *Appl. Opt.* **1**, 759 (1962); D. E. Burch, W. L. France, and D. Williams, *Appl. Opt.* **2**, 585 (1962); **3**, 55 (1964).

[91] D. Williams, D. C. Wenstrand, R. J. Brockman, and B. Curnutte, *Mol. Phys.* **20**, 769 (1971).

[92] L. D. Tubbs and D. Williams, *J. Opt. Soc. Amer.* **62**, 284, 423 (1972); **63**, 859 (1973).

[93] W. S. Benedict, R. Hērman, G. E. Moore, and S. Silverman, *Can. J. Phys.* **34**, 850 (1956).

[94] P. W. Anderson, *Phys. Rev.* **76**, 471, 647 (1949).

[95] C. J. Tsao and B. Curnutte, *J. Quant. Spectrosc. Radiat. Transfer* **2**, 41 (1962).

† See Chapter 4.2.

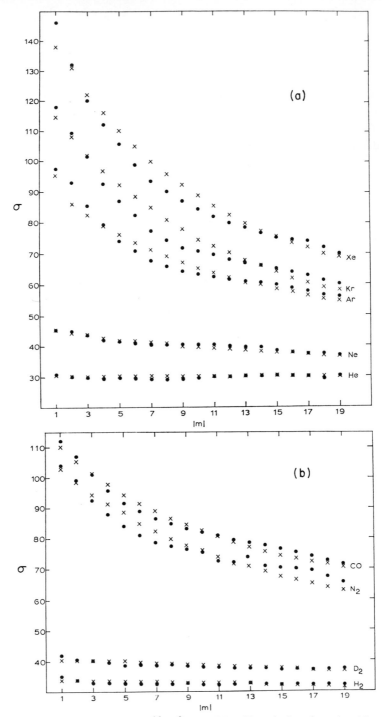

FIG. 10.   The cross sections in 10^{-16} cm² for collisional broadening of rotational lines in the vibrational bands of CO. The quantity $|m|$ plotted on the abscissa is equal to the rotational quantum number J in the lower state of the CO molecule involved in the collision; the cross sections σ between the absorbing molecule (CO) with the broadening molecule are indicated in the figure. Observed values are shown by filled circles; computed values by crosses [from D. Williams, D. C. Wenstrand, R. J. Brockman, and B. Curnutte, *Mol. Phys.* **20**, 769 (1971)].

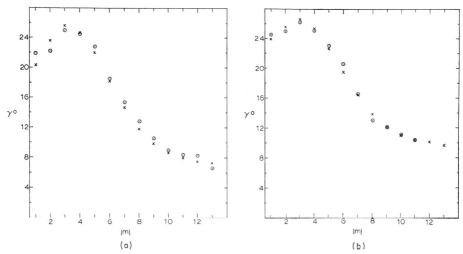

FIG. 11. The half-widths γ^0 at atmospheric pressure of self-broadened rotational lines in (a) the fundamental $v = 0 \rightarrow 1$ and (b) first overtone $v = 0 \rightarrow 2$ bands of HCl. The value $|m|$ is equal to the rotational quantum number J of the HCl molecule in its initial level; values of γ^0 are given in units of 0.01 cm^{-1}. Observed values are shown as circles; calculated values by crosses [from D. Williams, D. C. Wenstrand, R. J. Brockman, and B. Curnutte, *Mol. Phys.* **20**, 769 (1971)].

At pressures lower than 1 atm the frequency and duration of the collisions causing the broadening of spectral lines are such that an absorbing molecule spends only a small fraction of its life in actual collision; the influence of the collisions thus represents a relatively small time-averaged effect. In gases at much higher pressures or in liquids, molecular collisions produce larger time-averaged effects as a result of molecular distortions. As discovered independently by Welsh and his associates[96] and by Oxholm and Williams,[97] the nonpolar molecules N_2 and O_2 can absorb infrared radiation in the vicinity of their vibrational fundamentals as a result of *enforced dipole radiative processes*[5,6] involving molecules in collision; no rotational line structure in the N_2 and O_2 bands was observed. Later Welsh and his colleagues[98] in a series of brilliant experiments on H_2, HD, and D_2 elucidated the details of the phenomena involved; they showed that the spectral transmittance in a path length l by compressed samples of these gases at pressure p is given by the expression $T(v) = \exp[-\beta(v)p^2l]$ under conditions in which the absorber thickness w is proportional to pl and the binary collision frequency is also proportional to p. At still higher pressures, they observed

[96] F. H. Crawford, H. L. Welsh, and J. L. Locke, *Phys. Rev.* **75**, 1607 (1949); **76**, 580 (1949).
[97] M. L. Oxholm and D. Williams, *Phys. Rev.* **76**, 151 (1949).
[98] H. L. Welsh, *et. al.*, *Can. J. Phys.* **37**, 362, 1249 (1959).

the influence of triple collisions by observing additional absorption described by a smaller additional exponential proportional to $p^3 l$. Because of the wide spacings between the rotational energy levels of H_2, HD, and D_2, Welsh and his colleagues were able to observe greatly pressure-broadened rotational lines in the fundamental vibration band and also to observe lines in the pure rotational spectrum; they noted that the selection rules for rotational transitions were $\Delta J = 0, \pm 2$ rather than those involved in usual dipole radiative transitions. A satisfactory theoretical treatment of the observed phenomena has been given by Welsh's colleague van Kranendonk.[99] The pressure-induced pure rotational spectrum of N_2 in the far infrared has been observed[100] but no resolved rotational lines were observed at the high pressures involved.

A series of interesting studies of enforced absorption has been conducted by Ketelaar.[101] In some of these he employed mixtures of nonpolar gases such as N_2 and H_2 and was able to observe absorption bands due to simultaneous transitions in which the frequency of the absorbed radiation is the sum of the fundamental vibration frequencies of the separate H_2 and N_2 molecules.[102] In such cases, the collision pair $N_2–H_2$ must be considered as a single system interacting with the electromagnetic field!

Another type of induced absorption has been discovered by Welsh and his associates, who reported infrared absorption by a mixture of He and Ar at high pressure.[103] The absorption consists of a broad structureless band with a maximum near 200 cm^{-1} in the far infrared and is caused by the interaction of radiation with the dipole moment induced in He–Ar collision pairs. Since the absorbed electromagnetic energy goes directly into increased translational kinetic energies of the He and Ar atoms following collision, the phenomenon is called *translational absorption*. It is possible in all collisions between spherical molecules of different polarizabilities and is also observed when nonspherical molecules like H_2 collide, since the polarizability of such molecules depends on the orientations of the molecules during collision. The translational absorption band of H_2 overlaps the pressure-induced rotational band of H_2; the total region of absorption is called the *translation–rotation spectrum*. The theoretical interpretation of translational absorption has been presented by van Kranendonk and by Levine and Birnbaum.[104]

[99] J. van Kranendonk and R. Bird, *Physica* **17**, 953, 968 (1952); J. van Kranendonk and Z. J. Kiss, *Can. J. Phys.* **37**, 1187 (1959).

[100] H. A. Gebbie, N. W. B. Stone, and D. Williams, *Mol. Phys.* **6**, 215 (1963).

[101] J. P. Colpa and J. A. A. Ketelaar, *Mol. Phys.* **1**, 14, 343 (1958).

[102] F. N. Hooge and J. A. A. Ketelaar, *Physica* **23**, 423 (1957).

[103] D. R. Bosomworth and H. P. Gush, *Can. J. Phys.* **43**, 729, 751 (1965).

[104] H. B. Levine and G. Birnbaum, *Phys. Rev.* **154**, 86 (1967).

Perhaps the ultimate in molecular interactions are those in which colliding molecules undergo chemical reactions to form one or more molecules of a different compound. Many years ago it was demonstrated that infrared spectroscopy could be used to trace the course of slow reactions like the formation of an organic acid from its anhydride[105] or the hydrolysis of a salt.[106] Pimentel[107] has recently modified a conventional infrared spectrograph employing quantum detectors by mounting the Littrow mirror on a rapidly rotating shaft and has used it to follow chemical reactions with reaction times in the millisecond range and below. This development has not only been of importance in helping to extend the study of fast chemical reactions into the time domain but has been important in the identification of many short-lived intermediate reaction products.

4.1.7. Vibrational Spectra of Larger Polyatomic Molecules

The rotational lines in the spectra of many large polyatomic molecules are so closely spaced that they cannot be resolved with existing instruments; in the spectra of these molecules in the liquid state any rotational structure tends to be "washed out" as a result of collisions with neighboring molecules; free rotational motion of such molecules does not occur in the solid state. The normal vibration bands of such molecules are typically broad but devoid of observable fine structure.

Studies of the spectra of large polyatomic molecules have nonetheless given valuable information regarding their vibrational energies, the forces holding the atoms in their equilibrium positions, and the intramolecular motions of atomic groups experiencing hindered rotation. The general methods employed in the analysis of *vibrational spectroscopy* have been described in the book by Herzberg[6] and in more recent books.[108-111] A detailed review of the results obtained with many molecules of different types has been given in a recent book by Sverdlov et al.[112] along with a lengthy

[105] E. K. Plyler and E. S. Barr, *J. Chem. Phys.* **3**, 679 (1935); **4**, 90 (1936).

[106] D. Williams, *J. Amer. Chem. Soc.* **62**, 2442 (1940).

[107] G. C. Pimentel, *Eur. Congr. Mol. Spectrosc. 8th, Copenhagen*, pp. 563–569. Butterworth, London and Washington, D.C., 1965.

[108] E. B. Wilson, Jr., J. C. Decius, and P. C. Cross, "Molecular Vibrations." McGraw-Hill, New York, 1955.

[109] J. Charette, "Theory of Molecular Structure." Van Nostrand-Rheinhold, Princeton, New Jersey, 1966.

[110] N. B. Colthup, L. H. Daly, and S. E. Wiberly, "Introduction to Infrared and Raman Spectroscopy." Academic Press, New York, 1964.

[111] L. J. Bellamy, "The Infrared Spectra of Complex Molecules," 2nd ed. Methuen, London, 1958.

[112] L. M. Sverdlov, M. A. Kovner, and E. P. Krainov, "Vibrational Spectra of Polyatomic Molecules." Wiley, New York, 1974.

discussion of the treatment of molecular vibration problems by group-theoretical methods; a recent parallel treatment of the closely related subject of Raman spectroscopy has been given in a recent book by Dollish et al.[113] Since the spectra of large molecules has been treated exhaustively in various books, we shall not attempt further discussion.

4.1.8. Molecules in Condensed Phases

A molecule of a gas in its ground electronic state has translational, rotational, and vibrational degrees of freedom. In the liquid and solid states its translational and rotational motion is no longer free and its vibrational motion is modified to some extent by its neighbors. Although the fundamental vibrational frequencies are changed, changes are usually not so large as to make the vibrational spectrum so different as to make the spectrum of the molecule unrecognizable. The spectrum of water in vapor, liquid, and solid phases has been studied intensively and has characteristics that serve to illustrate some of the spectroscopic effects accompanying changes of phase.

The absorption spectrum of the water vapor molecule is characterized by the following fundamental vibration bands: v_1 at 3652 cm^{-1}, v_2 at 1595 cm^{-1}, and v_3 at 3756 cm^{-1}, each of which is characterized by the irregularly spaced rotational lines characteristic of an asymmetric top. Its pure rotation spectrum consists of many lines irregularly spaced throughout the far infrared with a group of extremely intense lines in the 80–300 cm^{-1} region. No absorption is involved in its free translational motion.

The absorption spectrum of water in the liquid state is shown in Fig. 12, which shows the spectral transmittance of a 10-μm layer of the liquid. The v_1 and v_3 fundamentals in the vapor spectrum are replaced by a broad region of general absorption centered at 3400 cm^{-1}; the v_2 fundamental, which is a rather broad band in the vapor spectrum is replaced by a narrow band near 1640 cm^{-1} in the liquid. The absorption spectrum in the far infrared is dominated by a broad absorption band with a maximum near 685 cm^{-1}; by comparison of the position of this band with the position of the corresponding band in D_2O it has been possible to attribute the strong band to a lattice motion involving the *hindered rotation* or *libration* of an H_2O molecule in the field of its neighbors. Similarly, the weaker sharp band appearing in the spectrum near 190 cm^{-1} can be attributed to the *hindered translation* of an H_2O molecule in the field of its neighbors. There are, of course, lattice modes of lower frequency but these do not contribute importantly to the infrared absorption. The motions of hindered rotation and translation of water molecules have been discussed briefly by Curnutte

[113] F. R. Dollish, W. G. Fateley, and F. F. Bentley, "Characteristic Raman Frequencies of Organic Compounds." Wiley, New York, 1974.

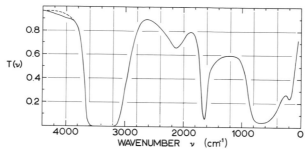

FIG. 12. The calculated values of fractional spectral transmission $T(\nu)$ of a 10-μm layer of water in the infrared. Effects due to reflection and absorption by cell windows and due to interference in the thin water layer have been eliminated [from C. W. Robertson, B. Curnutte, and D. Williams, *Mol. Phys.* **26**, 183 (1973)].

and Williams,[114] who give references to various original papers; more extensive discussions of the water spectrum can be found in Franck's recent treatise.[115] The weaker absorption bands at 2120 and 3950 cm^{-1} in Fig. 12 can be attributed to combinations involving the fundamentals and the lattice modes.[116]

The infrared spectrum[117] of ice at $-170°C$ has an appearance similar in many respects to that of water as shown in Fig. 12. The strong band corresponding to water vapor fundamentals ν_1 and ν_3 appears near 3251 cm^{-1}; the band corresponding to ν_2 appears near 1600 cm^{-1}; the major librational peak has its maximum near 830 cm^{-1} and the peak due to hindered translation appears at 220 cm^{-1}. There is evidence of band structure in the ice spectrum that is "smeared out" in the spectrum of liquid water. The general spectral features of water and ice have corresponding counterparts in the spectra of the liquid and solid phases of other compounds with small molecules. Considerations of hindered molecular motions in glasses have been given recently by Dean.[118]

Little quantitative information regarding the intensities of the strong water bands in the vicinity of 3400 cm^{-1} could be obtained from a spectral scan like that given in Fig. 12, since the 10-μm layer of water would be essentially opaque in that spectral region. In other spectral regions information regarding band intensities would be limited by the difficulties in preparing the required thin layers of uniform thickness, in making corrections

[114] B. Curnutte and D. Williams, "Structure of Water and Aqueous Solutions" (W. A. P. Luck, ed.), pp. 208–218. Verlag Chem./Phys. Verlag, Weinheim, 1974.

[115] F. Franck (ed.), "Water: A Comprehensive Treatise." Plenum Press, New York, 1972.

[116] C. W. Robertson, B. Curnutte, and D. Williams, *Mol. Phys.* **26**, 183 (1973).

[117] J. E. Bertie, H. J. Labbé, and E. Whalley, *J. Chem. Phys.* **50**, 4501 (1971).

[118] P. Dean, *Rev. Mod. Phys.* **44**, 127 (1972).

for reflection and absorption by the cell windows, in taking account of small amounts of stray radiation transmitted by a spectrograph, and in making corrections for interference effects in the thin absorbing layer. Robertson[119] has been able to handle these experimental difficulties by means of a specially designed cell providing a wedge-shaped absorbing film ranging in thickness from 20 μm at one edge to near-optical contact at the other; the cell is aligned interferometrically so that its thickness can be measured at various points along the wedge. The Lambert absorption coefficient at a given frequency can be determined by measuring spectral transmittance as the cell is translated laterally through the beam. Values of $\alpha(v)$ for water as determined with cells of this type are shown in Fig. 13; experimental uncertainties are discussed in the original papers.[116, 119]

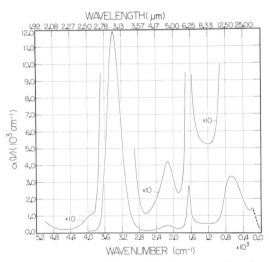

FIG. 13.   The Lambert absorption coefficient $\alpha(v)$ of water as a function of frequency v in reciprocal centimeters in the infrared [from H. D. Downing and D. Williams, *J. Geophys. Res.* **80**, 1656 (1975)].

For certain purposes, including Mie calculations of scattering, the complete statement of the optical properties of a medium involves not only a knowledge of its Lambert absorption coefficient but a knowledge of the real and imaginary parts of the medium's complex index of refraction $\hat{N}(v) = n(v) + ik(v)$. The imaginary part $k(v)$, called the *absorption index*, is related to the Lambert coefficient $\alpha(v)$ by the relation $k(v) = \lambda\alpha(v)/4\pi = \alpha(v)/4\pi v$ with v in reciprocal centimeters. Once this is known, the refractive index $n(v)$ can be determined from accurately measured values of spectral

119 C. W. Robertson and D. Williams, *J. Opt. Soc. Amer.* **61**, 1316 (1971).

reflectance $R(v)$ at near-normal incidence from the familiar relation

$$R(v) = \{[n(v) - 1]^2 + k^2(v)\}/\{[n(v) + 1]^2 + k^2(v)\}. \quad (4.1.8)$$

The values of $n(v)$ and $k(v)$ for water are given in the plots in Figs. 14 and 15, respectively.

We note that Kramers–Kronig techniques[120] can be employed to obtain both $n(v)$ and $k(v)$ from spectral reflectance $R(v)$ alone and to obtain $n(v)$ from measured values of $\alpha(v)$; the subtractive Kramers–Kronig relations[121] offer certain advantages when spectral measurements have been made over limited spectral ranges and abrupt truncation of the Kramers–Kronig integrals is necessary. The optical constants $n(v)$ and $k(v)$ plotted in Figs. 14 and 15 are based on reflection and absorption measurements taken over a wide spectral interval extending from the near ultraviolet to the microwave region; the uncertainty bars in these plots represent maximum differences between values determined directly from Eq. (4.1.8) and values based on Kramers–Kronig analysis alone.[122] The small uncertainties reveal the importance of Kramers–Kronig relations that can be used effectively since the development of high-speed digital computers.

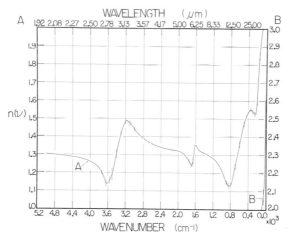

FIG. 14. The real part $n(v)$ of the complex index of refraction $\hat{N}(v) = n(v) + ik(v)$ of water in the infrared. The uncertainty bars represent the spread between $n(v)$ values obtained by direct measurements of reflection and absorption and $n(v)$ values obtained by Kramers–Kronig analyses based on the same data [from H. D. Downing and D. Williams, J. Geophys. Res. 80, 1656 (1975)].

[120] M. Gottlieb, J. Opt. Soc. Amer. 50, 343 (1960).
[121] R. K. Ahrenkiel, J. Opt. Soc. Amer. 61, 1651 (1971).
[122] H. D. Downing and D. Williams, J. Geophys. Res. 80, 1656 (1975).

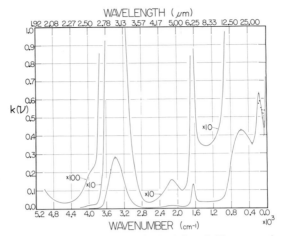

FIG. 15.   The absorption index $k(v)$ of water in the infrared. The uncertainty bars represent the spread between values of $k(v) = \lambda\alpha(v)/4\pi$ based on absorption measurements and values of $k(v)$ based on a Kramers–Kronig analysis based on reflection measurements [from H. D. Downing and D. Williams, *J. Geophys. Res.* **80**, 1656 (1975)].

A different approach to quantitative *intensity spectroscopy* involves attenuated total reflection (ATR) techniques and was first employed by Fahrenfort[123] and Harrick.[124] ATR measurements have been made by B. L. Crawford and his colleagues to determine $n(v)$ and $k(v)$ for many liquids. Various experimental arrangements used in ATR studies have been outlined by Stewart.[13]

4.1.9. Applications to Astronomy†

Although the value of infrared spectroscopy in laboratory studies of chemistry and molecular structure has been widely recognized for many years, its applications to astronomy and to space studies are less well known. With modern techniques it has become possible to obtain high-resolution spectra of planets from ground-based observatories and from space probes; these studies have added to our knowledge of absorbing gases in planetary atmospheres and have given information regarding the temperatures of planetary atmospheres from estimates of the line intensities in the observed bands. Other studies have provided information regarding the materials

[123] J. Fahrenfort, *Spectrochim. Acta* **17**, 698 (1961).
[124] N. J. Harrick, *J. Phys. Chem.* **64**, 1110 (1960); *Appl. Opt.* **4**, 1664 (1965); *Anal. Chem.* **36**, 188 (1964).

---

† See also Vol. 12A (Astrophysics), Parts 9–11.

present in planetary cloud covers. It should be mentioned that observations of spectra of CN molecules in space were used to give 3 K as the temperature of the interstellar medium more than 20 years before Dicke's microwave radiometer measurement gave an identical result, which has recently been acclaimed as evidence for the "big bang theory" of the origin of the universe!

4.2. Far-Infrared and Submillimeter-Wave Regions [†][*]

4.2.1. Introduction

The far infrared forms the transition between the infrared and the microwave regions. The point where the infrared becomes "far" is not well defined; it may be taken somewhere around 20-μm wavelength. The regular microwave region starts at about 2 mm, and so the far infrared extends roughly over the two decades from 20 μm to 2 mm in wavelength, or from 500 to 5 cm^{-1}, or 15 THz to 150 GHz in the wavenumber and frequency scales, respectively.

The submillimeter-wave region is in fact the same region, but viewed from the other end and with an emphasis on the region around 1 mm; say 2 to 0.2 mm, or in round numbers on the frequency scale 100–1000 GHz. This way of describing the region of present interest suggests that it has no characteristic features by itself. Although there are effects and methods encountered specifically in this part of the spectrum, it is indeed mainly defined by the inapplicability of standard methods used in the adjacent regions. It has been, consequently, rather inaccessible for quite a long time.

Extension of microwave techniques to wavelengths of a millimeter and less entails severe problems. Other transmission methods can be substituted for the characteristic waveguide systems, but alternatives for sources and detectors are not easily found. The usual electron tube oscillators and semiconductor devices can no longer be used at frequencies around 100 GHz. Submillimeter radiation can be generated with special types, but the efficiency decreases rapidly with increasing frequency and near-infrared frequencies cannot be attained. Much the same holds for point-contact diodes used as detectors.

At the short wavelength side, the infrared, the situation is similar. There are no purely optical problems in entering the far infrared; the main difficulty involves again the radiation source. In the near infrared, thermal sources approaching blackbody characteristics are conveniently applied to provide the continuous background spectrum required in absorption spectroscopy. Toward longer wavelengths, however, blackbody emission decreases rapidly

---

†See also Vol. 10 ("Physical Principles of Far-Infrared Radiation") by L. C. Robinson.

* Chapter 4.2 is by D. Oepts.

according to Planck's radiation law. In the region of present interest, the Rayleigh–Jeans approximation to the Planck expression applies and the radiation flux emitted per unit area in the wavenumber interval $(\sigma, \sigma + d\sigma)$ by a source of temperature T is

$$B(\sigma)\, d\sigma = 2\pi c k T \sigma^2\, d\sigma, \tag{4.2.1}$$

where c is the velocity of light in vacuum and k is Boltzmann's constant. Thus the emitted power in a wavenumber interval of fixed width decreases quadratically with decreasing wavenumber or increasing wavelength. In going from, say, 1 μm or 10,000 cm^{-1} in the near infrared to 100 μm or 100 cm^{-1} in the far infrared, the source power density drops by a factor of 10^4. We also see from (4.2.1) that the far-infrared emission from a source with a temperature around 1000 K, as the usual Globar or Nernst infrared sources, is not much stronger than the emission from the spectrometer itself, and its surroundings, at about 300 K. In the case of emission spectra the situation usually is still worse. With a hotter source, the far-infrared signal increases in proportionality with T, while the total radiance from a blackbody is, according to Stefan's law, proportional to T^4, and so the radiation at shorter wavelengths increases very much more. This adds to the problem of detecting a very weak signal the problem of rejecting an unwanted signal that can be orders of magnitude larger.

The source most commonly employed in the far infrared is the high pressure mercury arc with quartz envelope. At intermediate wavelengths most of the radiation comes from the redhot envelope, while for λ larger than about 100 μm, the arc itself contributes appreciably and increases the efficiency.[1]

With the extremely small signal powers available, detectors[2,3] with highest possible sensitivity are required. This is again a problem, due to the small energy of infrared photons. Thermal detectors, which sense the heat generated by the incident radiation, are frequently employed. An extensively used detector is the Golay cell,[4] which operates near the theoretical limit of sensitivity for a room-temperature detector. This performance is achieved by a built-in optical amplifier system. The incident radiation heats a tiny volume of gas and the resulting pressure increase deforms a thin reflecting membrane in the wall of the gas cell. This deformation modulates an auxiliary light beam which in turn produces a variation in the output of a photodetector. When equipped with a diamond window, the Golay cell has almost

[1] R. Papoular, *Infrared Phys.* **4**, 137–147 (1964).

[2] E. H. Putley, *J. Sci. Instrum.* **43**, 857–868 (1966).

[3] R. K. Willardson and A. C. Beer (eds.), "Semiconductors and Semimetals," Volume 5, Infrared Detectors. Academic Press, New York, 1970.

[4] M. J. E. Golay, *Rev. Sci. Instrum.* **18**, 357–362 (1947).

flat response from the visible down to millimeter wavelengths. A disadvantage is its rather long time constant, in the order of 30 msec. Much faster, but somewhat less sensitive, are the pyroelectric detectors,[5] in which a temperature change due to incident radiation induces a potential difference between the surfaces of a ferroelectric material. In bolometers, the temperature change of the detecting element is measured by the change in its electrical resistance. At cryogenic temperatures these form sensitive and reasonably fast detectors. Liquid helium-cooled carbon[6] and germanium[7] bolometers are frequently used.[8] Highly sensitive but complicated in operation is the superconducting bolometer described by Bloor et al.[9] Cooled GaAs and InSb provide very fast and sensitive photodetectors. Frequently employed types are known as Kinch–Rollin[10] and Putley[11] detectors.

A complication in far-infrared spectroscopy forms the strong and complex pure rotational absorption spectrum of atmospheric water vapor. This is usually eliminated by evacuating the whole instrument.

Given only very weak sources and relatively insensitive detectors, far-infrared spectroscopists are forced to make an as efficient use of the available radiation energy as possible. In this respect, new methods have offered improvements over conventional spectrometers. In practically all spectroscopic instruments different wavelengths are distinguished by dividing the radiation beam into two or more parts, introducing a phase difference between them, and observing the interference pattern obtained on recombining the beams.[12] When the phase differences introduced are not equal for different points of the source or entrance aperture, the interference pattern becomes blurred and the resolving power is decreased. It was shown by Jacquinot[13] that this effect is more serious when the interfering beams are obtained by a spatial division than it is when the division is made with a partially reflecting beamsplitter. In the former case the dependence of the phase differences on source position is of first order, while in the latter case the dependence can be of second or even higher order. Allowing the same loss in resolving power, a grating spectrometer therefore requires a slit with smaller area than the entrance aperture of a beamsplitter instrument and consequently can pass less radiation energy. This is known as the *Jacquinot*

[5] J. Cooper, *J. Sci. Instrum.* **39**, 462–472 (1962).

[6] W. S. Boyle and K. F. Rogers, Jr., *J. Opt. Soc. Amer.* **49**, 66–69 (1959).

[7] F. J. Low, *J. Opt. Soc. Amer.* **51**, 1300–1304 (1961).

[8] G. Dall'Oglio, B. Melchiorri, F. Melchiorri, and V. Natale, *Infrared Phys.* **14**, 347–350 (1974).

[9] D. Bloor, T. J. Dean, G. O. Jones, D. H. Martin, P. A. Mawer, and C. H. Perry, *Proc. Roy. Soc.* **A260**, 510–522 (1961); D. H. Martin and D. Bloor, *Cryogenics* **1**, 159 (1961).

[10] M. A. Kinch and B. V. Rollin, *Brit. J. Appl. Phys.* **14**, 672–676 (1963).

[11] E. H. Putley, *J. Phys. Chem. Solids* **22**, 241–247 (1961).

[12] H. A. Gebbie, *Appl. Opt.* **8**, 501–504 (1969).

[13] P. Jacquinot, *J. Opt. Soc. Amer.* **44**, 761–765 (1954).

advantage or *étendue*[†] *advantage* of beamsplitter or amplitude division instruments.

Scanning spectrometers as usually employed in the near infrared are inefficient for another reason. By measuring only a small range of wavelengths, one spectral element, at a time, most of the already feeble energy is wasted. The total measuring time is used more effectively by using a multiplex system. In such a system, the detector receives signals simultaneously from different spectral elements, these signals being coded in such a way that their separate contributions to the total signal can be reconstructed afterwards. Fellgett[14] has shown that the multiplexing of N spectral elements into one signal can give a gain of a factor of $N^{1/2}$ in the signal-to-noise ratio of the spectrum, provided that the noise is independent of the signal. He also indicated methods to achieve a suitable coding by means of an interferometer. In two-beam interferometric spectrometers based on this principle, the spectrum is coded as an interferogram and is reconstructed by a Fourier transformation. These instruments possess both the Jacquinot advantage and the *Fellgett* or *multiplex advantage*.

When the width $\delta\sigma$ of a spectral element is kept constant, the required resolving power $R = \sigma/\delta\sigma$ decreases at lower wavenumbers and frequently spectrometer slits can be set wide enough so that the detector area becomes the limiting factor in the étendue. In that case, the Jacquinot advantage of interferometers is of no importance. To a lesser extent this also holds for the Fellgett advantage. Additional advantages, however, such as wide spectral range, absence of stray light and filtering problems, still apply in the low wavenumber region. The Fourier transform spectroscopy method has therefore become widely used throughout the far-infrared and submillimeter-wave regions. A large part of the present chapter, Section 4.2.4, is devoted to this subject. Far-infrared spectroscopy with techniques adapted from microwave and near-infrared methods is considered in less detail, and less usual techniques are described briefly. Some applications are discussed in Section 4.2.6. Both techniques and applications of far infrared and submillimeter-wave spectroscopy have been treated in several textbooks to which we refer the reader for more details.[15-19]

[14] P. Fellgett, *J. Phys. Radium* **19**, 187–191 (1958).

[15] G. W. Chantry, "Submillimeter Spectroscopy." Academic Press, New York, 1971.

[16] M. F. Kimmit, "Far-Infrared Techniques." Pion, London, 1970.

[17] D. H. Martin (ed.), "Spectroscopic Techniques for Far Infrared, Submillimeter and Millimeter Waves." North-Holland Publ., Amsterdam, 1967.

[18] K. D. Möller and W. G. Rothschild, "Far-Infrared Spectroscopy." Wiley, New York, 1971.

[19] L. C. Robinson, "Physical Principles of Far-Infrared Radiation," Methods of Experimental Physics, Volume 10. Academic Press, New York, 1973.

[†] The étendue of a radiation beam is the product of cross-sectional area and solid angle, cf., W. H. Steel, *Appl. Opt.* **13**, 704 (1974).

4.2.2. Microwave and Laser Methods

Spectroscopy at submillimeter wavelengths employing methods similar to those used in the microwave region is possible, in principle, but still rare in practice, due to technical difficulties.[20] Klystron or magnetron oscillators are not available for wavelengths less than 2 mm. A microwave-type source that can still be used at wavelengths below 1 mm is the backward wave oscillator or carcinotron.[21, 22] Operation at $\lambda = 0.4$ mm has been reported.[23] A frequently used method to produce submillimeter waves consists in generating higher harmonics of the radiation from a standard millimeter-wave klystron. A nonlinear element, usually a point-contact diode, is placed in a waveguide containing the fundamental wave, and the resulting components of higher frequency are coupled into a smaller waveguide (see Fig. 1). This

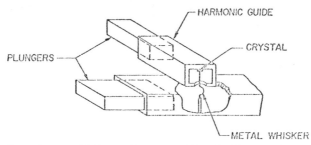

Fig. 1. Crossed-waveguide harmonic generator [from Robinson, *Advan. Electron. Electron Phys.* **26**, 171–215 (1969)].

technique is used very successfully by Gordy and co-workers[24] down to $\lambda = 0.37$ mm.[25] Instead of a point-contact diode, Froome[26] has used a small high-pressure arc discharge as the nonlinear element and he has been able to detect harmonics up to the 29th from a 35-GHz klystron.

Various types of solid-state microwave oscillators exist and some of these can be used in the millimeter region.[20] Extension to submillimeter wavelengths is difficult due to inherent frequency limitations and impedance

[20] J. Fox (ed.), "Submillimeter Waves" (*Proc. Symp. N.Y., 1970*). Polytechnic Press, Brooklyn, New York, 1971.

[21] T. Yéou, *Proc. Congr. Int. Tubes Hyperfréquences, 5th* (Microwave Tubes), p. 151–156. Dunod, Paris and Academic Press, New York, 1965.

[22] E. Glass and F. Cross, *Proc. Eur. Microwave Conf., Brussels, 1973*, C.12.4 (1973).

[23] Y. Couder and P. Goy, *in* "Submillimeter Waves" (*Proc. Symp. N.Y., 1970*) (J. Fox, ed.), pp. 417–430. Polytechnic Press, Brooklyn, New York, 1971.

[24] W. C. King and W. Gordy, *Phys. Rev.* **93**, 407–412 (1954).

[25] P. Helminger, F. C. De Lucia, and W. Gordy, *Phys. Rev. Lett.* **25**, 1397–1399 (1970).

[26] K. D. Froome, *Nature (London)* **193**, 1169–1170 (1962); *Proc. Int. Conf. Quantum Electron., 3rd, Paris, 1963*, 1527–1539 (1964).

matching problems. A Josephson junction oscillator has been shown to produce radiation in the 1–0.3-mm region.[27] Use of higher harmonics from Gunn diodes may present another possibility.[28]

When a tunable coherent radiation source and sufficiently fast detectors are available, heterodyne detection becomes possible, which can give considerable increase in signal-to-noise ratio over conventional power detection.

Microwave methods have the advantages of high resolution and frequency accuracy, high sensitivity, and no need for a monochromator or similar instrument. These advantages can be obtained in the submillimeter-wave region when suitable sources and detectors are available, but the shorter wavelengths and higher frequencies present some problems. Microwave waveguide systems cannot be scaled down to submillimeter wavelengths as mechanical tolerances and wall losses become prohibitive. Alternative transmission systems exist,[29] of which quasi-optical systems, light pipes, and oversized waveguides are commonly used.

A given relative frequency accuracy corresponds at higher frequencies to a larger absolute frequency uncertainty. In particular, when high harmonics are used, the fundamental frequency should be extremely stable to obtain a final accuracy comparable to that usual in microwave work. An advantage of the increased frequency variation of high harmonics is the increased tunability; the region between successive harmonics can be covered with a relatively small variation of the fundamental frequency, so that a wide frequency range becomes available.

When direct frequency measurement is not possible, optical methods have to be used to determine the wavelength of the radiation. Michelson and Fabry–Perot interferometers as well as grating spectrometers can be applied for this purpose. A Fabry–Perot interferometer can also be employed as a resonant cavity to contain the sample of material under study.[30] An example of a semiconfocal resonator used for determining the wavelength of the source and for measuring the sample absorption is shown in Fig. 2. The bolometer radiation detector is placed in the cavity as well, and the whole assembly is mounted in a liquid helium bath. Tuning of the cavity is achieved by translating the upper mirror and entrance cone. This part is interchangeable to accommodate for a broad wavelength range.

[27] R. K. Elsley and A. J. Sievers, *IEEE Trans. Microwave Theory Tech.* **MTT-22**, 1117 (Abstr.) (1974).

[28] L. F. Eastman, *in* "Submillimeter Waves" (*Proc. Symp. N.Y., 1970*) (J. Fox, ed.), pp. 41–51. Polytechnic Press, Brooklyn, New York, 1971.

[29] D. J. Kroon and J. M. van Nieuwland, *in* "Spectroscopic Techniques for Far Infrared Submillimeter and Millimeter Waves" (D. H. Martin, ed.), Chapter 7. North-Holland Publ., Amsterdam, 1967.

[30] E. P. Valkenburg and V. E. Derr, *Proc. IEEE* **54**, 493–498 (1966).

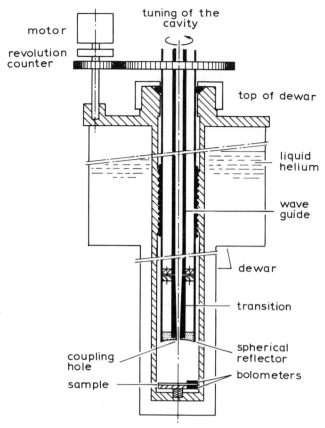

FIG. 2. Tunable Fabry–Perot cavity with transition cone adaptable for wavelengths from 9 to 0.5 mm [from Y. Couder and P. Goy, *in* "Submillimeter Waves" (*Proc. Symp. N.Y., 1970*) (J. Fox, ed.), pp. 417–430. Polytechnic Press, Brooklyn, New York, 1971].

A molecular-beam maser oscillator and amplifier operating at frequencies up to 317 GHz ($\lambda \approx 1$ mm) has been constructed by De Lucia.[31] The absence of Doppler and collision broadening makes highly accurate frequency measurements possible and has resulted in extremely accurate values of molecular-rotational constants.

With all submillimeter sources mentioned above, the radiated power rapidly decreases with increasing frequency. Still, even when the power seems negligible in comparison with usual microwave levels, one may have a powerful source as compared with a broad-band far-infrared source such as the mercury arc. The power emitted, for instance, by 1 cm² of the arc, into

[31] F. C. De Lucia and W. Gordy, *in* "Submillimeter Waves" (*Proc. Symp. N.Y., 1970*) (J. Fox, ed.), pp. 99–114. Polytechnic Press, Brooklyn, New York, 1971.

an $f/5$ cone, in a bandwidth of 3 GHz centered at 300 GHz, i.e., in a spectral element of 0.1 cm^{-1} width at 10 cm^{-1}, lies in the order of 10^{-9} W.

Much more powerful sources form submillimeter or far-infrared lasers. Several molecular gas lasers are known, providing emission at a large number of wavelengths throughout the region.[32] To date, they have not found very widespread use in spectroscopy due to their limited tunability. Investigations of lines almost coinciding with laser lines and of the laser molecules themselves can be made. Spectra that can be swept across the source frequency, for instance by applying an external magnetic field, and also isolated points of continuous spectra can be observed with fixed frequency sources.

More promising, however, for spectroscopic applications is the use of lasers for the indirect generation of tunable far-infrared radiation through some nonlinear process. By mixing two ruby laser beams in a LiNbO$_3$ crystal, tunable difference-frequency radiation has been obtained in the range 20–38 cm^{-1} and tuning from 0 to 50 cm^{-1} is expected to be possible.[33] Mixing of two CO$_2$ lasers has resulted in an almost continuous range of frequencies between 5 and 140 cm^{-1}.[34] Radiation tunable from 50 to 150 cm^{-1} with peak powers of 3 W has been obtained by utilizing Raman scattering and parametric generation in LiNbO$_3$ irradiated by a Q-switched ruby laser.[35] Far-infrared radiation has also been generated by mixing dye laser beams.[36] These fields of research are in a state of rapid development[37] and may drastically change the present situation in far-infrared spectroscopy.[38]

4.2.3. Far-Infrared Grating Spectroscopy

The dispersion of radiation with different wavelengths into different directions with the aid of a diffraction grating is a celebrated procedure in the optical regions and can also be used in the far infrared.[39] Tolerances on shape, surface finish, etc. are proportional to the wavelength and present no problem in this region. Plane reflection gratings are universally applied,

[32] L. C. Robinson, "Physical Principles of Far-Infrared Radiation," Methods of Experimental Physics, Vol. 10, Section 2.5. Academic Press, New York, 1973.

[33] D. W. Faries, P. L. Richards, Y. R. Shen, and K. H. Yang, *Phys. Rev.* **3A**, 2148–2150 (1971).

[34] R. L. Aggarwal, B. Lax, H. R. Fetterman, P. E. Tannenwald, and B. J. Clifton, *J. Appl. Phys.* **45**, 3972–3974 (1974).

[35] B. C. Johnson, H. E. Puthoff, J. SooHoo, and S. S. Sussmann, *Appl. Phys. Lett.* **18**, 181–183 (1971).

[36] K. H. Yang, J. R. Morris, P. L. Richards, and Y. R. Shen, *Appl. Phys. Lett.* **23**, 669–671 (1973).

[37] *Int. Conf. Submillimeter Waves their Appl., 1st., IEEE Trans. Microwave Theory Tech.* **MTT-22**, 981–1120 (1974).

[38] H. Yoshinaga, *Progr. Opt.* **11**, Chapter 2 (1973).

[39] F. Kneubühl, *Appl. Opt.* **8**, 505–519 (1969).

frequently with large dimensions (e.g., 28×35 cm^2 [40]) to maximize the amount of radiation reaching the detector. The theoretically possible resolving power is in general not attained due to the necessity of using wide slits to obtain sufficient signal. The Czerny–Turner or the Ebert–Fastie monochromator arrangements are usually employed. The latter mount in particular allows long slits to be used without prohibitive aberrations.[41] As the detector element usually has circular shape, an additional "image slicer" optical system is needed to couple a long exit slit to the detector.[42] A grating monochromator design with particular emphasis on the matching of the apertures of its optical components has been given by Sesnic.[43] Light pipes are frequently used to transmit radiation from the exit slit into a dewar containing a low-temperature detector. In order to discriminate the monochromated source radiation from the thermal radiation of the monochromator itself, a mechanical chopper modulating the source radiation and lock-in detection are commonly used. Frequently, the chopper blades are made of a material transparent for visible and near-infrared radiation so that in principle only the far infrared is modulated and detected.

The distribution of the radiation diffracted by a grating over the different orders is determined by the shape of the grooves. Gratings used in the far infrared are always of the echelette type (Fig. 3). The diffracted flux has a

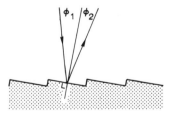

FIG. 3.   Echelette grating.

maximum in the "blaze" direction $\phi_1 = \phi_2$ corresponding to specular reflection from each groove facet. The grating constant and blaze angle are commonly chosen so as to concentrate the desired radiation in the first order. The blaze condition $\phi_1 = \phi_2$ and the grating equation (Section 3.4.3.3) with $n = 1$, are simultaneously satisfied only for a single wavelength λ_b. Away from λ_b the grating efficiency diminishes, but reasonable efficiency is obtained from about $\lambda = \frac{2}{3}\lambda_b$ to $\frac{3}{2}\lambda_b$ and a single grating can be used for one wavelength or wavenumber octave. The effect of radiation of the desired wave-

[40] P. L. Richards, *J. Opt. Soc. Amer.* **54**, 1474–1484 (1964).
[41] W. G. Fastie, *J. Opt. Soc. Amer.* **42**, 647–651 (1952).
[42] W. Benesch and J. Strong, *J. Opt. Soc. Amer.* **41**, 252–254 (1951).
[43] S. S. Sesnic, *IEEE Trans. Nucl. Sci.* **NS-18**, 365–375 (1971).

length being diffracted into unwanted orders is sufficiently controlled in this way, but the reverse effect: unwanted wavelengths being diffracted in higher orders into the same direction as the required wavelength, is by no means eliminated. For $\lambda = \lambda_b/2$, $\lambda_b/3$, etc., both the blaze condition and the grating equation are again satisfied. As the source intensity increases strongly toward shorter wavelengths, the rejection of higher orders forms a serious problem in far-infrared grating spectroscopy. A combination of filters is usually employed to achieve adequate suppression of higher orders without too much attenuation in the first order. Figure 4 shows an example of a far-infrared grating spectrometer with different filter arrangements. Various

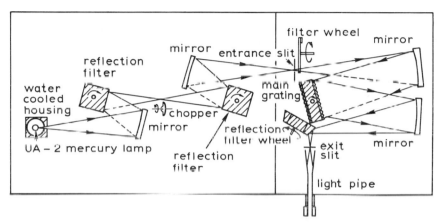

FIG. 4. Far-infrared grating spectrometer for the range 20–1600 μm with various filter arrangements to eliminate higher grating orders [from I. T. Silvera and G. Birnbaum, *Appl. Opt.* **9**, 617–625 (1970)].

types of filters are in use and some will now be discussed briefly. An extensive bibliography on filters is included in the work of Kneubühl.[39]

(a) *Simple absorption filters.* Most of the visible and near-infrared radiation can be removed by absorption and scattering in carbon black, commonly applied in the form of black polyethylene foil. In low-temperature detector systems the unwanted thermal radiation is blocked by cold windows of quartz, sapphire, or alkali halides. The sharpening of absorption profiles occurring at low temperatures can aid in effective filtering.

(b) *Reststrahlen filters.* The high reflection coefficients of ionic crystals in the far infrared have already been used by the earliest workers in this region, and alkali halide reststrahlen reflection filters are still widely used in grating spectrometers.

(c) *Yoshinaga filters.* Polyethylene sheets containing dispersed absorbing materials can be used as long-wavelength transmission filters. By a suitable choice of filling materials, a wide range of filters with different absorption regions can be constructed.[44]

(d) *Filter gratings.* A grating reflects wavelengths longer than the grating constant into the zero order, while shorter wavelengths are diffracted away mainly in the blaze direction. When using gratings as reflection filters in this way, polarization effects can form a problem. Transmission filter gratings can be made by molding polyethylene in an echelette shape. Crossed grooves on either side of the sheet are useful.

(e) *Mesh filters.* Metal meshes transmit and scatter short-wavelength radiation, while wavelengths larger than the mesh spacing are effectively reflected. The filter profile is steeper than with other filters and any cuton wavelength can be obtained by choosing the appropriate mesh spacing. Still more valuable are the "complementary" or "capacitive" meshes developed by Ulrich.[45] These can be used in transmission and combined to form interference filters.[46]

In the visible and near-infrared regions, double-beam spectrophotometers are extensively used in which the sample absorptance is measured by an optical null method. This method is difficult to apply in the far infrared, due to thermal radiation from sample and reference cells, etc. Double-beam systems exist, however, in which sample and reference signals are measured independently with a four-phase chopping system.[47, 48] A commercial far-infrared grating instrument is available from the Perkin–Elmer Corp. Other manufacturers have, at least for $\sigma \lesssim 200 \text{ cm}^{-1}$, switched to interferometers.

4.2.4. Fourier Transform Spectroscopy

The multiplex method employing Fourier coding by means of a two-beam interferometer will be discussed in some detail in this section. The following notations and definitions related to Fourier transforms will be used:

$$\mathscr{F}f(x) = \int_{-\infty}^{\infty} f(x)e^{2\pi ixy}\, dx$$
$$= \int_{-\infty}^{\infty} f(x)\cos 2\pi yx\, dx + i \int_{-\infty}^{\infty} f(x)\sin 2\pi yx\, dx$$
$$= \mathscr{F}_c f + i\mathscr{F}_s f,$$
$$\mathscr{F}^{-1}F(y) = \int_{-\infty}^{\infty} F(y)e^{-2\pi ixy}\, dy.$$

[44] N. G. Baldecchi and B. Melchiorri, *Infrared Phys.* **13**, 189–211 (1973).

[45] R. Ulrich, *Infrared Phys.* **7**, 37–56 (1967).

[46] G. D. Holah and J. P. Auton, *Infrared Phys.* **14**, 217–229 (1974).

[47] M. J. French, D. E. H. Jones, and J. L. Wood, *J. Phys. E* (*Sci. Instrum.*) **2**, 664–672 (1969).

[48] T. M. Hard and R. C. Lord, *Appl. Opt.* **7**, 589–598 (1968).

For real even functions, the sine transform is zero, and $\mathscr{F} = \mathscr{F}_c$. Frequent use will be made of the convolution theorem

$$\mathscr{F}(fg) = \mathscr{F}f \otimes \mathscr{F}g \qquad \text{and} \qquad \mathscr{F}(f \otimes g) = \mathscr{F}f\mathscr{F}g,$$

where $\otimes$ denotes the convolution product

$$f \otimes g(x) = \int_{-\infty}^{\infty} f(x')g(x - x')\, dx' = \int_{-\infty}^{\infty} f(x - x')g(x')\, dx'.$$

The essentials of Fourier transform spectroscopy have been treated in almost every detail by Connes.[49] Comprehensive accounts have also been given by Gebbie and Twiss,[50] Mertz,[51] Vanasse and Sakai,[52] Richards,[53] and others. The book by Bell[54] presents a detailed treatment of theory and techniques and includes an extensive bibliography. Introductory articles as well as specialized contributions are found in the proceedings of the 1970 Aspen conference.[55]

4.2.4.1. **Principle of the Method.** When two beams of radiation obtained from the same source are recombined, the radiant flux in the resulting beam is $I_t = I_1 + I_2 + I_{\text{int}}$ when the two beams have a flux I_1 and I_2, respectively, where I_{int} is due to interference between the beams. With a monochromatic source and parallel beams, as obtained for example in a Michelson interferometer (see Fig. 5), the interference term is $I_{\text{int}} = 2(I_1 I_2)^{1/2} \cos \delta$, where δ is the phase difference between the beams.[56] When this phase difference is caused by a path difference x in vacuum, one has $\delta = 2\pi\sigma_0 x$, where σ_0 is the wavenumber of the radiation. When $I_1 = I_2$, the flux I_t as a function of x is

$$I_t(x) = 2I_1(1 + \cos 2\pi\sigma_0 x). \tag{4.2.2}$$

This is the interference function or interferogram. For a nonmonochromatic source, each spectral element contributes to $I_t(x)$ with its own x-dependence according to $\cos 2\pi\sigma x$. The result is

$$I_t(x) = 2 \int_0^{\infty} B(\sigma)(1 + \cos 2\pi\sigma x)\, d\sigma, \tag{4.2.3}$$

where $B(\sigma)$ is the spectral density of the source radiation and $B(\sigma)\, d\sigma$ the

[49] J. Connes, *Rev. Opt.* **40**, 45–79, 116–140, 171–190, 231–265 (1961).

[50] H. A. Gebbie and R. Q. Twiss, *Rep. Progr. Phys.* **29**, 729–755 (1966).

[51] L. Mertz, "Transformations in Optics." Wiley, New York, 1965.

[52] G. A. Vanasse and H. Sakai, *Progr. Opt.* **6**, Chapter 7 (1967).

[53] P. L. Richards, *in* "Spectroscopic Techniques for Far Infrared, Submillimeter and Millimeter Waves" (D. H. Martin, ed.), Chapter 2. North-Holland Publ., Amsterdam, 1967.

[54] R. J. Bell, "Introductory Fourier Transform Spectroscopy." Academic Press, New York, 1972.

[55] G. A. Vanasse, A. T. Stair, Jr., and D. J. Baker (eds.), *Aspen Int. Conf. Fourier Spectrosc., 1970*, AFCRL–71–0019, Spec. Rep. 114 (1971).

[56] M. Born and E. Wolf, "Principles of Optics," Sect. 7.2. Pergamon, Oxford, 1959.

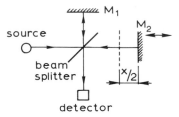

FIG. 5. Principle of Michelson interferometer.

flux in the wavenumber element $(\sigma, \sigma + d\sigma)$, corresponding to I_1 in (4.2.2) (see also Part 2, Section 2.2.1.7). Equation (4.2.3) is rewritten as

$$I_t(x) = 2 \int_0^\infty B(\sigma)\, d\sigma + 2 \int_0^\infty B(\sigma) \cos 2\pi\sigma x\, d\sigma = I_c + I(x). \quad (4.2.4)$$

In (4.2.3) only positive frequencies were assumed: $B(\sigma) \equiv 0$ for $\sigma < 0$. A more convenient relation between the spectrum and the interferogram is obtained by defining the even spectrum

$$B_e(\sigma) = \tfrac{1}{2}[B(\sigma) + B(-\sigma)]. \quad (4.2.5)$$

The x-dependent part of (4.2.4) is

$$I(x) = 2 \int_0^\infty B(\sigma) \cos 2\pi\sigma x\, d\sigma = \int_{-\infty}^\infty B_e(\sigma) \cos 2\pi\sigma x\, d\sigma. \quad (4.2.6)$$

This is the basic relation in Fourier spectroscopy: the varying part of the interferogram is the Fourier (cosine) transform of the even spectrum. The even spectrum is obtained from the interferogram by reversing the transform:

$$B_e(\sigma) = \int_{-\infty}^\infty I(x)e^{-2\pi i\sigma x}\, dx = \int_{-\infty}^\infty I(x) \cos 2\pi\sigma x\, dx. \quad (4.2.7)$$

Here, use is made of the fact that $I(x)$ is even; see, however, Section 4.2.4.5.

The normal spectrum $B(\sigma)$ is twice the even spectrum for positive frequencies. The spectrum will show a strong line at $\sigma = 0$ when it is calculated from the total interferogram $I_t(x)$ [Eq. (4.2.4)] instead of from $I(x)$. Although in principle this does not influence the remainder of the spectrum, it is desirable for practical reasons to remove the constant part I_c when the total interferogram has been measured. From (4.2.4) we see that $I_c = I(0) = \tfrac{1}{2}I_t(0)$, but this is true only when the two beams have equal flux as was assumed here. When $I_1 \neq I_2$, one has $I_c > \tfrac{1}{2}I_t(0)$.

In the following, by "interferogram" is usually meant the varying part $I(x)$ only. Examples of interferograms and associated spectra are shown in Fig. 6. Only one half of a symmetric interferogram is usually measured.

4.2.4.2. Resolving Power, Instrumental Function, Apodization. Just as in other spectrometers, the maximum path difference between interfering

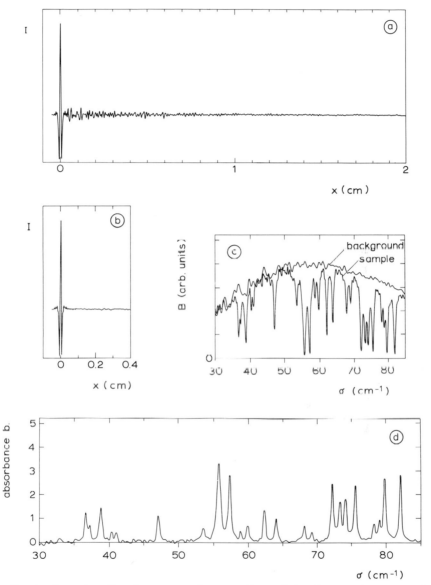

FIG. 6. Examples of interferograms and spectra: (a) sample interferogram (water vapor in 10 atm helium); (b) part of background interferogram (evacuated cell); (c) part of the spectra corresponding to (a) and (b); (d) absorbance spectrum $b(\sigma) = -\ln\{B(\sigma)/B_0(\sigma)\}$ obtained from the spectra in (c) [from D. Oepts, Thesis, Univ. of Amsterdam, 1972].

rays determines the resolving power in Fourier spectroscopy. When $I(x)$ is known for $-L \leqslant x \leqslant L$, one may compute, instead of the true $B(\sigma)$,

$$B'(\sigma) = \int_{-L}^{L} I(x) \cos 2\pi\sigma x \, dx.$$

Following Connes[49] we write this in the form

$$B'(\sigma) = \int_{-\infty}^{\infty} D_L(x) I(x) \cos 2\pi\sigma x \, dx, \qquad (4.2.8)$$

where D_L is the rectangular function of width $2L$

$$D_L = \begin{cases} 1, & |x| < L \\ \frac{1}{2}, & x = \pm L \\ 0, & |x| > L. \end{cases} \qquad (4.2.9)$$

Using the convolution theorem and the fact that D_L and $I(x)$ are even functions one finds

$$B'(\sigma) = B_e \otimes f_0(\sigma),$$

where $B_e(\sigma)$ is the (unknown) even spectrum and $f_0(\sigma)$ is the Fourier transform of $D_L(x)$:

$$f_0(\sigma) = 2L(\sin 2\pi\sigma L)/2\pi\sigma L = 2L \operatorname{sinc}(2\sigma L).$$

This $f_0(\sigma)$ is the instrumental function (or scanning function, slit function, spectral window, etc). An infinitely narrow spectral line at σ_0: $B(\sigma) = B\delta(\sigma - \sigma_0)$ will be reproduced in the calculated spectrum as $\frac{1}{2}Bf_0(\sigma - \sigma_0) + \frac{1}{2}Bf_0(\sigma + \sigma_0)$. The line centered at $-\sigma_0$ shows up due to the appearance of the even spectrum.

The half-width of the function f_0 (see Fig. 7) is approximately $1/2L$ and the limit of resolution or the width of a spectral element may thus be taken to be $\delta\sigma = 1/2L$.

4.2.4.2.1. APODIZATION. Frequently, $f_0(\sigma)$ is not considered a convenient instrumental function because of its large and slowly decaying sidelobes. It can be modified by the process of apodization. This is usually achieved by multiplying the interferogram by an apodizing function $A(x)$ varying smoothly from $A(0) = 1$ to $A(\pm L) = 0$. The spectrum resulting from the new interferogram is found by substituting $A(x)$ for $D_L(x)$ in (4.2.8), and so the new instrumental function $f_1(\sigma)$ is the Fourier transform of $A(x)$ and can be given a convenient shape by suitable choice of $A(x)$.[57] As $f_1(\sigma)$ is always broader than $f_0(\sigma)$, one pays for this improvement with resolution.

The instrumental function is observed only with narrow spectral lines; for lines with a width $\gtrsim 1/L$ the sidelobes of f_0 are smoothed out in the

[57] A. S. Filler, *J. Opt. Soc. Amer.* **54**, 762–767 (1964).

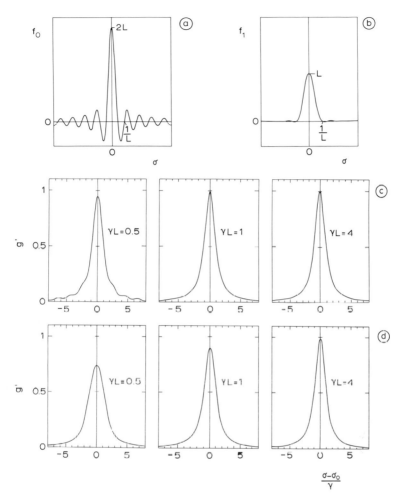

FIG. 7.   (a) Instrumental function $f_0(\sigma) = 2L\,\mathrm{sinc}(2\sigma L)$; (b) instrumental function obtained with apodization according to $A(x) = D_L(x)\cos^2(\pi x/2L)$; (c) convolution of a Lorentz line shape (semihalf-width γ) with the instrumental function $f_0(\sigma)$ for three values of L; (d) the same as (c), but with the instrumental function $f_1(\sigma)$ of (b).

convolution with the line shape (see Fig. 7). The Fourier transform of a line with a width γ—defined in some reasonable way—has a width of order $1/\gamma$. This Fourier transform is the contribution of the line under consideration to the interferogram, according to (4.2.6). Thus the part of $I(x)$ with $|x| < 1/\gamma$ contains practically all the information pertaining to this line and truncating the interferogram at a length $L > 1/\gamma$ does not much influence the observed line shape. On the other hand, when $I(x)$ is multiplied by an apodizing function that is different from unity in the region $|x| < 1/\gamma$, the Fourier transform of the line shape is altered, resulting in a distorted line. This has been verified quantitatively for Lorentz lines and associated absorption line shapes.[58, 59]

4.2.4.2.2. INCREASED RESOLUTION. One might wonder whether it is not possible to apply some inverse apodization to increase the resolution beyond the value set by L and f_0. Indeed, as the interferogram is an analytic function, it can in theory be extrapolated indefinitely. The problem is analogous to that of image restoration or superresolution in optical imaging,[60, 61] and the same methods apply.[62] Useful applications in Fourier transform spectroscopy will be extremely rare, however, as a significant gain in resolution is obtained only at the expense of an enormous increase in the noise.[63]

4.2.4.2.3. THE USE OF SIGNATURES. The periodic structure of a rotational spectrum causes a pattern of repeated features in the interferogram known as *signatures*[64, 65] (see Fig. 8). A similar structure appears with channel spectra caused by interference in plane parallel samples.[66] All information relevant to the rotational or channel structure is contained in the signatures. Important data can be obtained from studying the signatures without needing a Fourier transformation.[64, 65] For a detailed knowledge of the spectrum the Fourier transformation remains necessary, but it is possible to take advantage of the particular shape of the interferogram. By taking $I(x) = 0$ between the signatures, these parts of the interferogram do not contribute to the noise, and by rapidly passing over these parts in the measurement of the interferogram, more time can be spent at the relevant points. The spectrum obtained from the "edited" interferogram becomes at each σ a weighted average over $B(\sigma)$ values at corresponding points in successive lines. When the spectrum consists of m lines at equal distances d with shapes varying only gradually over the spectrum, interferogram slices of width $2l = 4/md$ will suffice to obtain an almost undistorted spectrum given by[67]

$$B'(\sigma) = (4/m) \sum_{n=-\infty}^{\infty} \text{sinc}(4n/m)B_e(\sigma - nd).$$

[58] J. M. Dowling, *J. Quant. Spectrosc. Radiat. Transfer* **9**, 1613–1627 (1969).

[59] A. Lightman, *Infrared Phys.* **11**, 125–127 (1971).

[60] J. L. Harris, *J. Opt. Soc. Amer.* **54**, 931–936 (1964).

[61] C. W. McCutchen, *J. Opt. Soc. Amer.* **57**, 1190–1192 (1967).

[62] B. R. Frieden, *Progr. Opt.* **9**, Chapter 7 (1971).

[63] A. M. Despain and J. W. Bell, *Aspen Int. Conf. Fourier Spectrosc., 1970* (G. A. Vanasse et. al.; eds.), pp. 397–406, AFCRL–71–0019 (1971).

[64] T. Williams, *J. Opt. Soc. Amer.* **50**, 1159–1167 (1960).

[65] J. M. Dowling, *J. Opt. Soc. Amer.* **54**, 663–667 (1964).

[66] E. V. Loewenstein and D. R. Smith, *Appl. Opt.* **10**, 577–583 (1971).

[67] D. Oepts, Thesis, Univ. of Amsterdam (1972).

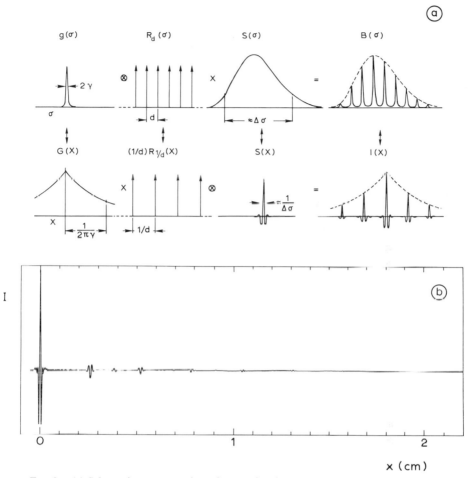

FIG. 8.   (a) Schematic representation of a rotational spectrum and its Fourier transform, illustrating the formation of signatures in $I(x)$; (b) actual interferogram showing (absorption) signatures [from D. Oepts, Thesis, Univ. of Amsterdam, 1972].

As only a fraction $4/m$ of the interferogram contributes to the noise, the signal-to-noise ratio will be increased by a factor $(m/4)^{1/2}$, while the measurement time is reduced by almost the factor $4/m$. Alternatively, without affecting the signal-to-noise ratio, a gain in measurement time of $(4/m)^2$ can be obtained. In practice, some time will be spent in passing the uninteresting interferogram parts; and in choosing the width of the parts to be used, possible broadening of the signatures, e.g., due to centrifugal distortion, has to be taken into account as well as the shape of the background spectrum.

Editing of interferograms can also be used (with care, of course!) to remove undesirable features from the spectrum such as channel structure caused by cell windows, or effects of unwanted reflections.

4.2.4.3. Sampling of the Interferograms. When, as is usual, the Fourier transform (4.2.7) is to be calculated with a digital computer, the interferogram has to be sampled at discrete points. This does not affect the spectrum obtained, provided that the sampling interval is properly chosen. When $I_k = I(x_k)$ are the interferogram values measured at equidistant points $x_k = kh$, the sampled interferogram can be represented by

$$I^h(x) = h \sum_k I_k \delta(x - x_k) = I(x)h \sum_k \delta(x - x_k) = I(x)hR_h(x), \quad (4.2.10)$$

where $R_h(x)$ is the Dirac comb

$$R_h(x) = \sum_{n=-\infty}^{\infty} \delta(x - nh)$$

with the property $\mathscr{F}R_h = (1/h)R_{1/h}(y)$. Using the convolution theorem, the Fourier transform of $I^h(x)$ becomes

$$\begin{aligned} B^h(\sigma) = \mathscr{F}I \otimes \mathscr{F}hR_h &= B_e \otimes R_{1/h} \\ &= \int_{-\infty}^{\infty} B_e(\sigma') \sum_{-\infty}^{\infty} \delta(\sigma - \sigma' - (n/h)) \, d\sigma' \\ &= \sum_{-\infty}^{\infty} B_e(\sigma - (n/h)). \end{aligned}$$

Thus $B^h(\sigma)$ is the sum of the even spectrum B_e and higher orders $B_e(\sigma \pm 1/h)$, $B_e(\sigma \pm 2/h)$, etc. The same effect occurs in other spectrometers employing discrete steps of path difference between interfering rays. Due to the appearance of the even spectrum, each order is accompanied by a mirror image in the present case, and the free spectral range is $\Delta\sigma = 1/2h$.

In the far infrared one generally works in the zero order and overlap does not occur when wavenumbers $\sigma > \sigma_m = 1/2h$ are filtered out. With $h \leqslant 10 \ \mu m$ the whole region up to $\sigma_m = 500 \ cm^{-1}$ can be covered without problems.

The overlap effect is also called "aliasing" and the fact that a function whose Fourier transform is limited to a range $\Delta\sigma$ is completely defined by its values sampled at distances $1/2\Delta\sigma$ (the Nyquist interval) is known as the *sampling theorem*.

4.2.4.3.1. SAMPLING ERRORS. In practice, samples are not taken exactly at the points $x_k = kh$ as was previously assumed. Incorrect zero point and periodic or random variations in the sampling interval h can occur. The zero-point error causes samples to be taken at $x_h = kh + \varepsilon$; the effects and correction of this error are considered in Section 4.2.4.5.1. Variations in h are analogous to ruling errors in a diffraction grating and produce the same effects: a progressive error causes displacement and broadening of spectral lines; a periodic error gives rise to ghost lines; and random errors cause background noise or stray light. The accuracy needed to make these effects negligible is much higher in a Fourier spectrometer than in an equivalent grating instrument.[49] Fortunately, this problem is considerably reduced in the far infrared as the errors that can be tolerated are proportional to the wavelength. Periodic errors are likely to occur when the interval h is determined by the rotation of a micrometer screw. If ε_1 is the amplitude of a sinusoidally varying error, the relative intensity of the nth ghost line accompanying a line at σ_0 is $J_n(\beta)$, where $\beta = 2\pi\varepsilon_1\sigma_0$ and J_n is a Bessel function of order n.[49] The amplitude of the noise caused by random errors is determined by $(\varepsilon_2\sigma)^2$, where ε_2 is the rms error in h.[68] A deviation of h from its nominal value causes a proportional error in the wavenumber scale. These effects are usually negligible.

4.2.4.4. Scanning and Modulation. The interferogram samples I_h can be obtained with continuous or stepwise scanning. In the first method, the path difference x is varied at a constant speed and samples are taken at appropriate times. In the second method, the path difference is kept constant for some time τ. During this time the signal is averaged or integrated. The resulting signal is recorded and the path difference increased by h in a time short compared with τ. Combination of both methods is also possible: scanning with variable speed[49] or averaging during continuous scanning.[51] The latter process always occurs to some extent in the continuous mode due to the finite response time of the measuring system.

A mechanical chopper is usually employed to modulate the radiation of the source with a suitable frequency to enable detection and subsequent amplification and phase-sensitive rectification.

4.2.4.4.1. RAPID SCANNING. No additional modulation is necessary when the path difference is varied with a speed high enough to obtain signal variations with frequencies in a suitable range with respect to the detector response time. In this rapid scanning or periodic method,[69] the scan is repeated many times in a sawtooth fashion and the successive interferograms can be co-added to increase the signal-to-noise ratio.[70] The absence of a chopper, which shuts off the source radiation half of the time, gives a gain in signal of a factor 2. Furthermore, the constant part of the total interferogram I_c in (4.2.4) is not modulated and the sensitivity to fluctuations in source intensity is greatly reduced (cf. Section 4.2.4.6).

[68] M. T. Surh, *Appl. Opt.* **5**, 880–881 (1966).
[69] L. Genzel, *J. Mol. Spectrosc.* **4**, 241–261 (1960).
[70] P. R. Griffiths, C. T. Foskett, and R. Curbelo, *Appl. Spectrosc. Rev.* **6**, 31–78 (1972).

4.2.4.4.2. PHASE MODULATION. The use of a chopper can be avoided in a slow scanning or aperiodic system also by the superposition of a small-amplitude oscillation on the slowly varying path difference. In a Michelson interferometer this can be achieved easily by vibration of the "fixed" mirror (M_1 in Fig. 5). To see the effect of this procedure we assume that the path difference is switched between $x - \frac{1}{2}\delta$ and $x + \frac{1}{2}\delta$. The difference signal measured by the detector gives a modified interferogram that can be written, using (4.2.6), as

$$
\begin{aligned}
I^{\mathrm{P}}(x) &= I(x - \tfrac{1}{2}\delta) - I(x + \tfrac{1}{2}\delta) \\
&= \int_{-\infty}^{\infty} B_{\mathrm{e}}(\sigma)\left[\cos\{2\pi\sigma(x - \tfrac{1}{2}\delta)\} - \cos\{2\pi\sigma(x + \tfrac{1}{2}\delta)\}\right] d\sigma \\
&= \int_{-\infty}^{\infty} 2B_{\mathrm{e}}(\sigma) \sin \pi\sigma\delta \, \sin 2\pi\sigma x \, d\sigma.
\end{aligned}
$$

Hence, the interferogram obtained with phase modulation is the sine transform of a modified spectrum

$$
B^{\mathrm{P}}(\sigma) = B_{\mathrm{e}}(\sigma) 2 \sin \pi\sigma\delta. \tag{4.2.11}
$$

The peak-to-peak modulation amplitude can be chosen as $\delta = 2h = 1/\sigma_m$ so that the first half-period of $\sin \pi\sigma\delta$ coincides with the free spectral region.

In practice, sinusoidal rather than square modulation is applied, with the result that the sine in (4.2.11) has to be replaced by the Bessel function $J_1(\pi\sigma\delta)$.[71]

The phase-modulation or internal-modulation method has the same advantages as rapid scanning: suppression of constant and low-frequency signals and no chopper loss. The modification of the observed spectrum presents no problem; in absorption measurements the additional σ-dependent factor drops out automatically, otherwise it can be divided out if anyway necessary.

4.2.4.5. Asymmetric Interferograms.   4.2.4.5.1. PHASE ERRORS. In the basic relation (4.2.6) it was assumed that all interferogram components have zero phase at $x = 0$. In a real instrument, frequency-dependent phase shifts can occur due to the frequency characteristics of the measuring system, or due to dispersion in the beamsplitter or in a sample placed in one of the interferometer arms. In that case we have

$$
\begin{aligned}
I^{\mathrm{a}}(x) &= 2 \int_{0}^{\infty} B(\sigma) \cos(2\pi\sigma x + \phi(\sigma)) \, d\sigma \\
&= \int_{-\infty}^{\infty} B_{\mathrm{e}}(\sigma) e^{i\phi(\sigma)} e^{2\pi i\sigma x} \, d\sigma. \tag{4.2.12}
\end{aligned}
$$

[71] J. Chamberlain, *Infrared Phys.* **11**, 25–55 (1971).

The last part of (4.2.12) is obtained with the definition $\phi(\sigma) = -\phi(-\sigma)$ for negative wavenumbers. The interferogram is now no longer symmetric—or antisymmetric with phase modulation—and its cosine, or sine, transform is not the desired spectrum. Even a small asymmetry of the interferogram causes important discrepancies between $B_e(\sigma)$ and the cosine transform of $I^a(x)$.[49, 72]

A special case of phase error occurs when the point of zero path difference has not been correctly determined so that instead of $I(x)$ the recorded interferogram is

$$I^a(x) = I(x + \varepsilon) = 2 \int_0^\infty B(\sigma) \cos(2\pi\sigma(x + \varepsilon)) \, d\sigma.$$

Clearly, this corresponds to a linear phase error $\phi(\sigma) = 2\pi\sigma\varepsilon$.

The even spectrum can be retrieved from (4.2.12) by complex Fourier transformation:

$$B_e(\sigma)e^{i\phi(\sigma)} = \mathscr{F}^{-1}I^a(x) = B^a(\sigma) = B_c{}^a(\sigma) - iB_s{}^a(\sigma), \qquad (4.2.13)$$

where $B_c{}^a$ and $B_s{}^a$ are the cosine and sine transforms of $I^a(x)$. We now have

$$B_e(\sigma) = |B^a(\sigma)| = (B_c{}^a(\sigma)^2 + B_s{}^a(\sigma)^2)^{1/2} \qquad (4.2.14a)$$

and

$$\phi(\sigma) = \arg B^a(\sigma) = -\arctan(B_s{}^a(\sigma)/B_c{}^a(\sigma)). \qquad (4.2.14b)$$

To obtain $B_e(\sigma)$ in this way, it is necessary to measure $I^a(x)$ for both positive and negative x and to compute both its cosine and sine transforms. In addition, the signal-to-noise ratio is reduced in comparison with the symmetric case and the noise always gives a positive contribution in (4.2.14a), leading to baseline uncertainty. A further complication can arise when the truncation or apodization is asymmetric with respect to the center of the interferogram.[73] These difficulties can be avoided when the instrumental phase errors are known. It is then possible to correct the distorted spectrum obtained from the asymmetric interferogram,[51, 74] or to correct the interferogram prior to transformation.[75] Both correction methods, known as the multiplicative and the convolution methods, use the fact that, apart from noise, the phase function $\phi(\sigma)$ is always a smooth function varying slowly over the spectrum. It is therefore possible to compute $\phi(\sigma)$ from only a small portion of the interferogram around $x = 0$. We shall describe the convolution method in some detail. We define

$$G(\sigma) = A(\sigma)e^{-i\phi(\sigma)}, \qquad (4.2.15)$$

[72] D. R. Bosomworth and H. P. Gush, *Can. J. Phys.* **43**, 729–750 (1965).
[73] W. H. Steel and M. L. Forman, *J. Opt. Soc. Amer.* **56**, 982–983 (1966).
[74] R. B. Sanderson and E. E. Bell, *Appl. Opt.* **12**, 266–270 (1973).
[75] M. L. Forman, W. H. Steel, and G. A. Vanasse, *J. Opt. Soc. Amer.* **56**, 59–63 (1966).

where $A(\sigma)$ is a real and even function to be specified later, and

$$F(x) = \mathscr{F}G = \int_{-\infty}^{\infty} G(\sigma)e^{2\pi i\sigma x}\,d\sigma. \qquad (4.2.16)$$

This $F(x)$ is real as a consequence of the even and odd properties of A and ϕ. The sampled version of $I^a(x)$ is written as [cf. (4.2.10)] $I^h(x) = I^a(x)hR_h(x)$, and a corrected interferogram is defined as

$$\hat{I}(x) = F \otimes I^h = h \sum_{k=-\infty}^{\infty} F(x - kh)I^a(kh). \qquad (4.2.17)$$

The corresponding spectrum is

$$\hat{B}(\sigma) = \mathscr{F}^{-1}(F \otimes I^hhR_h) = (\mathscr{F}^{-1}F)(\mathscr{F}^{-1}I^a \otimes \mathscr{F}^{-1}hR_h).$$

Using (4.2.13) and (4.2.16) this becomes

$$\hat{B}(\sigma) = A(\sigma)e^{-i\phi(\sigma)}(B_e(\sigma)e^{i\phi(\sigma)} \otimes R_{1/h}(\sigma)) = A(\sigma)B_e(\sigma). \qquad (4.2.18)$$

In the last step it has been assumed that both $B_e(\sigma)$ and $A(\sigma)$ are confined to the free spectral range $|\sigma| \leqslant 1/2h$ so that the higher orders have no effect. With $A(\sigma) = D_{1/2h}(\sigma)$, the rectangular function (4.2.9) of width $1/2h$, the spectrum obtained from the corrected interferogram $\hat{I}(x)$ is identical with $B_e(\sigma)$, and so $\hat{I}(x)$ is equal to the interferogram without phase errors.

In the special case $\phi(\sigma) = 2\pi\sigma\varepsilon$ we have $G(\sigma) = A(\sigma)e^{-2\pi i\sigma\varepsilon}$ and $F(x) = F_0(x - \varepsilon)$, where F_0 is the Fourier transform of $A(\sigma)$. With $A(\sigma) = D_{1/2h}(\sigma)$ this gives

$$F_0(x) = (1/h)\,\mathrm{sinc}(x/h)$$

and the corrected interferogram is

$$\hat{I}(x) = \sum_{k=-\infty}^{\infty} \mathrm{sinc}((x - \varepsilon - kh)/h)I^a(kh).$$

Remembering that the linear phase error means $I^a(x) = I(x + \varepsilon)$ and that the corrected interferogram is equal to $I(x) = I^a(x - \varepsilon)$, we can write this in the form

$$I^a(x') = \sum_{-\infty}^{\infty} \mathrm{sinc}((x'/h) - k)I^a(kh),$$

which is the explicit expression of the sampling theorem, showing how an unknown $I^a(x')$ value is expressed in the sampled values $I^a(kh)$.

With the symmetric or antisymmetric corrected interferogram $\hat{I}(x)$, the spectrum can be computed in the usual way from a single-sided interferogram.

Taking only a finite number of terms in sum (4.2.17) corresponds to truncating $F(x)$ and modifies $G(\sigma)$. This can result in spurious ripples in the interferogram and overlap effects. A large number of terms is required to avoid these effects when the rectangular function is used for $A(\sigma)$. Therefore, $F(x)$ is usually multiplied by an apodizing function instead of abruptly truncated. Alternatively, one can apodize F by using a suitable $A(\sigma)$, for instance one of the usual apodizing functions. The resulting multiplication (4.2.18) of the spectrum with an attenuating function presents no more a problem than it did with phase modulation. As the noise is attenuated as well, the signal-to-noise ratio is unaffected.

By confining $A(\sigma)$ to a region smaller than $1/2h$, the spectral bandwidth is reduced and coarser sampling is allowed for the corrected interferogram. This procedure is known as *mathematical filtering*.[76]

4.2.4.5.2. AMPLITUDE SPECTROSCOPY.[77, 78] Let a transparent nonreflecting sample be introduced into the fixed arm of a Michelson interferometer. When its refractive index does not depend on the wavenumber σ, and no absorption occurs, the only effect is to alter the optical path difference between the interfering beams, and the interferogram $I(x)$ is shifted along the x-axis. The amount of shift and the thickness of the sample then give us the refractive index.

When the radiation source of the interferometer is monochromatic, this method can still be used when dispersion is present to determine the refractive index at a single wavelength. With a polychromatic source, the shift of the interferogram maximum can be used to determine an average index of refraction for a dispersive sample. Dispersion causes the optical path length to differ for different σ and makes the interferogram asymmetric. (see Fig. 9) When absorption is present as well, the interferogram can be written as

$$I(x) = \int_{-\infty}^{\infty} B_e(\sigma) T(\sigma) e^{i\phi(\sigma)} e^{2\pi i \sigma x} \, d\sigma, \qquad (4.2.19)$$

where $B_e(\sigma)$ is the even spectrum of the source, $T(\sigma)$ the transmission factor of the sample, and $\phi(\sigma) = -2d[n(\sigma) - 1] + \phi_0(\sigma)$, where d is the thickness and $n(\sigma)$ the refractive index of the sample; $\phi_0(\sigma)$ represents residual phase errors not caused by the sample. Now, $T(\sigma)e^{i\phi(\sigma)}$ is the complex transfer function, and its Fourier transform $P(x)$ is the impulse response of the sample. Thus the interferogram can be written as

$$I(x) = I^b \otimes P(x),$$

where I^b is the interferogram without the sample. For reasons of causality $P(x)$ differs from zero only for $x \geqslant 0$ and features introduced in the interferogram by the presence of the sample are found at the positive x-side only.

[76] J. Connes and V. Nozal, *J. Phys. Radium* **22**, 359–366 (1961).
[77] E. E. Bell, *Infrared Phys.* **6**, 57–74 (1966).
[78] J. Chamberlain, J. E. Gibbs, and H. A. Gebbie, *Infrared Phys.* **9**, 185–209 (1969).

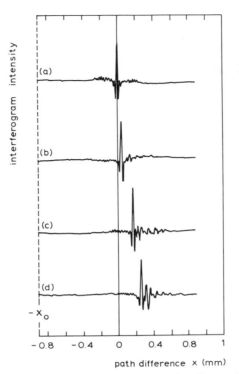

FIG. 9.   Interferograms obtained with dispersive samples of increasing thickness in the fixed arm of a Michelson interferometer: (a) no sample; (b) 14.5 μm; (c) 53.5 μm; (d) 88.0 μm [from J. Chamberlain, J. E. Gibbs, and H. A. Gebbie, *Infrared Phys.* **9**, 185–209 (1969), by permission of Pergamon Press].

Both the transmission and the phase spectrum of the sample can be determined by taking the complex Fourier transform of the asymmetric interferogram (4.2.19), and dividing by the background spectrum, which can also be complex due to $\phi_0(\sigma)$. The complex reflection spectrum of a solid or liquid can be obtained in the same way. Measurements of the difference in refractive index at both sides of an absorption line can be used to determine the line strength.[79]

Some problems can arise in the application of the amplitude spectroscopy method just described. The sample must be of interferometric quality with respect to homogeneity, thickness variations, and surface finish. Even in the far infrared this is not always easily achieved. Effects of (multiple) reflections at the sample faces or at cell windows have to be taken into account.[80]

[79] R. B. Sanderson, *Appl. Opt.* **6**, 1257–1230 (1967).

[80] J. Chamberlain, *Infrared Phys.* **12**, 145–164 (1972); D. D. Honijk, W. F. Passchier, and M. Mandel, *Physica* **64**, 171–188 (1973).

The x-scales in sample and background interferograms should coincide precisely.[81] When absorption is present, the ratio of the variable to the constant part of the interferogram [cf. (4.2.4)] is reduced. Phase modulation to suppress I_c is therefore advisable; apart from a constant phase shift $\pi/2$ this does not distort the phase spectrum.

4.2.4.6. Noise. 4.2.4.6.1. DETECTOR NOISE. Spontaneous thermal or electrical fluctuations in the radiation detection system form the most important source of noise in far-infrared work. A common method to reduce the influence of noise in a measurement consists in reducing the noise bandwidth by means of filters in the electronic circuit and reducing the signal bandwidth by slower scanning. This method also applies in Fourier transform spectroscopy, but there is a difference in comparison with more conventional systems. As the noise is Fourier transformed along with the signal, noise outside the range of signal frequencies of interest has no influence within this range.[†] In principle, the bandwidth or time constant used in recording the interferogram has no effect on the noise in the spectrum. This is no longer true, however, when the interferogram is sampled. If the noise spectrum extends to frequencies outside the free spectral range $\Delta\sigma = 1/2h$, noise is added in the form of overlapping higher orders. It is therefore advantageous to limit the noise bandwidth to $\Delta\sigma$ prior to sampling. Electrical filters for this purpose can cause nonlinear phase shifts. No noise enhancement due to sampling occurs in the step-and-integrate mode or in the continuous mode when the signal is integrated over the sampling interval. In the latter case, however, the signal spectrum becomes multiplied by $\text{sinc}(\sigma h)$, while the root mean square noise is unaffected when its bandwidth was initially much larger than $1/2h$ (or $1/2\tau$ in time–frequency units). When the integration is performed numerically with subsamples taken during the sampling time, the procedure is a form of mathematical filtering.

The relation between the noise in the interferogram and in the spectrum is given by Parseval's theorem

$$\int_{-\infty}^{\infty} |B_c(\sigma)|^2 \, d\sigma = \int_{-\infty}^{\infty} |I(x)|^2 \, dx. \tag{4.2.20}$$

If the interferogram is a sum of signal and noise parts, (4.2.20) applies for the noise and the signal separately. The mean square signal and mean square noise in the spectrum are therefore both equal to those in the interferogram. The actual signal-to-noise ratio (SNR) in the spectrum depends

[81] T. J. Parker, W. G. Chambers, and J. F. Angress, *Infrared Phys.* **14**, 207–215 (1974).

[†] Electrical signal frequencies f and wavenumbers σ are related by $f = v/\sigma$, where $v = h/\tau$ is the effective path difference scanning speed. We express signal and noise frequencies in wavenumber units with this conversion in mind.

on the shape of the spectrum. When the signal is concentrated in a few lines, the SNR in the spectrum will be much higher than in the interferogram; in absorption measurements with a broad background spectrum, the opposite can occur.

When the interferogram noise is limited to the free spectral range, (4.2.20) gives for the mean square noise amplitudes n_s^2 and n_1^2 in the spectrum and in the interferogram

$$2 \, \Delta\sigma \, n_s^2 = 2Lqn_1^2.$$

With apodization, one has $|A(x)I(x)|^2$ in (4.2.20) and q, the factor taking this into account, is

$$q = (1/2L) \int_{-L}^{L} |A(x)|^2 \, dx.$$

Thus for fixed observation time n_s is proportional to $L^{1/2}$. The signal initially increases proportional to L with increasing resolution, and optimum SNR is obtained when the interesting spectral detail is just resolved.[82] The factor $q^{1/2}$ due to apodization is usually somewhat larger than $\frac{1}{2}$, while the resolution is roughly halved. For unresolved lines, the SNR with apodization is therefore less than without.[49] This effect is eliminated when the sampling time per interferogram point is made proportional to $A(x)$ so that no time is spent in gathering information (and noise) that is to be thrown away afterwards.[83] Maximum SNR is obtained for isolated lines with apodization such that the instrumental function is the same as the true line shape, at the expense of doubling the line width.

When the measurement time T is increased, keeping L fixed, the relation between electrical noise frequencies and signal frequencies is altered; the noise bandwidth becomes relatively narrower and n_s decreases in proportion with $T^{-1/2}$,[49, 52] so that we have

$$n_s = \text{const} \times (qL/T)^{1/2}.$$

When the sampling time per point τ is kept constant, L becomes proportional to T and $n_s = \text{const} \times \tau^{-1/2}$, *independent of the resolution*. To double the resolution one just keeps the instrument scanning twice as long, everything else staying the same, including noise level and average signal. This is in sharp contrast with the case of a grating spectrometer where a factor of 16 increase in time is usually required,[40] and illustrates the increase of multiplex and étendue gain with increasing resolution.

4.2.4.6.2. MODULATION NOISE. Modulation or scintillation noise is caused by fluctuations in the intensity of the source or in the sensitivity of the detector. This noise is not additive but multiplicative and is not reduced

[82] J. M. Dowling, *Appl. Opt.* **6**, 1580–1581 (1967).
[83] M. F. A'Hearn, F. J. Ahern, and D. Zipoy, *Appl. Opt.* **13**, 1147–1157 (1974).

by multiplexing. The modulation of the interferogram by such fluctuations causes sidebands associated with each spectral line. As long as the relative fluctuations are small, this effect is negligible save for the sidebands pertaining to the line at $\sigma = 0$ caused by the constant interferogram part I_c (4.2.4). Although I_c is generally subtracted from the total interferogram before computing the spectrum, its variations remain and may be important in comparison with the varying signal part $I(x)$. This occurs especially in absorption measurements where the absorbed energy causing the true signal variations is a small fraction of the background energy which determines I_c.[84] This problem is absent in double-beam instruments.[85, 86] Modulation noise is also practically eliminated with rapid scanning and with phase modulation (Section 4.2.4.4).

4.2.4.6.3. DRIFT. Slow variations in the interferogram signal, with characteristic times in the order of the total measuring time, have almost no influence on the shape of the spectrum. Their effect can be considered as a variation of the apodizing function $A(x)$, resulting in a slightly distorted instrumental function.

4.2.4.6.4. DIGITIZING NOISE. The finite accuracy of the analog-to-digital conversion used in recording the interferogram adds random variations to the signal. Usually the dynamic range of the interferogram signal is large and accurate sampling is required. This is particularly the case when a long time constant is used in the continuous scanning mode so that the signal-to-noise ratio in the analog signal is high. In the integrating modes, a short time constant can be used in taking the subsamples while the total sampling time may be arbitrarily long. By performing the integration or averaging *after* the analog-to-digital conversion, the dynamic range of the measuring system is increased by the presence of noise.[87] The same effect occurs in the rapid scanning mode with co-adding of interferograms.[70]

4.2.4.6.5. PHOTON NOISE. A measurable far-infrared signal requires such a large number of photons that the statistical fluctuations in this number are always negligible in comparison with other noise sources.

4.2.4.7. Fourier Transformation. With a properly sampled, symmetric interferogram $I(x)$, the Fourier transformation (4.2.7) leading to the spectrum is reduced to

$$B_e(\sigma) = \sum_{k=0}^{N} I_k \cos 2\pi\sigma kh, \qquad (4.2.21)$$

with $I_k = I(kh)$ for $k \neq 0$ and $I_0 = \frac{1}{2}I(0)$. When phase modulation is used,

[84] G. Roland, *J. Phys. Radium* **28**, Supple. C2, 26–32 (1967).

[85] F. J. Ahern and C. Pritchet, *Appl. Opt.* **13**, 2240–2243 (1974).

[86] R. T. Hall, D. Vrabec, and J. M. Dowling, *Appl. Opt.* **5**, 1147–1158 (1966).

[87] J. Butterworth, D. E. MacLaughlin, and B. C. Moss, *J. Sci. Instrum.* **44**, 1029–1030 (1967).

the sine transform of an antisymmetric interferogram is computed analogously. In amplitude spectroscopy both the cosine and the sine transform must be computed from a two-sided interferogram, i.e., k runs from $-N$ to $+N$. It is advisable in this case to take the origin $x = 0$ not at the vacuum zero path difference position, but at the center of the shifted interferogram.[78]

Straightforward evaluation of (4.2.21) for N_B wavenumber points can take an excessive amount of computer time as both N_B and N can be large, typically 500–5000; even measurements with $N = 10^6$ have been made in the near infrared.[55]

The fast Fourier transform (FFT) method allows the transformation to be executed in a time proportional to $N \log N$ instead of NN_B.[88] A clear explanation of the principle of this method and details on its use have been given by Connes.[89] Standard FFT programs are available with most computer systems. The method is generally used with $N_B = N = 2^n$; when necessary the interferogram is extended with zeros to obtain a suitable N. A complex transform is produced from a two-sided complex input function. When only the cosine or sine transform is needed and one interferogram half is used, it is possible to transform two interferograms at the same time. One point is obtained per resolution element $\delta\sigma = 1/2L = 1/2Nh$ and interpolation is usually required in plotting the spectrum or accurately locating the position of a peak.

Other methods have been used that usually give more flexibility in output at the cost of longer computing time. When a special purpose computer is used, the transformation can often be made in real time. A method frequently used in that case is sometimes referred to as *Fourier synthesis* instead of analysis.[51] Each time that an interferogram sample becomes available a new term is added to sum (4.2.21) for all wavenumbers of interest. The development of the spectrum can be observed directly in this way.[90]

Analog methods for performing the Fourier transformation have been applied as well, especially with periodic scanning instruments. When frequency analysis is performed in real time with a single channel system, the multiplex advantage is lost, however.

4.2.4.8. Practical Realization.   4.2.4.8.1. INSTRUMENTAL LIMITATIONS. The real instrumental function differs from the theoretical shape in most spectrometers due to imperfections of the optical components or to the width of the slits used. With the exception of the equivalent of grating ruling errors (cf. Section 4.2.4.3.1), such limitations are far less important in Fourier transform spectroscopy.

[88] M. L. Forman, *J. Opt. Soc. Amer.* **56**, 978–979 (1966).

[89] J. Connes, in *Aspen Int. Conf. Fourier Spectrosc., 1970* (G. A. Vanasse *et al.*, eds.), pp. 83–115, AFCRL–71–0019 (1971).

[90] H. Yoshinaga *et al.*, *Appl. Opt.* **5**, 1159–1165 (1966).

For example, suppose that, due to misalignment or surface defects of a component, the path difference x is not equal for all rays in the beam, but varies from $x - d$ to $x + d$. One then obtains the interferogram

$$I^d(x) = \int_{x-d}^{x+d} I(x')\, dx' = \int_{-\infty}^{\infty} D_d(x - x')I(x')\, dx'$$

$$= D_d \otimes I(x). \qquad (4.2.22)$$

A homogeneous distribution of path differences, represented by $D_d(x)$ [see Eq. (4.2.9)] has been assumed here for simplicity; another one can easily be substituted for it. The spectrum obtained from (4.2.22) is

$$B^d(\sigma) = \mathscr{F}(D_d \otimes I) = \mathscr{F}D_d\mathscr{F}I = \operatorname{sinc}(\sigma d)B_e(\sigma).$$

The result is an attenuation increasing toward higher wavenumbers and no useful signal is obtained when the inaccuracy in x is in the order of the wavelength so that $\sigma d \approx 1$. The resolution, however, is the same as in $B_e(\sigma)$ and is still determined only by the available length of $I(x)$.

Stray light only contributes to the constant part I_c of the interferogram and has no effect on the spectrum.

The detection system should be sufficiently linear over the usually large dynamic range of the interferogram to prevent distortions of the spectrum.[91] Some saturation in the central peak $I(0)$ causes a subtraction of a constant from all spectral intensities. Special procedures (e.g., "chirping") are sometimes applied to reduce the effects of nonlinearity.[92]

Two optical arrangements to produce the interferogram are commonly used, the Michelson and the lamellar grating interferometer.

4.2.4.8.2. MICHELSON INTERFEROMETER. The principle of the Michelson interferometer arrangement is illustrated in Fig. 5. Mostly, one actually uses the Twyman–Green arrangement with collimated beams.

Uncoated thin polyester foil (Mylar or Melinex) is usually applied for the beamsplitter. This makes a compensator plate unnecessary. Due to interference in the foil, the efficiency η of the beamsplitter varies roughly as $\sin^2(\text{const } \sigma d)$, where d is the thickness; in general more than half the incident flux is reflected back toward the source. Usually d is chosen so that the first maximum of η is at the center of the σ-region to be studied. Polarization properties of the radiation can be studied by using polarizing beamsplitters.[93]

For rays making an angle ϕ with the axis, the path difference is $x' = x \cos \phi$. The use of an extended entrance aperture causes an increasing

[91] K. Hemmerich, W. Lahmann, and W. Witte, *Infrared Phys.* **6**, 123–128 (1966).
[92] Th. P. Sheahen, *J. Opt. Soc. Amer.* **64**, 485–494 (1974); *Appl. Opt.* **13**, 2907–2912 (1974).
[93] D. H. Martin and E. Puplett, *Infrared Phys.* **10**, 105–109 (1970).

variation in x and, unlike the case of a constant spread in x, this limits the attainable resolution. When Ω is the solid angle of the beams, one has $\delta\sigma_{min} = \sigma\Omega/2\pi$.[49] In addition, one obtains $B_e(\sigma - (\sigma\Omega/4\pi))$ instead of $B_e(\sigma)$. In the far infrared these effects are usually very small;[94] they are sometimes compensated for by field widening procedures in the near infrared.[95] In that region, effects of mirror tilt are frequently also compensated by using corner reflectors or "cat's eyes."

A micrometer screw and stepping motor usually provides a suitable scanning system, but at high wavenumbers more accuracy is again required. Interferometric control can be applied as in grating ruling engines.

A widely used instrument is the modular cube interferometer developed at NPL (Fig. 10).[96] This instrument can be used in the range $2-700 \text{ cm}^{-1}$, with maximum resolution $1/2L = 0.05 \text{ cm}^{-1}$.[94] It is produced commercially by Grubb Parsons & Co Ltd. Newcastle upon Tyne, England. One of the accessories enables phase modulation to be performed. Comparable interferometers are produced by Beckmann–R.I.I.C. Ltd., London, England. Suitable computer systems can be supplied as well by both firms. Systems with real-time computation and facilities for absorption and reflection measurements are marketed by Coderg, Clichy, France, and by Polytec, Karlsruhe, West Germany. A complete rapid scanning system is available from Digilab Inc., Cambridge, Massachusetts.

4.2.4.8.3. LAMELLAR GRATING INTERFEROMETER.[97] Two beams of radiation with a variable path difference can be obtained with a lamellar grating as sketched in Fig. 11. The beamsplitter efficiency problems of a Michelson interferometer are absent with this method. Aperture limitations are more stringent, but in the low wavenumber region these are usually no problem. The efficiency decreases when radiation reflected in the first grating order reaches the detector, as these partly cancel the interferogram signal due to an additional phase difference π. This occurs when $\sigma > \sigma_c$, with $\sigma_c = F/sd$.

Problems can arise at high resolution and low σ due to waveguide effects in the groove regions.[98] These effects can be reduced by increasing d, but this decreases σ_c.

A high-resolution lamellar grating interferometer has been constructed by Dowling et al.[86] A commercial instrument is produced by Beckmann–R.I.I.C. Ltd.[99]

[94] J. W. Fleming and J. Chamberlain, *Infrared Phys.* **14**, 277–292 (1974).

[95] W. H. Steel, in *Aspen Int. Conf. Fourier Spectrosc., 1970* (G. A. Vanasse *et. al.*, eds.), pp. 43–53, AFCRL–71–0019 (1971).

[96] G. W. Chantry, H. M. Evans, J. Chamberlain, and H. A. Gebbie, *Infrared Phys.* **9**, 85–93 (1969).

[97] J. Strong and G. A. Vanasse, *J. Opt. Soc. Amer.* **50**, 113–118 (1960).

[98] A. Wirgin, *Opt. Commun.* **7**, 205–210 (1973).

[99] R. C. Milward, *Infrared Phys.* **9**, 59–74 (1969).

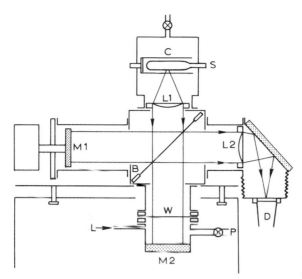

FIG. 10. NPL cube interferometer adapted for measurements in the asymmetric mode on liquid samples . S: mercury arc source, C: chopper, L1, L2: lenses, B: beamsplitter, M1, M2: mirrors, D: detector, W: window, L: liquid filling tube, P: dry air inlet [from J. Chamberlain, J. E. Gibbs, and H. A. Gebbie, *Infrared Phys.* **9**, 185–209 (1969), by permission of Pergamon Press].

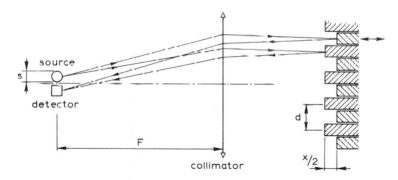

FIG. 11. Principle of lamellar grating interferometer.

A lamellar grating instrument is mechanically more complex than a Michelson; amplitude spectroscopy is not possible, and phase modulation difficult to realize. Due to its higher efficiency at low wavenumbers, however, it may still be preferred over a Michelson instrument. An existing grating monochromator can sometimes relatively easily be modified into a lamellar grating interferometer.

4.2.4.9. Final Remarks. The Fourier transform method has had its major impact in the infrared regions, but it can also be applied in other regions. Even in circumstances where the Jacquinot and the Fellgett advantages do not apply, the possibility to attain very high wavenumber accuracy and resolution, the so-called *Connes advantage*, can make it an attractive method.

Fourier transform spectroscopy is also successfully applied in nuclear magnetic resonance (NMR) measurements. Considerations on multiplex advantage, apodization, phase errors, etc., apply just as in the infrared. In the version using noise excitation,[100] the cross-correlation function of input and sample signal corresponds to the interferogram $I(x)$, with the path difference x replaced by the delay time t. The signal obtained with the more usual pulse method[101] gives directly the impulse response of the sample, corresponding to $P(x)$ in Section 4.2.4.5. Besides the signal-to-noise or sensitivity gain, multiplexing has an advantage in high-resolution NMR in the study of time-varying spectra, as the required resolution determines a minimum observation time per spectral element.

Although the Fourier transform method is now well established and generally applied, people not acquainted with it frequently seem to have difficulty in really accepting its reliability and usefulness. The following are examples of often-raised objections:

1.   Unless you manipulate your spectrum by apodization, negative intensities can appear (Fig. 7(a)). This gives me little confidence in the results and I prefer a system where such unrealistic things do not occur.

2.   With apodization and things like that, you can obtain any spectrum you like; how am I to believe that I get the real one?

3.   In normally scanning a spectrum, it is easy to see when anything goes wrong and one can repeat the measurement or ignore a disturbed or noisy part of the spectrum. With an interferogram I have no idea about the reliability of the results and furthermore, a single erroneous point or noise spike can spoil the whole spectrum.

4.   All your computations make the relation between the measurement and the spectrum very indirect and although you might get accurate line positions, I feel unsure about intensities or the shape of broad bands.

As regards (1), the occurrence of a negative signal can be compared with overshoot in an underdamped recording system scanning a narrow line. It has nothing to do with fundamentally incorrect operation, but is due to the inability to respond accurately to a signal requiring higher resolution than is used. Apodization can reduce the effect just as heavier damping of the recorder.

(2)   It is equally possible with a conventional spectrometer to distort the spectrum by widening the slits, increasing the scanning speed, or filtering

[100] R. R. Ernst, *J. Magn. Resonance* **3**, 10–27 (1970).
[101] R. R. Ernst and W. A. Anderson, *Rev. Sci. Instrum.* **37**, 93–102 (1966).

the signal, yet one has reasonable confidence in a spectrum recorded under appropriate conditions. Quite the same is true in the Fourier spectroscopy case.

(3)   A gross error or excessive noise peak in the interferogram at some point $x = a$ can be identified not only in the interferogram, but also afterwards in the spectrum, as it gives rise to a contribution of the form $A \cos(2\pi\sigma a)$ throughout the spectrum. From the period and amplitude of this signal, the position and magnitude of the error can be found and the interferogram inspected and corrected. In practice it usually suffices to replace an erroneous point, caused, for instance, by a punching error in a paper tape record, by an average interferogram value. Errors that are too small to be directly recognized, are indeed spread all over the spectrum but they remain small and each point in the spectrum is a result of all interferogram points in which the signal contributes coherently while noise adds incoherently. For each spectral element, signal and noise are in effect averaged over the full measurement time; this is just the multiplex advantage.

(4)   The Fourier transformation is a perfectly reliable and reversible operation and there is no information lost in an uncontrollable way in the transformation or additional operations. Of course, quantitative work needs separate or simultaneous (double-beam) measurement of a reference or background spectrum to eliminate the effects of the spectral response profile of the detector, window absorptions, instrumental transmission factors, etc., just as with a conventional spectrometer. With the necessary precautions, however, line intensities and shapes of broad spectra can very well be determined quantitatively (see, e.g., Section 4.2.6.2).

The time delay between the measurement of the interferogram and the availability of the spectrum can be a disadvantage of Fourier spectroscopy. With the widespread availability of computing facilities and the development of small and inexpensive computers, this seldom forms a problem nowadays.

4.2.5. Other Methods

4.2.5.1. Introduction. Various methods that have been applied or proposed to increase the signal-to-noise ratio or to simplify the equipment in comparison with conventional spectrometers, will be summed up in this section. Not all of these methods have actually been used in the far infrared, but as most are rather closely related, it seems appropriate to include them here.

The selective detection of a small wavelength portion out of a polychromatic beam can be performed by filtering or modulation techniques. In an instrument of the first type an arrangement is made to reject the unwanted parts of the spectrum before they can reach the detector, by means of absorption, reflection, or refraction. In instruments of the second type, signals from different spectral elements are modulated with different frequencies or different amplitudes. Different spectral elements already correspond, of course, to different frequencies, but no detector is able to follow these; a time-averaged radiation power is always observed, and modulation of this power in a specific and detectable way is required.

4.2.5.2. Filter Methods.   4.2.5.2.1. SELECTIVE DETECTORS. A detector never gives equal responses for all wavelengths, and sometimes the sensitivity is sufficiently peaked so that the detector itself acts as a filter to isolate a—usually rather broad—spectral band. A rough spectrum of a source can be obtained by looking at it with a set of detectors having different sensitivity profiles (as do our eyes).[102] Narrow-band tunable detectors may become available in the future so that higher resolution can be obtained with this method.[103]

4.2.5.2.2. RESTSTRAHLEN FILTERS. Selective reflection from reststrahlen crystals can be used to select parts of the far-infrared region.[104]

4.2.5.2.3. MESH FILTERS. Metal meshes can be used as low-pass reflection and high-pass transmission filters, and a combination of these gives a useful bandpass filter.[105]

4.2.5.2.4. INTERFERENCE FILTERS. With metal meshes and their complementary structures, interference filters can be assembled having a narrow transmission band.[46, 106] In a two-element filter the passband can be tuned by varying the separation of the elements.[107]

4.2.5.2.5. FABRY–PEROT INTERFEROMETER. Metal meshes or metal mirrors with a small coupling hole are used as reflectors in submillimeter-wave Fabry–Perot (Section 3.4.3.5) interferometers.[30] With high reflectivities one obtains high-Q resonators with narrow transmission peaks. Scanning of the spectrum is relatively simple and can be performed at high speed.[108] The narrow free spectral range often makes additional wavelength selection necessary. The instrument is particularly useful in combination with coherent sources.

4.2.5.2.6. FRESNEL ZONE PLATE. The focal length of a Fresnel zone plate depends on the wavelength and a simple spectrometer making use of this effect has been realized.[109]

[102] K. Shivanandan, J. R. Houck, and M. O. Harwit, *Phys. Rev. Lett.* **21**, 1460–1462 (1968); *Astrophys. J.* **157**, L245–248 (1969).

[103] P. L. Richards and S. A. Sterling, *Appl. Phys. Lett.* **14**, 394–396 (1969).

[104] W. G. Mankin, *Infrared Phys.* **13**, 333–336 (1973).

[105] S. F. Nee, and A. W. Trivelpiece, *Rev. Sci. Instrum.* **44**, 916–917 (1973).

[106] M. M. Pradhan, *Infrared Phys.* **11**, 241–245 (1971); **12**, 263–266 (1972).

[107] R. Ulrich, K. F. Renk, and L. Genzel, *IEEE Trans. Microwave Theory Tech.* **MTT-11**, 363–371 (1963).

[108] D. S. Komm, R. A. Blanken, and Ph. Brossier, *Appl. Opt.* **14**, 460–464 (1975).

[109] R. H. Wright and P. N. Daykin, *Nature (London)* **189**, 212 (1961).

4.2.5.2.7. FOCAL ISOLATION. The strong wavelength dependence of the focal length of a quartz lens in the far infrared has been used in early work in this region[110] to isolate parts of the spectrum. More recent proposals for a spectrometer with longitudinal dispersion employ a modulation method.

4.2.5.3. Modulation Methods.[†]   4.2.5.3.1. GOLAY (STATIC) MULTISLIT SPECTROMETER. A method to increase the étendue of a normal spectrometer without reducing the resolving power has been developed by Golay.[111] The entrance and exit slits are replaced by identical quasi-random arrays of slits. The wavelength for which the entrance array is imaged in register with the exit one is fully transmitted, while for wavelengths for which the image is translated due to the spectrometer dispersion by more than the width of a single slit, the transmission averaged over the array is only one half. By using two complementary patterns and chopping the exit beam between the two, an ac signal is obtained for the selected wavelength while others give no modulation. Alternatively, separate detectors can be used at the complementary exits and the difference signal recorded while the spectrum is (rapidly) scanned.

4.2.5.3.2. GIRARD GRILLE SPECTROMETER. Following the Golay multislit principle, Girard has realized a high-resolution infrared spectrometer.[112] His entrance and exit "grilles" are in the form of sets of hyperboles; the nontransmitting parts of the exit grille are reflective and a difference signal of transmitted and reflected beams is detected.

4.2.5.3.3. LONGITUDINAL DISPERSION SPECTROMETER. In the proposed instruments an entrance grid is imaged onto a corresponding exit grid for a specific wavelength. For other wavelengths the image is out of focus and translation of one of the grids in its plane modulates the selected wavelength only.[113]

4.2.5.3.4. SISAM. The "spectromètre interférentiel à sélection par l'amplitude de modulation," devised by Connes,[114] employs a Michelson interferometer setup to modulate selectively radiation dispersed by a grating (see Fig. 12). For a wavelength λ_s such that $2d \sin \theta = n\lambda_s$, the gratings function as well-adjusted mirrors in a Michelson interferometer and variation of the path difference between the two arms of the interferometer modulates the signal at the detector (cf. Section 4.2.4.1). For a different wavelength,

[110] H. Rubens and R. W. Wood, *Phil. Mag.* **21**, 249–261 (1911).

[111] M. J. E. Golay, *J. Opt. Soc. Amer.* **41**, 468–472 (1951).

[112] A. Girard, *Appl. Opt.* **2**, 79–87 (1963); *J. Phys. Radium* **24**, 139–141 (1963).

[113] J. F. James and R. S. Sternberg, *J. Phys. Radium Suppl. C2* **28**, 326–329 (1967); W. Witte, *Infrared Phys.* **13**, 285–300 (1973).

[114] P. Connes, *J. Phys. Radium* **19**, 215–222 (1958).

†For a recent review see M. Harwit and J. A. Decker, Jr., *Progr. Opt.* **12**, Chapter 3 (1974).

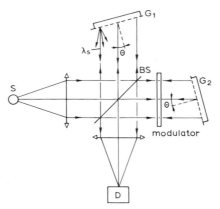

FIG. 12.   Principle of SISAM spectrometer.

the gratings act as tilted mirrors producing a fringe pattern which, on modulating the path difference, moves across the field of view of the detector without producing a net modulation of the signal. The spectrum is scanned by rotating both gratings. The resolving power is the same as would be obtained with one of the gratings in a normal Littrow mount using infinitely narrow slits, while the useful étendue is the same as for other beamsplitter instruments possessing the Jacquinot advantage. Signals of different grating orders are modulated with different frequencies and are easily distinguished. A high-resolution near-infrared SISAM has been described recently.[115]

4.2.5.3.5. MOCK INTERFEROMETER. This instrument, devised by Mertz,[51] is a Fourier spectrometer in which the specific modulation of each spectral element is not obtained by interferometric means but by a nonuniformly rotated grid in the entrance and exit focal plane of a conventional spectrometer. It possesses multiplex, stray light and étendue advantage just as an interferometer, but does not eliminate higher orders. A high-resolution instrument of this type for use in the visible and near infrared has been constructed by Ring and Selby.[116]

Selective sinusoidal modulation in the exit plane of a spectrometer with a single entrance slit, as proposed by Golay[117] and Fellgett,[14] has been obtained also with the aid of a Girard grille[118] and with an encoder disk.[119]

[115] P. Pinson, *Appl. Opt.* **13**, 1618–1620 (1974).

[116] M. J. Selby, *Infrared Phys.* **6**, 21–32 (1966); J. Ring and M. J. Selby, *Infrared Phys.* **6**, 33–43 (1966).

[117] M. J. E. Golay, *J. Opt. Soc. Amer.* **39**, 437–444 (1949).

[118] A. Girard, in *Aspen Int. Conf. Fourier Spectrosc., 1970* (G. A. Vanasse *et al.*, eds.), pp. 425–428, AFCRL–71–0019 (1971).

[119] J. F. Grainger, J. Ring, and J. H. Stell, *J. Phys. Radium Suppl. C2* **28**, 44–52 (1967).

4.2.5.3.6. HADAMARD SPECTROMETER. In this multiplex instrument a coding mask is again used at the exit of a standard spectrometer and the modulation is obtained by translating the mask in steps over the spectrum. At each position of the mask, the detector signal is a linear combination of the signals from the different spectral elements. The coefficients in this combination are either 1 or 0 as the mask is either open or opaque at each spectral element. This simplifies the decoding of the signal, and a further simplification is obtained by employing a particular coding sequence based on Hadamard matrices.[120] When the entrance slit of the spectrometer is also replaced by a coding mask, an étendue advantage is obtained in addition to the multiplex one. It is then necessary to translate both masks separately but not all possible combinations of positions need be realized.[121]

A multiplex attachment to a conventional spectrometer as used in the mock interferometers and Hadamard spectrometers requires in general less mechanical precision than an interferometer for the same spectral region. The attainable resolution is determined by the spectrometer, the fineness of the coding masks, and possible aberrations in the optical system. Reduced dynamic range of the coded signal can be an advantage of such systems.

4.2.5.3.7. HETERODYNE SPECTROSCOPY. The most direct way to obtain selective modulation is to combine the radiation signal at the detector with radiation from a coherent local oscillator. When both signals are spatially coherent over the detector area, the detected power will be modulated with frequencies $\omega_d = \omega_{sig} - \omega_{loc}$. Even with the fastest detectors, however, manageable ω_d's are extremely small compared with ω_{sig}, and tuning of ω_{loc} over the frequencies of interest is required. This still severely limits the applicability of this method which allows extremely high resolution to be obtained and gives a considerable increase in signal-to-noise ratio as compared with conventional detection.[122]

4.2.6. Applications

4.2.6.1. Introduction. The energies of far-infrared photons correspond to thermal energies at temperatures ranging from a few Kelvin to slightly above room temperature. Accordingly, the kind of transitions that can be studied in this spectral region have to do with rather weak forces or comparatively large masses. Forces within molecules give rise mainly to near-infrared frequencies, but some weak bonding forces, bending and torsion

[120] R. N. Ibett, D. Aspinall, and J. F. Grainger, *Appl. Opt.* **7**, 1089–1093 (1968).

[121] M. Harwit, P. G. Phillips, L. W. King, and D. A. Briotta, Jr, *Appl. Opt.* **13**, 2669–2674 (1974).

[122] M. C. Teich, *Proc. IEEE* **56**, 37–46 (1968); *in* "Semiconductors and Semimetals," Volume 5 (R. K. Willardson and A. C. Beer, eds.), Chap. 9. Academic Press, New York, 1970.

effects, etc., can be studied in the far infrared. The effects of the motions of molecules as individuals in gases and liquids and of collective motions in crystals also appear in this region. Resonant behavior of charged particles in gaseous plasmas and in semiconductors and metals can be observed at submillimeter frequencies.

The far-infrared region is of astronomical interest in connection with the cosmic background radiation, the existence of strong sources of radio and submillimeter-wave emission, and the presence of various molecular species in interstellar space.[†]

The possibility of realizing high-capacity communication links with submillimeter waves requires the study of atmospheric transmission properties in this region.

We shall not attempt to review all fields of application in detail; instead some examples will be given to illustrate the use of far-infrared spectroscopy methods in the study of physical problems. For more extensive surveys we refer the reader to the existing textbooks on the subject.[15–19, 123]

4.2.6.2. Gases. The pure rotational absorption by polar molecules in gases has been a favorite object for study in far-infrared spectroscopy from the beginning of measurements in this region. In dilute gases, the pure rotational spectrum consists of narrow lines whose positions and intensities give information on the structure of molecules (see Section 4.1.5). Interactions between the gas molecules cause random perturbations in the rotational energy levels, with the result that the lines are broadened. By using the expression for the transition probability and transforming to the Heisenberg representation,[124] or else by applying linear response theory,[125] the expression for the absorption coefficient, which relates the observed spectrum with the molecular motions, is found to be (cf. Chapter 2.4)

$$a(\omega) = (2\pi^2 n\mu^2/3\varepsilon_0 hc)\omega(1 - e^{-\hbar\omega/kT})f(\omega), \qquad (4.2.23)$$

where $f(\omega)$, called the *reduced* or *normalized spectrum*, is the Fourier transform of the dipolar autocorrelation function

$$f(\omega) = (1/2\pi) \int_{-\infty}^{\infty} \langle \mathbf{u}(0) \cdot \mathbf{u}(t) \rangle e^{-i\omega t} \, dt, \qquad (4.2.24)$$

where $\mathbf{u}(0)$ is a unit vector having the direction of the molecular dipole moment at $t = 0$, $\mathbf{u}(t)$ the corresponding vector at time t, and $\langle \mathbf{u}(0) \cdot \mathbf{u}(t) \rangle$ the equilibrium ensemble average of their scalar product.

[123] A. Finch, P. N. Gates, K. Radcliffe, F. N. Dickson, and F. F. Bentley, "Chemical Applications of Far-infrared Spectroscopy." Academic Press, New York, 1970.
[124] R. G. Gordon, *Advan. Magn. Res.* **3**, 1–43 (1968); *J. Chem. Phys.* **43**, 1307–1312 (1965).
[125] B. J. Berne and G. D. Harp, *Advan. Chem. Phys.* **17**, 63–227 (1970).

[†] See also Vol. 12A ("Astrophysics").

At not too high densities, the spectrum consists of independent lines and the correlation function can likewise be split into parts

$$f(\omega) = \sum_m d_m^2 f_m(\omega - \omega_m)$$

and

$$\langle \mathbf{u}(0) \cdot \mathbf{u}(t) \rangle = \sum_m d_m^2 \langle \mathbf{u}(0) \cdot \mathbf{u}(t) \rangle_m = \sum_m d_m^2 e^{i\omega_m t} \phi_m(t).$$

By writing $\langle \mathbf{u}(0) \cdot \mathbf{u}(t) \rangle_m$ in this way, the coherent part of its time development, due to the free rotation, is given by $e^{i\omega_m t}$, while $\phi_m(t)$ describes the decay of correlation between $\mathbf{u}(0)$ and $\mathbf{u}(t)$ due to random perturbations by other molecules. For sufficiently low pressures, the impact approximation can be made: the perturbations consist of separate independent collisions of negligible duration. This leads, independent of any model for the effects of individual collisions, to an exponential decay for ϕ_m, which results in a Lorentzian reduced line shape[124, 126]

$$\phi_m = e^{-|t|/\tau_m} \quad \text{and} \quad f_m(\omega - \omega_m) = (1/\pi\tau_m)/[(\omega - \omega_m)^2 + (1/\tau_m^2)].$$

The collision frequencies and collision cross sections associated with the decay times τ_m can thus be determined from the widths of the rotational lines and compared with calculations using a specific model for the collision dynamics. As the effect of collisions differs for the various rotational states, the line width depends on the line number m.

Due to the finite duration of real collisions, the impact approximation result for ϕ_m does not describe its short time behavior correctly. Accordingly, the true line shape will differ from a Lorentzian, especially in the far wings as short times correspond to high frequencies.

With increasing densities, the lines start to overlap and it is then not always possible to separate the spectrum into independent lines. Coupling between the lines can change the overall shape of the spectrum.[124] For the continuous spectrum obtained at high densities, a short-time expansion of $\langle \mathbf{u}(0) \cdot \mathbf{u}(t) \rangle$ can be used.[124]

An example of the study of pressure effects on rotational spectra will be discussed in some detail. The pure rotational spectrum of carbon monoxide has been observed at pressures between 1 and 200 atm in pure CO and in CO diluted in helium.[67] The Fourier transform method was used, employing an early NPL–Grubb–Parsons interferometer with mercury arc source, Golay detector, and Mylar beamsplitter. Gas cells of different lengths, closed with crystalline quartz windows, were placed between the interferometer part and the detector. Some spectra obtained at pressures from 2 to 20 atm are

[126] R. Kubo, in "Fluctuation, Relaxation, and Resonance in Magnetic Systems" (D. ter Haar, ed.), pp. 23–68. Oliver and Boyd, Edinburgh, 1961.

illustrated in Fig. 13. At the lower pressures one has the typical linear rotator spectrum with practically equidistant lines. The line widths increase with increasing pressure and around 40 atm the lines were found to merge into a continuum. Widths and strengths of partly overlapping lines were determined by numerically fitting a sum of Lorentz lines to the observed spectra. Examples of fitted spectra together with observed ones are shown in Fig. 14. A good fit could be obtained up to 30 atm with Lorentz lines having width and strength proportional to the gas density. Thus coupling effects due to line overlap were found insignificant. Some indications were found for deviations from the Lorentz shape in the wings of the lines, but as appreciable deviations are to be expected only for distances of more than about 10 cm^{-1} from the line center, these effects are difficult to measure.

Results of measurements at pressures between 50 and 200 atm are given in Fig. 15. The absorption coefficient per unit absorber density was found to be independent of the pressure apart from the high-wavenumber region in the pure CO case. The increasing absorption in that region can be interpreted as pressure-induced pure rotational absorption (cf. Section 4.1.6.1). The calculated spectra in Fig. 15 were obtained by taking for the reduced spectrum the classical probability distribution of rotational frequencies for free linear rotators. Apparently, the influence of intermolecular interactions is still small on the relatively short time scale involved here.

4.2.6.3. Liquids and Dielectric Solids.[†]

The rotational motion of molecules causes absorption of radiation in polar liquids as it does in gases. Evidence for relatively free rotation has been observed in the spectra of polar molecules dissolved in nonpolar liquids,[127, 128] but generally the rotation is strongly perturbed by the surrounding molecules and the motion is of a random nature rather than being a free rotation. The correlation function $\langle \mathbf{u}(0) \cdot \mathbf{u}(t) \rangle$ then does not contain a separable coherent part and when the correlation time of the perturbations is short enough, it will show an exponential decay.[126, 129] As a result the reduced spectrum (4.2.24) will become a Lorentzian centered at $\omega = 0$, and the absorption coefficient becomes

$$a(\omega) = \frac{2\pi^2 n\mu^2}{3\varepsilon_0 hc} \; \omega(1 - e^{-\hbar\omega/kT}) \frac{\tau/\pi}{\omega^2\tau^2 + 1} \approx \frac{n\mu^2}{3\varepsilon_0 ckT} \frac{\omega^2\tau}{\omega^2\tau^2 + 1}.$$

$$(4.2.25)$$

The last part is obtained by using $\hbar\omega \ll kT$. This type of absorption is known as *Debye relaxation* and can be observed in the radio-frequency to submillimeter regions (cf. Section 4.4.3). The behavior of real liquids is only

[127] D. Datta and G. M. Barrow, *J. Chem. Phys.* **43**, 2137–2139 (1965).
[128] R. M. van Aalst and J. van der Elsken, *Chem. Phys. Lett.* **23**, 198–199 (1973).
[129] H. Shimizu, *J. Chem. Phys.* **43**, 2453–2465 (1965).

[†] See Chap. 4.4 for further treatment of relaxation phenomena.

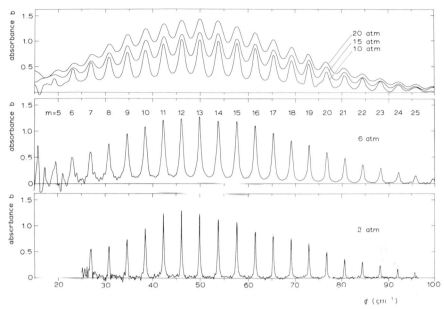

FIG. 13.   Examples of carbon monoxide absorbance spectra at different pressures [from D. Oepts, Thesis, Univ. of Amsterdam, 1972]

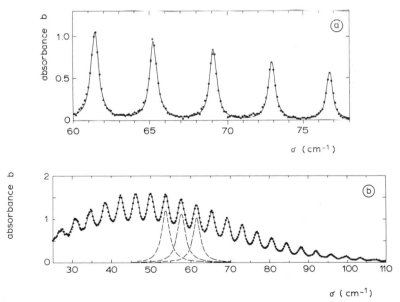

FIG. 14.   Examples of the fitting of synthetic spectra (solid lines) to observed CO spectra (points). (a) Lines at 5 atm showing little overlap; (b) overlapping lines at 20 atm. Three lines of the fitted spectrum are shown separately [from D. Oepts, Thesis, Univ. of Amsterdam, 1972].

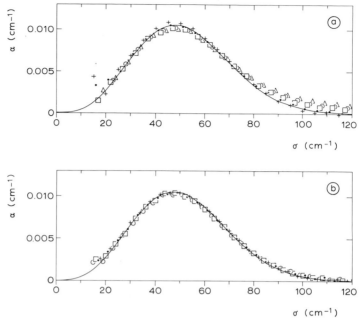

FIG. 15. Absorption coefficient spectra of carbon monoxide at different pressures. (a) Pure CO with pressures of 46(+), 70($\bullet$), 95 ($\square$), 144($\bigcirc$), and 195($\triangle$) atm. The solid line is the calculated value with $\mu_D = 0.116$. (b) CO-helium mixture containing 10% CO at pressures of 95(+), 128($\odot$), 168($\square$), and 186($\bigcirc$) atm. The solid line is the calculated value with $\mu_D = 0.117$ [from D. Oepts, Thesis, Univ. of Amsterdam, 1972].

approximately described by (4.2.25). Deviations occur for instance at high frequencies as the exponential form for the correlation function is incorrect for short times where inertial effects dominate in the motion of $\mathbf{u}(t)$.

Not only absorption of radiation is observed in liquids and solids, but also refraction and reflection. The radiation field acting on the molecules in a dense medium is not just the incident field, but the fields caused by surrounding molecules have to be taken into account and the interaction between matter and radiation is more complicated than in gases. The response of the material can be described by a complex dielectric constant $\varepsilon(\omega) = \varepsilon'(\omega) - i\varepsilon''(\omega)$, or by the associated complex refractive index. For example, the expression for the Debye absorption is frequently written as[130]

$$\frac{\varepsilon(\omega) - \varepsilon(\infty)}{\varepsilon(0) - \varepsilon(\infty)} = \frac{1}{1 + i\omega\tau}. \tag{4.2.26}$$

[130] V. V. Daniel, "Dielectric Relaxation," Academic Press, New York, 1967.

The absorption is determined mainly by the imaginary part of ε; in gases, where the real or dispersive part is essentially unity, the absorption coefficient is equal to $a(\omega) = \frac{1}{2}\omega\varepsilon''(\omega)$ so that (4.2.26) agrees with (4.2.25). The real and imaginary parts of ε can be solved from (4.2.26) and when ε'' is plotted against ε' a semicircle is obtained. By plotting experimental values for ε' and ε'' in the same way, known as a *Cole–Cole plot*, deviations from the simple Debye behavior appear as deviations from the semicircular arc. With the amplitude spectroscopy technique (Section 4.2.4.5) both the real and imaginary parts of the refractive index can be determined.

In addition to absorption associated with rotational molecular motion, both polar and nonpolar liquids show far-infrared absorptions than can be associated with vibrational and translational motions (cf. Section 4.1.4). In this respect, the behavior of liquids shows a remote similarity to that of solids and the spectra have some resemblance with those obtained in the solid form of the same substance.

The spectrum of liquid chloroform has been investigated at NPL and Nancy[131] over the extremely broad frequency range 0–4200 GHz, using interferometers for the microwave-, millimeter-, and submillimeter-wave regions. For the submillimeter part of the spectrum an NPL Grubb–Parsons cube interferometer as illustrated in Fig. 10 was employed. A liquid-helium cooled InSb Putley detector was used in the 4–36 cm^{-1} region and a Golay cell for the higher wavenumbers. The Cole–Cole plots obtained at two temperatures are shown in Fig. 16, and Fig. 17 gives the absorption coefficient and the real part of the refractive index separately. Deviations from the Debye behavior including additional absorption at intermediate frequencies and high-frequency rolloff are apparent.

In crystalline solids, far-infrared absorption can occur due to dipolar motions under the influence of intermolecular or intramolecular forces. A well-known example of the first type is the lattice vibration in alkali halide crystals, which can be considered as a vibration of the two ionic sublattices relative to each other. The strong coupling of this vibration with the radiation field gives rise to the very strong absorption and high reflectivity in the far infrared exhibited by these crystals and put to use in reststrahlen filters.

Up to now, we have considered only electric dipole interactions of the sample material and the radiation field. Magnetic interaction is also possible although this is usually very much weaker. Magnetic resonance associated with paramagnetic ions in a crystal lattice can be used to obtain information about the environment of these ions. The transition frequencies involved are determined by an external magnetic field and this is usually chosen such that

[131] J. Goulon, J. L. Rivail, J. W. Fleming, J. Chamberlain, and G. W. Chantry, *Chem. Phys. Lett.* **18**, 211–216 (1973).

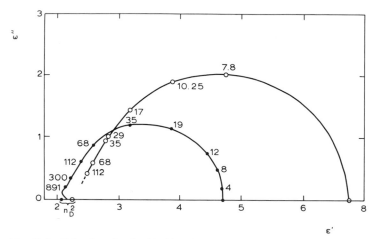

FIG. 16. Cole–Cole plots for liquid chloroform at 25(●) and −60(○) °C. The numbers against the points refer to the frequencies in gigahertz [from J. Goulon, J. L. Rivail, J. W. Fleming, J. Chamberlain, and G. W. Chantry, *Chem. Phys. Lett.* **18**, 211–216 (1973), by permission of North-Holland Publ.].

resonance is observed in the microwave region. Sometimes it is advantageous to use submillimeter waves, for instance to study transitions between component levels already split up by the internal electrostatic crystal field. Zero-field splitting and transitions in high magnetic fields have been studied with the use of Fourier spectroscopy.[132, 133]

In antiferromagnetic substances, the ordered spin systems can perform collective motions and these spin waves or magnons can be excited by radiation of the proper frequency, somewhat as a magnetic analog of lattice vibrations or phonons. Characteristic frequencies lie in the submillimeter-wave region and so submillimeter-wave spectroscopy methods are essential for the study of this phenomenon. Antiferromagnetic resonance (AFMR) has been observed in the far infrared for the first time with a grating spectrometer[134] and has subsequently been studied with Fourier transform methods.[135]

4.2.6.4. Semiconductors and Metals. The absorption of far infrared radiation by semiconductors can be applied to the study of lattice vibrations just as in isolators. In addition, however, the free charge carriers couple with the radiation field and give the material interesting far-infrared properties.

[132] J. C. Hill and R. G. Wheeler, *Phys. Rev.* **152**, 482–494 (1966).

[133] C. C. Brackett, P. L. Richards, and W. S. Caughey, *J. Chem. Phys.* **54**, 4383–4401 (1971).

[134] R. C. Ohlman and M. Tinkham, *Phys. Rev.* **123**, 425–434 (1961); D. Bloor and D. H. Martin, *Proc. Phys. Soc.* **78**, 774–776 (1961).

[135] P. L. Richards, *Phys. Rev.* **138**, A1769–1775 (1965).

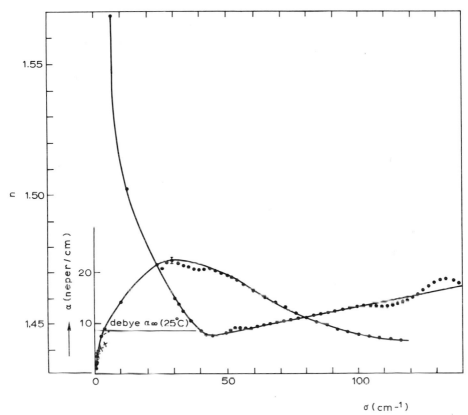

FIG. 17. Absorption coefficient α and refractive index n of liquid chloroform in the milli-meter-wave regions at 25($\bullet$) and $-60(\times)^{\circ}$C [from J. Goulon, J. L. Rivail, J. W. Fleming, J. Chamberlain, and G. W. Chantry, *Chem. Phys. Lett.* **18**, 211–216 (1973), by permission of North-Holland Publ.].

The study of these properties is of interest among other reasons in view of their application in the development of detectors, mixers, and sources for this spectral region. For instance, photoconductivity effects have been studied in various materials and now widely used detectors have resulted.[136]

A phenomenon that can be studied with advantage in the submillimeter region is cyclotron resonance absorption. It presents an example of a mea-surement where a fixed frequency source can be employed expediently. A charged particle moving freely in a magnetic field describes a circular motion around the lines of force, with angular frequency $\omega_c = eB/m$. The electrons and holes in a semiconductor are not free, but move in the periodic potential

[136] E. H. Putley, *Phys. Status Solidi* **6**, 571–614 (1964).

of the lattice. This can be taken into account by replacing m with an effective mass m^*. Charge carriers with different effective masses can be present, and m^* can also be anisotropic.[137] These effects can be studied by observing the absorption occurring when ω_c for any m^* coincides with the frequency of the incident radiation. The cyclotron motion of the charge carriers is damped by their interaction with the lattice, with a characteristic time τ. In order to observe clear resonances, $\omega_c \tau > 1$ should hold. Quantum effects in the spectrum can be observed when $\hbar \omega_c > kT$. These conditions can be satisfied by using high magnetic fields so that ω_c falls in the submillimeter-wave region, and by keeping T low.

Several materials have been studied by Button and co-workers using molecular lasers as the radiation source.[138] A schematic diagram of their setup is given in Fig. 18. The transmission of a sample of the material in a

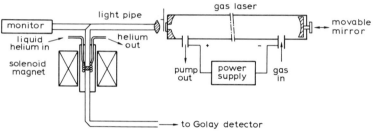

FIG. 18.   Diagram of laser spectrometer system used by Button *et al.* [from K. J. Button and B. Lax, *in* "Submillimeter Waves" (*Proc. Symp. N.Y.*, 1970) (J. Fox, ed.), Polytechnic Press, Brooklyn, New York, 1971].

cryostat is observed while the magnetic field strength is varied to scan the spectrum. One of the results is reproduced in Fig. 19. The complicated spectrum of transitions in the valence band illustrates the usefulness of low temperatures. At 200 K, sufficient electrons are excited into the conduction band to observe their cyclotron resonance as well. With weaker fields and microwave frequencies this peak would have been unobservable.

The high density of free electrons in metals prevents the propagation of far-infrared radiation in the material. The refractive index has a large imaginary part and the radiation is almost perfectly reflected. Under special conditions, however, cyclotron resonance can be observed. When the magnetic field is parallel to the surface of the material and the cyclotron radius

[137] J. M. Ziman, "Principles of the Theory of Solids," Cambridge Univ. Press, London and New York, 1964.

[138] K. J. Button, H. A. Gebbie, and B. Lax, *IEEE J. Quant. Electron.* **QE–2**, 202–207 (1966); K. J. Button and B. Lax, *in* "Submillimeter Waves" (*Proc. Symp. N.Y. 1970*) (J. Fox, ed.), pp. 401–416. Polytechnic, Brooklyn, New York, 1971.

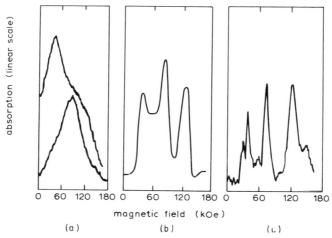

FIG. 19 Cyclotron resonance spectrum of holes in pTe at different temperatures, with conduction electron resonance at 200 K; $\lambda = 337$ μm and $H \perp c$. (a) T $\approx$ 200 K (upper curve), T $\approx$ 100 K (lower curve); (b) T $\approx$ 30 K; (c) T $\approx$ 10 K [after K. J. Button and B. Lax, in "Submillimeter Waves" (*Proc. Symp. N.Y., 1970*) (J. Fox, ed.). Polytechnic Press, Brooklyn, New York, 1971].

is larger than the skin depth to which the radiation penetrates, the gyrating electrons will cross the skin layer periodically during a fraction of the cyclotron period. Provided again that $\omega_c \tau > 1$, radiation polarized perpendicular to the magnetic field will be absorbed at the frequency ω_c and its harmonics. An example of such absorption, known as *Azbel'–Kaner resonance*, is given in Fig. 20. In these measurements use was made of a carcinotron source and a semiconfocal resonator as described in Section 4.2.2.

The energy gap between superconducting and normal electron states in superconductors corresponds to submillimeter wavelengths and becomes apparent in the submillimeter-wave absorption properties of superconductors, as observed, for instance, by Richards and Tinkham[139] with a grating spectrometer. Fourier transform spectroscopy has also been used for measurements in this field.

The properties of Josephson junctions are of considerable interest in view of their possible application to the detection, generation, and mixing of submillimeter waves.[103, 140]

4.2.6.5. Plasmas. The properties of electromagnetic wave propagation in a plasma[141] depend critically on a parameter called the *plasma frequency*, given by

$$\omega_p^2 = n_e e^2 / \varepsilon_0 m_e,$$

[139] P. L. Richards and M. Tinkham, *Phys. Rev.* **119**, 575–590 (1960).
[140] B. N. Taylor, *J. Appl. Phys.* **39**, 2490–2502 (1968).
[141] G. Bekefi, "Radiation Processes in Plasmas." Wiley, New York, 1966.

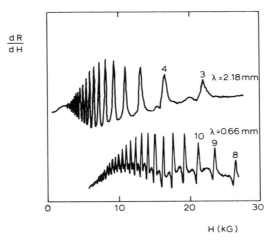

FIG. 20.   Azbel'–Kaner resonance in copper, observed with a carcinotron source [from Y. Couder and P. Goy, *in* "Submillimeter Waves" (*Proc. Symp. N.Y., 1970*). (J. Fox, ed.). Polytechnic Press, Brooklyn, New York, 1971].

where n_e is the number density of the electrons. For gaseous plasmas encountered in gas discharge physics, fusion research, and astrophysics, values of n_e are often such that ω_p corresponds to far-infrared wavelengths. This also occurs in semiconductors.

For a collisionless plasma the refractive index is given by

$$n(\omega)^2 = 1 - (\omega_p{}^2/\omega^2).$$

For frequencies less than ω_p, n becomes purely imaginary and incident waves do not propagate into the plasma but are totally reflected. For waves with $\omega > \omega_p$ the plasma is transparent and the refractive index can be measured to obtain ω_p and the electron density. Visible radiation can be used for this purpose, but in the far infrared the effect is larger and interferometers in this region are less sensitive to vibrations and misalignment. Molecular lasers can be used to advantage for this purpose.[142]

When the electron collision frequency is not negligible, strong absorption occurs for radiation frequencies near ω_p. The presence of magnetic fields further complicates the behavior; due to cyclotron resonance effects the transmission properties come to depend on the direction of propagation and state of polarization. In this connection, Faraday rotation and related effects can serve to probe plasma conditions.

[142] R. W. Peterson and F. C. Jahoda, *Appl. Phys. Lett.* **18**, 440–442 (1971); P. Brossier and R. A. Blanken, *IEEE Trans. Microwave Theory Tech.* **MTT-22**, 1053–1056 (1974); D. Véron, *IEEE Trans. Microwave Theory Tech.* **MTT-22**, 1117 (abstr) (1974).

A hot plasma emits as well as absorbs, refracts, and scatters radiation. The emitted radiation is important for the energy balance of the plasma and is of interest in the study of various physical processes in the plasma. An important source of radiation is the bremsstrahlung produced through the acceleration of electrons in the field of the ions. The emitted power per frequency or wavenumber interval is proportional to $n_e^2 T_e^{-1/2}$ and, in the far infrared, independent of the frequency as long as the blackbody level is not reached. Near ω_p where absorption is high, the plasma can become optically thick and emit like a blackbody. For lower frequencies, in the region of high reflectivity, the emissivity drops to very low values.[143] An example of such plasma radiation is found in the mercury arc, which is widely used as a source of far-infrared and submillimeter-wave radiation.[144]

In addition to bremsstrahlung by individual electrons, so called *suprathermal radiation* can arise from nonlinear interactions involving longitudinal waves which occur in the plasma at the plasma frequency. In turbulent plasmas this radiation can be orders of magnitude more intense than the regular thermal bremsstrahlung.[145]

Another type of radiation that is important in the millimeter- and submillimeter-wave regions is cyclotron radiation associated with the electron motion in strong magnetic fields. In first approximation, the electrons simply radiate with the cyclotron frequency $\omega_c = eB/m_e$. Higher harmonics of ω_o are present when the electron velocities v are such that v/c is not negligible. For relativistic electrons, $v \approx c$, the radiation is usually called *synchrotron radiation*; the maximum intensity is then found in the harmonics near $\omega = \frac{1}{2}\omega_c/(1 - (v/c)^2)$. This type of radiation can be responsible for a large part of the energy loss from the plasma. The discrete harmonic frequencies are broadened by various mechanisms and only the first few lines of a cyclotron resonance spectrum will be resolved. Measurements of such radiation from a laboratory plasma have been made in the 2.5–50 cm^{-1} region using a grating spectrometer.[146]

In plasma physics experiments the properties of the radiation source usually vary both in space and time. This adds new problems to those always present in submillimeter-wave spectroscopy. The use of long time constants to improve the signal-to-noise ratio is clearly impossible. Very rapid scanning, or multichannel operation, or shot-by-shot scanning in pulsed experiments has to be applied. The use of multiple detection systems is unattractive

[143] M. F. Kimmit, A. C. Prior, and V. Roberts, *in* "Plasma Diagnostic Techniques" (R. H. Huddlestone and S. L. Leonard, eds.), Chapter 9. Academic Press, New York, 1965.

[144] G. W. Chantry, "Far-Infrared Techniques," p. 209. Pion, London, 1970.

[145] C. Chin-Fatt and H. R. Griem, *Phys. Rev. Lett.* **25**, 1644–1646 (1970).

[146] S. Sesnic, A. J. Lichtenberg, A. W. Trivelpiece, and D. Tuma, *Phys. Fluids* **11**, 2025–2031 (1968).

due to their cost and complexity. Rapid scanning Fabry–Perot[147] and Fourier spectroscopy instruments have been devised and shot-by-shot measurements have been made both with gratings and Fourier methods.[148] Putley detectors are commonly used in view of their extremely fast response and high sensitivity.

The attainment of high resolution is impossible in these applications and fortunately usually not necessary. Therefore relatively broad-band filters are sometimes preferred over more elaborate monochromator systems. Cyclotron and suprathermal radiation spectra have been observed in this way.[149]

4.2.6.6. Astronomy and Atmospheric Physics.[†] The application of far-infrared spectroscopy in astronomy is hindered by the strong atmospheric absorption. Over most of the region, measurements are possible only with the use of rockets, balloons, or high-altitude aircraft to bring the instrument above or high up in the atmosphere. Due to their simplicity and high detection efficiency, Michelson interferometers are preferred for this purpose.[55]

The weakness of the radiation received from astronomical objects favors the use of Fourier transform spectroscopy in the near infrared as well. In that region, much higher resolving powers are usually applied than in the far infrared and the multiplex and étendue advantages are fully effective. The most impressive achievement of Fourier transform spectroscopy is still the measurement of planetary spectra with high resolution by the pioneers in this field, J. and P. Connes.[150]

Atmospheric absorption is of interest in meteorologic and environmental studies and also in view of possible submillimeter communication systems. Due to the long absorption paths involved and the presence of pressure broadening, there is considerable absorption between the pertinent rotational lines, and both strength and shape of the lines are of interest. Measurements have been made with Michelson interferometers using molecular lasers as sources[151] and also with the sun as the radiation source.[152] High-resolution laboratory measurements of water vapor absorption have also been made with the use of Froome-type harmonic generators. The arrangement used in

[147] R. A. Blanken, Ph. Brossier, and D. S. Komm, *IEEE Trans. Microwave Theory Tech.* **MTT-22**, 1057–1061 (1974).

[148] A. E. Costley, R. J. Hastie, J. W. M. Paul, and J. Chamberlain, *Phys. Rev. Lett.* **33**, 758–761 (1974).

[149] S. F. Nee, A. W. Trivelpiece, and R. E. Pechacek, *Phys. Fluids* **16**, 502–508 (1973).

[150] J. Connes and P. Connes, *J. Opt. Soc. Amer.* **56**, 896–910 (1966); J. Connes, P. Connes, and J. P. Maillard, "Atlas des spectres dans le proche infrarouge de Vénus, Mars, Jupiter et Saturne." Editions du C.N.R.S., Paris, 1969.

[151] J. R. Birch, W. J. Burroughs, and R. J. Emery, *Infrared Phys.* **9**, 75–83 (1969).

[152] J. E. Harries and P. A. R. Ade, *Infrared Phys.* **12**, 81–94 (1972).

[†] See also Vol. 12A and 12B ("Astrophysics").

such an investigation[153] is shown in Fig. 21. The absorption cell has a length of 6 m and is periodically and automatically replaced by a similar reference cell.

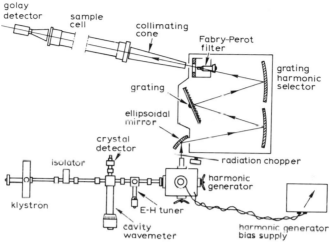

FIG. 21.   Arrangement for submillimeter-wave atmospheric absorption measurements using an arc harmonic generator [from R. Emery, *Infrared Phys.* **12**, 65–79 (1972), by permission of Pergamon Press].

An interesting application of interferometric spectroscopy has been proposed by Dick and Levy.[154] In their correlation interferometer the interferogram of a remote source is periodically scanned and correlated with a reference interferogram obtained with a known amount of a sample gas. The path difference is scanned by means of a rotating plate and the reference interferogram is recorded on magnetic material rigidly fixed to the rotating plate so that a fixed phase relation is guaranteed. Simultaneous detection of different gas species is possible in principle.

[153] R. Emery, *Infrared Phys.* **12**, 65–79 (1972).

[154] R. Dick and G. Levy, in *Aspen Int. Conf. Fourier Spectrosc., 1970* (G. A. Vanasse *et al.*, eds.), pp. 353–360. AFCRL–71–0019 (1971).

4.3. Microwave Region[†][‡][*]

4.3.1. Introduction

Microwave spectroscopy as a research field is rapidly reaching maturity. The challenge for the future is to find new areas of science where the experience and technology developed during the past 30 years or so can have a truly useful impact.

The aim of the present chapter is to discuss that technology as it is commonly practiced in spectroscopic laboratories today, and to look briefly at a few new developments in instrumentation that show promise for the future. It is difficult to cover all aspects of the laboratory art in a single chapter. We will therefore limit ourselves to a few seasoned examples of practical techniques that clearly promise expanded usefulness in future applications.

The available microwave literature is rich indeed. The heritage of microwave instrumentation that followed from the radar developments of the early 1940's is well summarized in the excellent texts of Gordy *et al.*,[1] Strandberg,[2] and Townes and Schawlow[3] to name a few. Recent books by Wollrab,[4] Ingram,[5] Sugden and Kenney,[6] and Gordy and Cook[7] give excellent accounts of the details of the theoretical and experimental aspects

[1] W. Gordy, W. V. Smith, and R. F. Trambarulo, "Microwave Spectroscopy." Wiley, New York, 1953. (Republication, Dover, New York, 1966.)

[2] M. W. P. Strandberg, "Microwave Spectroscopy." Methuen, London, 1954.

[3] C. H. Townes and A. L. Schawlow, "Microwave Spectroscopy." McGraw-Hill, New York, 1955.

[4] J. E. Wollrab, "Rotational Spectra and Molecular Structure." Academic Press, New York, 1967.

[5] D. J. E. Ingram, "Spectroscopy at Radio and Microwave Frequencies." Butterworth, London and Washington, D.C. (2nd ed., Plenum Press, New York, 1967.)

[6] T. M. Sugden and C. N. Kenney, "Microwave Spectroscopy of Gases." Van Nostrand–Reinhold, Princeton, New Jersey, 1965.

[7] W. Gordy and R. L. Cook, "Microwave Molecular Spectra." Wiley (Interscience), New York, 1970.

[†] Contribution of the National Bureau of Standards. Not subject to copyright.

[‡] Certain commercial instruments and products are identified in this paper for completeness of the experimental discussion. In no case does such identification imply a recommendation or endorsement by the National Bureau of Standards.

[*]Chapter 4.3 is by Donald R. Johnson and Richard Pearson, Jr (NRC–NBS Postdoctoral Research Associate 1974–1976).

of microwave spectroscopy not treated at length in the present work. Carrington's new book[8] summarizes a wide variety of modern microwave systems used in recent experiments. Many reviews of the original microwave research literature are also available to compliment the texts referenced above. One of the more recent of these reviews by Morino and Saito[9] gives a particularly good sampling of references to the significant developments in the field.

In an earlier volume of this "Methods of Experimental Physics" treatise, Lide[10] gave particular emphasis to the interpretation of microwave spectra and discussed in detail the molecular information that can be obtained. It is our intention to build on that work in the present chapter. We shall therefore look to that earlier part of this same treatise for definitions of terms, units, etc.

A glance at tabulations of microwave spectral measurements, such as the NBS Monograph 70 Series,[11] clearly indicates that the most intensively explored region of the microwave spectrum lies between 8 and 40 GHz. Historically speaking, microwave hardware was first developed in the region 2–40 GHz for applications in military radar. Much of this hardware has been made available for spectroscopic applications through a variety of commerical and surplus sources. Reflex klystrons, developed during this same period, provided relatively inexpensive and good quality sources of spectrally pure microwave radiation. Above 8 GHz the hardware components were broad-banded and of convenient size for laboratory work. Dimensional tolerances were not particularly critical below 40 GHz so that homemade components could be fabricated without special techniques. The introduction of the Stark-modulated spectrograph by Hughes and Wilson[12] in 1946 tended to produce additional concentration of activities in this region because their septum-type absorption cell worked best above 8 and below 40 GHz. With the increased sensitivity of the Stark-modulated system, many polar molecules were found to have detectable absorptions between low-lying rotational levels in this region and much useful spectroscopy has been done there.

As will be seen in the ensuing sections, developments in instrumentation have come a long way during the past three decades. Commerical spectrometers have been developed that employ backward wave oscillators and broad-banded components so that entire waveguide bands can now be swept

[8] A. Carrington, "Microwave Spectroscopy of Free Radicals." Academic Press, New York, 1974.

[9] Y. Morino and S. Saito, Microwave spectra, in "Molecular Spectroscopy: Modern Research" (K. Narahari Rao and C. Weldon Mathews, eds.). Academic Press, New York, 1972.

[10] D. R. Lide, Jr., Microwave spectroscopy, in "Molecular Physics" (D. Williams, ed.), Methods of Experimental Physics, Vol. 3, 2nd ed., Part A, Chapter 2.1. Academic Press, New York, 1974.

[11] Microwave Spectral Tables, Nat. Bur. Std. U.S. Monograph 70, Vol. I, 1964; Vol. III, 1969; Vol. IV, 1968; Vol. V, 1968.

[12] R. H. Hughes and E. B. Wilson, Jr., *Phys. Rev.* **71**, 562L (1947).

without tuning. Four waveguide bands cover the entire range 8–40 GHz and, with computer control, the data-taking process is almost completely automatic.

The millimeter-wavelength region of the spectrum has also seen extensive development over the years although at a somewhat slower pace than the "conventional region." Gordy and his collaborators[13] pioneered much of the early work in the millimeter-wave region using reflex klystrons with harmonic generation techniques. Recently, Helminger et al.[14] extended these techniques to a measurement of the $J = 67 \leftarrow 66$ transition of $^{16}O^{12}C^{32}S$ at 813,353.706 MHz. Krupnov and his collaborators[15] in the U.S.S.R. have recently explored the possibility of using special submillimeter-backward-wave oscillators as spectroscopic sources at very short wavelengths. Employing acoustical detection techniques, they have developed a new type of broadbanded spectrometer which has been used to observe molecular spectra of frequencies above 1 THz (1 THz = 10^{12} Hz).

4.3.2. Sources

If one single characteristic sets the microwave region apart from other regions of spectral investigation, it is the availability of tunable and essentially monochromatic sources of radiation at microwave frequencies. Typical microwave spectroscopic experiments require a source stability of a few parts in 10^7 and power levels in the 10-mW range. Instruments intended for scanning over many spectral features also require that the radiation source be easily varied in frequency. Specialized experiments sometimes place much more stringent requirements on source stability or power. Double-resonance experiments, for example, often require output power levels of 1 W or more. A large variety of oscillator devices has been developed to meet these requirements and each type has its own distinct character that offers both advantages and disadvantages for spectroscopic applications.

The noise generated by the source oscillator off the main carrier frequency can be a limiting factor in the base sensitivity of the entire spectrometer system. Leskovar[16] has investigated both the amplitude-modulated (AM) noise and the frequency-modulated (FM) noise off a carrier frequency of 10 GHz for a variety of current state-of-the-art sources. Some preliminary results of his investigations are summarized in Figs. 1 and 2, and provide a comparison of the relative performances of the various devices. It should be

[13] W. Gordy, *Pure Appl. Chem.* **11**, 403 (1965).

[14] P. Helminger, F. C. De Lucia, and W. Gordy, *Phys. Rev. Lett.* **25**, 1397 (1970).

[15] A. F. Krupnov and A. V. Burenin, *in* "Molecular Spectroscopy: Modern Research" (K. Narahari Rao, ed.), Vol. 2. Academic Press, New York, 1976.

[16] B. Leskovar, Lawrence Berkeley Lab., Berkeley, California, private communication.

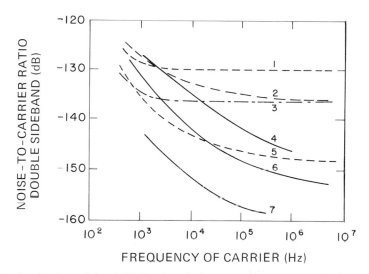

FIG. 1. Amplitude-modulated (AM) noise of microwave sources at a carrier frequency of 10 GHz and a bandwidth of 100 Hz. The sources shown are: (1) Si IMPATT, $Q = 500$; (2) backward wave oscillator; (3) GaAs IMPATT, $Q = 500$; (4) varactor harmonic generator, (5) GaAs Gunn, $Q = 500$; (6) tunable reflex klystron; (7) experimental two-cavity klystron. (B. Leskovar, Lawrence Berkeley Lab., Berkeley, California, private communication).

emphasized that these plots compare noise characteristics for only one center or carrier frequency near 10 GHz. As this center frequency is increased, the noise performance of all of the devices listed in Figs. 1 and 2 deteriorates at approximately logarithmic rates and their relative positions change somewhat. At frequencies above 40 GHz, noise in solid-state sources appears to increase more rapidly with frequency than it does in klystrons.

Reflex klystrons remain the most widely used oscillators for microwave spectroscopy. They compare favorably with other devices in both AM and FM noise characteristics, have good reliability, and offer output power levels from 10 mW to more than 1 W. At a given frequency, lower power tubes tend to have lower Q cavities and higher stability, and are thus better suited to most spectroscopic applications. Klystrons are characteristically narrow-banded devices which can be mechanically tuned approximately $\pm 5\%$ of their center frequency and swept electronically over a limited region ranging from 5 to 300 MHz depending on the individual tube characteristics. This rather limited tuning range for an individual tube is not a serious handicap for many spectroscopic applications. Continuous coverage over a wide spectral region can, however, require a significant collection of tubes. At the National Bureau of Standards a set of 19 reflex klystrons with carefully selected center frequencies is used to provide fundamental power throughout

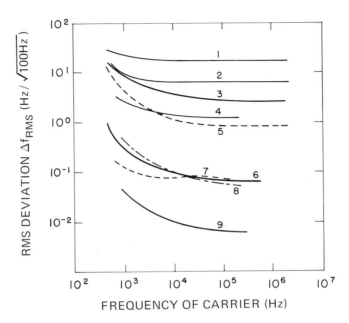

FIG. 2. Frequency-modulated (FM) noise of microwave sources at a carrier frequency of 10 GHz and a bandwidth of 100 Hz. The sources shown are: (1) Si IMPATT, $Q = 500$; (2) GaAs IMPATT, $Q = 500$; (3) GaAs Gunn, $Q = 240$; (4) reflex klystron; (5) GaAs Gunn, $Q = 2300$; (6) GaAs Gunn, $Q = 23,000$; (7) cavity stab. refl. kly., $Q = 8000$; (8) two-cavity klystron; (9) cavity stab. two-cav. kly., $Q = 23,000$ (B. Leskovar, Lawrence Berkeley Lab., Berkeley, California, private communication).

the region from 4 to 132 GHz with a minimum amount of overlap between tubes.

Reflex klystrons often tend to be extremely sensitive to thermal changes and mechanical vibrations. Significant improvements in both source stability and noise can be achieved with any reflex klystron by providing a firm mount and adequate cooling. Typically, forced air cooling has been used in most spectroscopic applications, often with less than satisfactory results. Some improvement has been obtained with tubes operated on water-cooled heat sinks. The thermal gradients can, however, be large and the tube life adversely affected. A simple solution which is both convenient and inexpensive is to operate the tube submerged in a bath of oil. Figure 3 illustrates an oil bath design that has been used successfully at NBS for many years. The flange at the output of the klystron has been fitted with a small O-ring and sealed to a short piece of waveguide that passes through the side of the can. About 3 gal. of SAE #20 nondetergent motor oil produces an excellent heat sink which

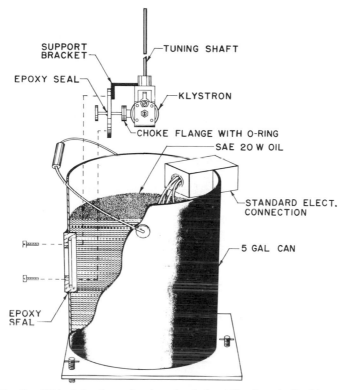

FIG. 3. Oil bath for thermal and mechanical stabilization of reflex klystrons.

stabilizes by radiative cooling, so that the can surface is only slightly warm to the touch after 20 or 30 min of operation. Rubber-coated "meter lead" does not deteriorate on contact with the oil and allows electrode potentials of up to 3000 V to be safely applied. Typical reflex klystrons operating in oil baths of this type achieve a short-term frequency stability approaching 1 part in 10^7 which is adequate for most spectroscopic applications without additional electronic stabilization.

Backward wave oscillators (BWO) have undergone extensive development during the last few years and are now available from a wide variety of commercial sources. In spectroscopic applications they can offer several distinct advantages. BWO's are inherently broad-banded devices that can be electronically swept in frequency over entire waveguide bands. When coupled to suitably broad-banded absorption cells and hardware they are particularly well suited for low-resolution spectral scanning. The average output power level of a BWO can be reasonably high over the entire band and is therefore capable of being electronically leveled for applications that require constant power.

Although the BWO tested for curve 2 of Fig. 1 represents typical operation at a single-carrier frequency, it is symtomatic of a common difficulty with most BWO's throughout the microwave spectral region. Without additional electronic stabilization, they tend to be too noisy for high-resolution spectroscopic applications. Phase locking a broad-banded oscillator introduces significant complications which tend to offset the characteristics which make this class of devices attractive for use as microwave sources.

Several manufacturers are now marketing integral BWO−power supply−sweeper packages that automatically scan an entire waveguide band. At least one company presently offers a device capable of oscillation well beyond 100 GHz.[17] Experimental BWO's have already been successfully employed by Krupnov and his collaborators[15] in broad-banded spectroscopic experiments above 1 THz. Again, the simplicity of wide-range electronic tuning appears to be a decisive factor in the choice BWO's as sources of radiation in these submillimeter-wave experiments. The introduction of superconducting materials for solenoids and the improvement of materials for permanent magnets will allow higher-density electron beams to be focused and may make commerical submillimeter-wave BWO's practical in the near future.

Solid-state oscillators are rapidly becoming practical for many phases of laboratory spectroscopy. The potential of such devices as IMPATT (*imp*act ionization *a*valanche *t*ransit *t*ime) diodes and Gunn effect oscillators have been extensively investigated over the last few years. Epitaxial silicon IMPATT diodes have good power capabilities but, as can be seen from Figs. 1 and 2, they are generally noisy devices because the avalanche breakdown is a noisy phenomenon. Gallium−arsenide IMPATTs show improved noise characteristics but generally do not have the power capabilities of the silicon devices. IMPATTs using other material such as germanium may offer potential for improved performance. An improved version called a TRAPATT (*t*rapped *p*lasma *a*valanche-*t*riggered *t*ransit) diode uses an alternating cycle between two avalanche modes and shows significant promise for improved efficiency. IMPATT diode devices are commerically available to at least 40 GHz and prototype devices have operated beyond 100 GHz. If future developments improve the noise characteristics of these avalanche devices, their large power capabilities and wide range of oscillation may find many practical uses in spectroscopy.

The Gunn effect in its simplest form is an oscillation that results when a dc voltage above a certain threshold is applied across a junctionless bulk sample of the base semiconductor material. A region of high field strength travels

[17] F. Gross, *Microwave J.* **14** (4), 55 (1971).

through the material from one ohmic contact to the other. As one field peak drains away at the anode, a new field peak builds up behind it. The Gunn effect is therefore an electron transfer effect in the bulk material and does not have associated with it the noise inherent in the avalanche processes. An excellent discussion of the design parameters that determine the output frequency of a Gunn effect oscillator has been presented by Sweet,[18] and we refer the reader to this source for details.

From Figs. 1 and 2 it is clear that Gunn effect oscillators compare favorably in both AM and FM noise with reflex klystrons. In addition, they are sufficiently broad-banded that they compete favorably with backward wave oscillators. Both varactor diode and YIG[†] tuning are now in common use with cavity-stabilized Gunn oscillators. Commerical oscillators are now available to 40 GHz with output power levels of at least several milliwatts. The state of the art limit for Gunn oscillators appears to be around 60 GHz; but other bulk property devices, currently under investigation, show promise of higher-frequency operation.

Solid-state devices in general require only low-voltage power supplies and therefore lend themselves nicely to applications such as chemical monitoring where weight and bulk must be kept at a minimum.

4.3.3. Detectors

Molecular resonances in the microwave region generally absorb very small amounts of power. In a representative laboratory experiment the magnitude of the signal due to molecular absorption is on the order of

$$P_{sig} \lesssim 10^{-4} P_0,$$

where P_0 is the incident microwave power. In spite of these intrinsically small signals, the availability of very sensitive detectors and sophisticated signal-processing techniques makes it possible to build practical spectrometers of remarkable sensitivity.

Ideally a microwave detector should noiselessly convert all of the signal information into a voltage that could be amplified and displayed for measurement. Practical detectors, however, generate a significant amount of noise when microwave power is incident upon them and rarely approach 25% signal conversion efficiency. Modulation and synchronous detection are almost universally used to reduce the noise in the recovered signal, but detector performance is still the limiting factor that determines the ultimate sensitivity of most spectrometers. For this reason, advances in detector

[18] A. A. Sweet, *Elec. Des. News* **21**, 40 (May 5, 1974).

[†] YIG (single-crystal yttrium iron garnet) is widely employed as a magnetic tuning material in microwave oscillators and filters.

technology can be expected to have a larger impact on the capabilities of microwave spectroscopy than any other single area of development.

This section will compare the characteristics of some of the many types of microwave detectors available. Diode detectors will be treated in considerable detail since they are widely used and readily available on the commercial market. Bolometer and acoustic methods of detecting microwave spectra will be treated briefly. Finally, we will look at such possibilities as using low noise microwave amplifiers to recover weak signals.

There are many kinds of diode devices that can be used to detect microwave radiation[19-22] but only two, the point-contact diode and the Shottky barrier diode, have found wide use in spectroscopic laboratories. Both of these types of diodes, in common with many other detection devices, exhibit a noise spectrum with three basic components. Assume the microwave power at a frequency v is producing a current I in a detector device. Let P_n be the noise power measured in a fixed bandwidth displaced from v by a frequency f. For sufficiently large f, thermal and shot noise predominate and give a "white noise" spectrum. Flicker noise predominates for small values of f and the noise power here takes the form $P_n \approx I^2/f$. The magnitude of the flicker or $1/f$ noise as it is commonly called varies significantly from diode to diode, but some general comments can still be made. Sometimes the frequency on which the signal information is carried can be chosen sufficiently high that the thermal and shot noise dominate. This is commonly the case for heterodyne detection where f is the intermediate frequency and can be as high as several GHz. Source modulation such as the type developed by Törring[23] can operate at frequencies of 5 MHz or more. Generally speaking, direct molecular modulations such as Stark, Zeeman, or double-resonance modulation must operate at frequencies of 100 kHz or lower. In these applications $1/f$ noise can be a serious problem and great care should be taken in choosing a detector to minimize the effect. Weinreb and Kerr[24] have demonstrated significant improvements in both flicker and shot noise performance for microwave detectors by using high intermediate frequencies and carefully selecting the detecting element. With cryogenic cooling they have obtained additional reduction in detector noise of a factor of 2 or more.

Another important consideration in detector selection is the conversion efficiency or ratio of detected signal power to input signal power. Typical

[19] H. C. Torrey and C. A. Whitmer, "Crystal Rectifiers." McGraw-Hill, New York, 1948.

[20] C. A. Burrus, *Proc. IEEE* **54**, 575 (1966).

[21] A. M. Cowley and H. O. Sorenson, *IEEE Trans. Microwave Theory Tech.* **MTT-14**, 588 (1966).

[22] H. A. Watson (ed.), "Microwave Semiconductor Devices and Their Circuit Applications." McGraw-Hill, New York, 1969.

[23] T. Törring, *J. Mol. Spectrosc.* **48**, 148 (1973).

[24] S. Weinreb and A. R. Kerr, *IEEE J. Solid State Circuits* **SC-8**, 58 (1973).

conversion efficiency for most devices is below 25%. Therefore, it is important that the impedance of the detector be carefully matched to the amplifier system so that as much of the detected signal power is transferred as possible. The typical impedance of a self-biased microwave detector diode is in the range 100–1000 Ω generally decreasing with increasing bias current. Transformer coupling has proved to be an efficient and practical solution in most applications. With these general comments in mind, we will now take a more detailed look at the specific devices that are in common use and briefly comment on those that show future promise.

Point contact diodes have historically been the most successfully employed detectors for microwave spectroscopy. The diode junction is established by making a firm contact between a semiconductor material and a finely pointed metal wire. Tungsten on silicon is probably the most widely used combination. These devices have a characteristically low junction capacitance which allows their operation at very high frequencies.

The technology for producing good working junctions has evolved over many years by cut and try empirical methods. Commercial diodes in convenient packages are available from several manufacturers for applications ranging from very low frequencies to well into the millimeter region. Unfortunately, the characteristics vary significantly from diode to diode. It is quite difficult to choose individual diodes for optimum performance without actually trying them on a spectrometer. For high-frequency applications this can be an expensive proposition. The common low-frequency cartridge-type diodes (1N21, 1N23, 1N78, 1N26, 1N53) are, however, relatively inexpensive and can be used as both detectors and mixers. Several methods of mounting the cartridge diodes in waveguide configurations have been discussed in the literature[1-3] and commercial versions are available. Below about 30 GHz detector-mount design parameters are not particularly critical. Above 30 GHz, however, where the 1N53 is commonly used, many laboratories have experienced difficulties with the older mount design. An improved design offering increased flexibility in tuning is shown in Fig. 4. This mount uses cartridge-type 1N53 diodes and performs well from 30 to about 70 GHz. The diode is inserted in the holder at the bottom of the figure and pushed in until the mount pin assembly reaches the Teflon stop at the top of the figure. The diode center pin and the mount pin assembly are then fully engaged and maximum electrical contact is achieved. The diode and mount pin assembly then can be adjusted as a single unit in order to place the diode junction at the position of maximum signal. This procedure also allows the impedance between the waveguide and the coaxial diode cartridge to be matched. The critical dimensions for constructing this sliding pin mount in RG/97 waveguide are shown on the figure. All metal parts are gold plated over brass.

A similar mount in WR–42 waveguide (K-band) using selected 1N26 diodes has been successfully used in the 90–130 GHz region. For maximum

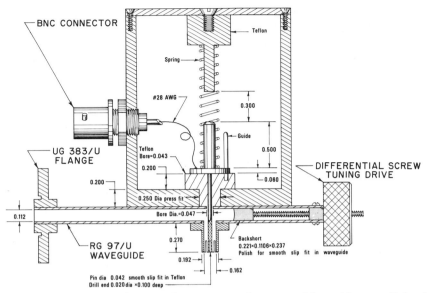

FIG. 4. Sliding-pin microwave detector mount for commercial cartridge-type diodes. The dimensions given are for a device to be operated in the 30–70 GHz region with 1N53 diode detectors (1 in. = 2.54 cm).

sensitivity it is essential to have adequate microwave power to self-bias the diodes ($\gtrsim 50$ μA). Tuning in this out-of-band configuration is far more critical than it is for lower frequences but is still less tedious than for most other millimeter-wave devices.

As mentioned earlier, the point contact diode is inherently a low-capacitance device which will work to very high frequencies. The high-frequency performance can, however, be limited very severely by the parasitic capacitance and inductance of the mount. One successful approach is to mount the diode directly in the waveguide forming the junction with a pointed tungsten whisker held in a micrometer-driven chuck.[25] Other designs have employed a diode and whisker permanently mounted in a short section of waveguide, called a wafer,[26] or on a quartz plate which is inserted into a slot in the waveguide.[27] Several commercial sources are now offering special point contact diode packages and mounts for millimeter wave applications.

Shottky barrier diodes (also called "hot carrier" diodes) are made by depositing a metal film on a small area of semiconductor, usually GaAs. A

[25] W. C. King and W. Gordy, *Phys. Rev.* **93**, 407 (1954).

[26] W. M. Sharpless, *Bell Syst. Tech. J.* **35**, 1385 (1956); **42**, 2496 (1963).

[27] J. W. Findlay, *Sky Telesc.* **48**, 352 (1974).

pointed wire is used to make electrical contact with the diode but is not involved in the rectification process. The properties of these devices are very uniform, and closely obey theoretical predictions. This is an important factor which has contributed significantly to the evolution of new applications for Shottky barrier devices. The excellent nonlinear characteristics of the Shottky devices make them well suited for applications as mixers and heterodyne detectors. Shottky barrier diodes are also reported to have lower $1/f$ noise than point contact devices, but their characteristics at low f do not appear to have been well studied. From all indications the GaAs Shottky barrier devices are among the most sensitive microwave detectors available at frequencies up to at least 160 GHz.[24, 27] Modern fabrication techniques have produced diodes smaller than 1 μm in diameter with very low junction capacitances.[28, 29] Tests indicate that these prototype devices are useful at frequencies as high as 200 GHz. As their technology evolves, these devices should certainly find wide application in laboratory spectroscopy.

At least two other semiconductor devices show some promise for microwave-detector applications. Tunnel diodes give better threshold sensitivity than other devices when biased near their current peak.[21] They generally have high junction capacitance and limited dynamic range. The $1/f$ characteristics are a bit better than point contact diodes but inferior to Shottky barrier devices. Back diodes are an improved version of tunnel diodes that do not require bias when operated as detectors. They appear to have very low $1/f$ noise and a sensitivity that competes directly with both the point contact and Shottky diode. Back diodes can be designed with good frequency response and should be given serious consideration in future detector applications.

Bolometers have also been used to detect microwave radiation and offer an entirely different set of characteristics from diode detectors. In its simplest form the bolometer detector is a fine wire in which the electrical resistance changes with temperature. Very sensitive microwave detectors can be made from Si[30] or InSb[13, 31, 32] cooled to 1.5 to 4.2 K. These devices respond to a broad range of source frequencies from the microwave into the far infrared, depending on the material used. The response time of bolometers is usually much slower than for diode detectors necessitating the use of modulation frequencies in the low audio range (100–400 Hz).

[28] G. T. Wrixon, *IEEE Trans. Microwave Theory Tech.* **MTT-22**, 1159 (1974).

[29] H. R. Fetterman, B. J. Clifton, P. E. Tannewald, C. D. Parker, and H. Penfield, *IEEE Trans. Microwave Theory Tech.* **MTT-22**, 1013 (1974).

[30] W. Steinbach and W. Gordy, *Phys. Rev. A* **8**, 1753 (1973).

[31] T. G. Phillips and K. B. Jefferts, *IEEE Trans. Microwave Theory Tech.* **MTT-22**, 1290 (1974).

[32] A. A. Penzias and C. A. Burrus, Millimeter-wavelength radio-astronomy techniques, *Annu. Rev. Astron. Astrophys.* **11**, 51 (1973).

Another technique which seems quite promising for millimeter and submillimeter wave applications is acoustic detection. The conversion of absorbed electromagnetic energy into sound waves was first reported by Bell[33] in a classic series of experiments carried out with mechanically chopped sunlight. Successful application of Bell's techniques to modern optical spectroscopy[34, 35] has been accomplished with condenser microphones.[36] In the microwave region the effect was first demonstrated with a gas-filled balloon placed in a modulated microwave field.[37]

Modern acoustical detectors have three principal distinguishing characteristics. Since the device is not a direct radiation detector it is useful at source frequencies from the microwave into the ultraviolet. In microwave experiments, higher source power can be used with acoustical detectors than is possible with most other detectors, and the devices are more rugged than solid-state millimeter wave detectors. The theoretical limit to acoustic detector noise is thermal fluctions in pressure from Brownian motion. In practice, instrumental noise and other absorption processes prevent the thermal limit from being reached. Krupnov and his co-workers[15] are presently using condenser microphones as microwave detectors well into the submillimeter region.

Another possibility for increasing sensitivity in recovering very weak signals is to amplify the signal at the microwave frequency before it is detected. Several types of low noise amplifiers are currently being used in spectrometers operating with radio telescopes. Parametric amplifiers appear to be among the best of the present generation of devices but tend to be narrow banded and are exceedingly expensive. As the technology of these devices evolves, they most certainly will be useful in laboratory spectroscopic systems.

In conclusion, it should be emphasized that many of the new detector devices which have been discussed in this section closely obey theoretical predictions. Significant advances can be anticipated in the future as a result of careful design engineering rather than the empirical cut and try methods used in the past.

4.3.4. Modulation

As mentioned earlier, modulation and synchronous detection are almost universally used to improve the sensitivity of microwave spectrometers. The general principles of modulation are discussed in detail in several of the

[33] A. G. Bell, *Proc. Amer. Ass. Advan. Sci.* **29**, 115 (1880); *Phil. Mag.* **11**, 510 (1881).

[34] L. B. Kreuzer, *J. Appl. Phys.* **42**, 2934 (1971).

[35] W. R. Harshbarger and M. B. Robin, *Accounts Chem. Res.* **6**, 229 (1973).

[36] G. M. Sessler and J. E. West, *J. Acoust. Soc. Amer.* **40**, 1433 (1966); *J. Audio Eng. Soc.* **12**, 129 (1964).

[37] W. D. Hershberger, E. T. Bush, and G. W. Leck, *RCA Rev.* **7**, 422 (1946).

general texts mentioned at the beginning of this chapter. For the purpose of the present section we shall rely on these earlier discussions for background and only briefly discuss the relative merits of the various modulation schemes. We shall also attempt to indicate the restrictions that a particular choice of modulation places on the other components of the spectrometer.

Modulation of a microwave absorption signal can be achieved in two different ways. The absorbing molecule can be affected directly by processes that shift the resonant energy levels (Stark or Zeeman effect) or by processes that regulate the population in those levels (saturation effects). Alternatively, the source of the microwave radiation can be varied systematically in frequency or amplitude. The net result in each of these cases is an absorption signal at the microwave detector that varies at the modulation rate. This signal can then be separated from the microwave background and processed with electronic techniques such as synchronous detection.

Stark-effect modulation has been by far the most popular technique and has been used on the most sensitive spectrometers. The Stark effect depends on the fact that the relative positions of the energy levels in polar molecules can be shifted by applying an external electric field. For most molecules the shift in resonant frequency of a given rotational transition depends on the square of the applied field. Typically for a molecule with a dipole moment of approximately 1 D, the shift in frequency, Δv in megahertz, is on the order of $10^{-6} E^2$ where E is in volts per centimeter. This means that an applied field of the order of 1000 V/cm is usually adequate for modulation purposes. The field is typically applied as a zero-based square wave with equal on and off times. Certain kinds of energy-level degeneracies produce much larger shifts in frequency which depend linearly on the applied field. Figure 5 illustrates this type of extremely simple linear dependence on the applied field for a degenerate pair of transitions in the S_2O molecule. Here, as in all spectra used for illustration in this chapter, the main absorption signal is up and the modulation-shifted signals down. The figure illustrates an important advantage of Stark-modulated spectra that comes from the fact that the applied field removes the usual degeneracies in the M_J sublevels. The frequency spacing and the relative intensities of the Stark shifted $\Delta M_J = 0$ spectra displayed in this figure clearly label the transition[7] involved as $J = 4 \leftarrow 3$. A particularly nice illustration of a typical pattern for second-order (E^2) Stark effect is shown on p. 472 of a recent review by Jones and Cook.[38] The reader is encouraged to compare the two figures. An additional bonus that occurs with Stark-effect modulation results from the fact that the magnitude of frequency shift that occurs with an applied electric field depends on the permanent electric dipole moment of the molecule. Precise values of the dipole moment are thus obtained directly from this technique.

[38] G. E. Jones and R. L. Cook, *CRC Criti. Rev. Anal. Chem.* **3** (4), 455–506 (1974).

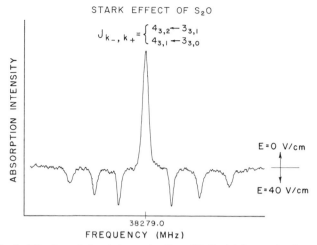

FIG. 5. Typical Stark-modulated absorption line of S_2O with first-order dependence on the applied electric field. The unshifted absorption signal is at the center of the figure with the Stark-shifted signals under $\Delta M_J = 0$ selection rules pointing down. The applied electric field is a 40 V/cm zero-based square wave and the modulated microwave signal is synchronously detected. The transition indicated arises from the degenerate pair $4_{32} \leftarrow 3_{31}$, $4_{31} \leftarrow 3_{30}$.

In order to produce the relatively high electric fields typically needed for Stark modulation it is necessary to use an absorption cell with closely spaced electrodes. Traditionally, septum-type absorption cells patterned after the Hughes–Wilson design[12] have been used. Because of the close proximity of the septum to the walls of the waveguide, these cells have large capacitance (~ 1000 pF). In addition, the dielectric material that is needed to provide electrical insulation between the septum and the waveguide often has disastrous effects on the microwave propagation characteristics. The large capacitance of these cells makes it very difficult to switch large electric fields on and off at rates much beyond 100 kHz. Stark systems are, therefore, generally limited to modulation frequencies where the $1/f$ noise of the microwave detector is a serious problem.

The propagation characteristics of a Stark spectrometer can be significantly improved by using a parallel plate absorption cell such as the one shown in Fig. 6. The parallel plate assembly itself does not exhibit a low-frequency transmission limit as does the conventional waveguide. By changing the waveguide and transition sections at the ends of the cell, a single plate assembly can be used over very wide frequency ranges. Cells of this type have been built with plate spacings from 1 to 10 mm and have been successfully used at frequencies from 2 to more than 200 GHz. The capacitance of these cells is generally smaller than the septum cells and can be made extremely small in the miniaturized versions used for millimeter wave applications.

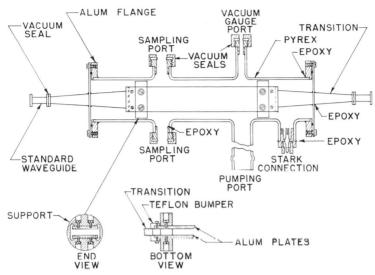

FIG. 6.   Parallel plate absorption cell. The drawing is to scale so that dimensions can be obtained from the broad face of the standard waveguide shown at either ends of the cell. The circular supports for the plates should be made from a low loss dielectric material such as Teflon.

Very uniform electric fields can be obtained with these cells if reasonable care is taken during construction to ensure that the plates are parallel. Surface finish on the plates is not particularly important, but it is most important that no spacers or other material be placed between the plates. Indeed, microwave transmission is significantly improved if clearance to external dielectrics of at least one half of the wavelength of the microwave radiation is left at the edges of the plates. Figure 6 is drawn to scale so that dimensions for a particular waveguide band can be approximately determined from the broad face dimension of the standard waveguide shown at the ends of the cell.

Parallel plate absorption cells can be designed for a wide range of special applications including high-temperature studies. Cooling these cells is difficult, but can be achieved by installing heat exchange coils on the back sides of the plates. The design shown in Fig. 6 is particularly convenient for dirty chemistry experiments as it can be easily disassembled for cleaning. One obvious disadvantage of these cells is that the volume of the cell tends to be quite large. This can be a serious handicap when precious samples are being studied.

The Zeeman effect, which depends on the fact that energy levels of molecules can be shifted by externally applied magnetic fields, can also be used as a modulation technique. Unfortunately most common molecules have closed-shell electronic ground state configurations and are therefore relatively insensitive to magnetic fields. The large field needed to modulate these molecules cannot be switched at rates fast enough to make the technique practical. There exists, however, a small class of molecules with open-shell electronic ground states that are quite sensitive to applied magnetic fields ($\Delta v \simeq 1$ MHz/G or $\sim 10^4$ MHz/T). A few of these such as O_2, NO, and NO_2 are permanent gases, but by far the largest number are molecules like OH, SO, and NCO which are very reactive and short lived. For this class of molecules, Zeeman modulation can be quite practical and in fact offers distinct advantages for the extremely reactive molecules. All glass absorption cells using free space[39] or dielectric rod[40, 41] propagation techniques can be used with the magnetic field applied from external coils. A significant refinement of the Zeeman modulation technique has been achieved by Radford[42] with his "field spinning" system. In this system the field is homogeneous and constant in amplitude, but its direction rotates with a frequency f. Woods[43] has recently reported excellent performance with this technique using cosine distribution cylindrical magnets.

It often happens that for large molecules with dense spectra or molecules in complicated spectroscopic systems, the usual molecular modulation techniques do not allow positive labeling of rotational quantum numbers. In most of these cases it is possible to estimate the structures of the molecules involved and predict their rotational energy level patterns. Figure 7 illustrates such a situation for the $CH_2CH_2C(CN)_2$ molecule.[44] The spectra displayed in the top right portion of the figure were taken with Stark modulation, but rotational quantum numbers cannot be positively attached to the individual transitions. In particular, the transitions labeled a, b, and c on the upper right-hand spectrum are all predicted to be in that general frequency region and to have about the same intensity. The spectrum illustrated in the lower right portion of Fig. 7 covers the same frequency region with the same absorbing gas sample, but was obtained with an entirely different technique called *modulated microwave double resonance*.[45, 46] In this double-resonance technique, the sample gas is irradiated simultaneously with two radiation fields. One is a strong saturating field near resonance between two energy

[39] R. Kewley, K. V. L. N. Sastry, M. Winnewisser, and W. Gordy, *J. Chem. Phys.* **39**, 2856 (1963).

[40] E. B. Brackett, P. H. Kasai, and R. J. Myers, *Rev. Sci. Instrum.* **28**, 699 (1957).

[41] D. R. Johnson and C. C. Lin, *J. Mol. Spectrosc.* **23**, 201 (1967).

[42] H. E. Radford, *Rev. Sci. Instrum.* **37**, 790 (1966).

[43] R. C. Woods, *Rev. Sci. Instrum.* **44**, 274 (1973).

[44] R. Pearson, Jr., A. Choplin, V. Laurie, and J. Schwartz, *J. Chem. Phys.* **62**, 2949 (1975).

[45] R. C. Woods, III, A. M. Ronn, and E. B. Wilson, Jr., *Rev. Sci. Instrum.* **37**, 927 (1966).

[46] O. L. Stiefvater, *J. Chem. Phys.* **62**, 233 (1974).

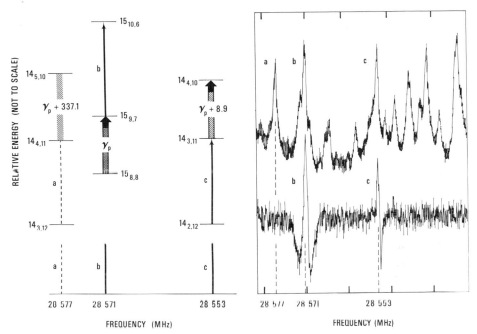

FIG. 7. Two forms of modulation were used to obtain microwave spectra of $\underline{CH_2CH_2C(CN)_2}$. The upper right-hand trace is a zero-based Stark-modulated spectrum. Directly below is a modulated microwave double-resonance spectrum observed by pumping at the frequency v_p (see text). Except for the type of modulation employed, both spectra were taken under essentially identical conditions. A schematic energy level diagram showing transitions which can be modulated by pump frequencies near v_p is shown on the left.

levels of the sample gas and the other is a much weaker field near resonance of another allowed transition in the gas. The weaker field is the usual microwave source radiation and is used to monitor the effect of the high-power pumping field. If the two transitions are chosen to have an energy level in common, as illustrated for the pump frequency v_p and the signal transition marked b on the left of Fig. 7, the molecular absorption seen at the signal frequency b will be characteristically modified when v_p is turned on. For this illustration the "fixed" pump frequency v_p is modulated at its source at the same rate as used in the Stark case. The three typical situations that can occur are illustrated in Fig. 7 as a, b, and c, in a scan along the signal frequency. At signal frequency a the fixed pump frequency is significantly off resonance with the corresponding possible pump transition, and no modulation occurs, so no double-resonance signal appears at a in the lower right-hand tracing. At signal frequency b the pump frequency is exactly on resonance with the corresponding possible pump transition, and the lower right-hand trace shows b fully modulated and split into two resonances when the pump is on.

At signal frequency c the pump is near, but not at resonance with the corresponding possible pump transition, so the lower right-hand tracing shows an asymmetric partially modulated signal.

It should be emphasized that this is a perfectly general technique which only requires that the two transitions are allowed by selection rules and that they share one level in common. The pump frequency can range from the radio-frequency region[47] through the microwave region.[44] Radio-frequency pump signals can generally be applied directly to the Stark electrode so that the same absorption cell can be used with both modulation techniques. Modulated microwave double resonance can be used in almost any kind of absorption cell that has adequate propagation characteristics at the two frequencies involved. The technique does increase the spectrometer complexity, but the sensitivity approaches that of Stark modulation and the advantages for assignment purposes are significant.

The alternative to molecular modulation that has traditionally been used in the millimeter region of the spectrum is direct modulation of the source radiation. This type of modulation produces derivative spectra when used with synchronous detection and does not allow easy identification of the rotational levels involved in the transition. The instrumentation is, however, quite straightforward and high-modulation frequencies can be used so that it is possible to operate out of the region limited by $1/f$ noise with most detectors. The most serious disadvantage that arises with this modulating scheme is that all variations in transmitted power will produce signals. Reflections anywhere in the waveguide system between the source and the detector can result in standing waves which produce large baseline variations in the synchronously detected signal.

Törring[23] has recently introduced a new spectrometer using a double-frequency source modulation technique that offers a significant suppression of the undesired background signals from the microwave transmission line. In this new technique, which Törring calls "saturation modulation," the microwave source is frequency modulated simultaneously at two frequencies f_1 and f_2. Both frequencies are large compared with the pressure-induced linewidth Δv of the molecular absorption, but the difference $f_2 - f_1$ is small compared with Δv. The molecular absorption acts as a nonlinear element causing harmonic distortions of the modulation envelope on the microwave carrier. After mixing with the original microwave carrier signal in the detector, new output signals appear at frequencies

$$(f_2 + f_1)/2 \pm n(f_2 - f_1)/2, \quad \text{where} \quad n = 3, 5, \ldots,$$

which are unique to the molecular absorption. Usually the lowest harmonic

---

[47] F. J. Wodarczyk and E. B. Wilson, *J. Mol. Spectrosc.* **37**, 445 (1971).

$n = 3$ gives the strongest signal and is selectively amplified. The background suppression is, however, often better with higher harmonics, so $n = 5$ is sometimes used.

Törring has shown that saturation-effect spectrometers can be built with a sensitivity, convenience, and simplicity comparable to Stark spectrometers. The most significant advantage that this new system offers over the Stark modulation system is in simplicity of absorption cell design and the elimination of problems with electrical breakdown from the applied Stark field. The new technique promises to be a useful tool, especially in cases where conventional modulation cannot be used.

4.3.5. Practical Spectrometers

Throughout the research literature one finds reference to the "conventional Stark-modulated spectrometer." In view of the wide range of options that have been discussed in the preceding sections it seems appropriate to attempt to describe the "conventional spectrometer" as it is currently being used in the not too modern laboratories where a good share of the spectroscopic work is still being carried out. This is not an easy task for, as Carrington[8] correctly points out, a microwave spectrometer is in many ways like an erector set and a certain amount of joy is derived from changing the configuration to meet the needs of a particular experiment.

The spectrometer system currently in use at NBS and illustrated in Fig. 8 will serve as a guide. We will attempt to point out the components in common use in many laboratories as well as a few of the little tricks of good practice that help to ease the experimental burden. As previously mentioned reflex klystrons are the "conventional source" in most laboratories, with BWO or Gunn oscillator sweepers becoming increasingly popular. In most applications where klystrons are still used, they are operated without phase locking.

It is desirable to provide some form of isolation between the microwave source and the rest of the system (the load) to prevent microwave power from being reflected back into the source. Without isolation and proper impedance matching, the load can significantly change or pull the frequency of oscillation. At low frequencies good quality broad-banded ferrite isolators are readily available at reasonable prices and are quite effective for this purpose. In the millimeter region, the available ferrite isolators are often impractical and $E-H$ tuners are used to match impedances in the transmission line in order to avoid frequency pulling. Immediately after the isolation, a directional coupler is usually inserted to tap off a small amount of power for frequency measurement purposes. If a wavemeter is to be used, it will interfere least with the main portion of the experiment if it is installed on the frequency measurement arm of the system. After the directional coupler it is usually necessary to employ an attenuator so that the microwave

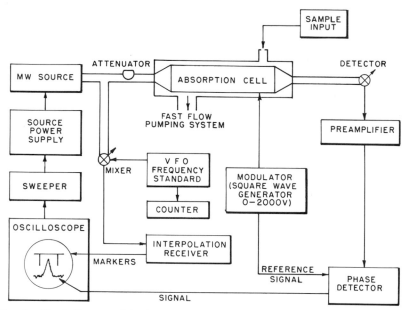

FIG. 8. Block diagram of the "conventional microwave spectrometer" currently in use at the National Bureau of Standards.

power level in the absorption cell can be reduced to a few milliwatts. Reflections and multimoding effects will be significantly reduced if the isolator, directional coupler, and attenuator are all of the same waveguide size as the source output. A properly matched transition section can then be used to connect the absorption cell.

For frequencies below 40 GHz septum-type waveguide absorption cells of the order of 3 m in length are still in common use. Parallel plate cells are gaining in popularity and are beginning to be more widely used for studies of stable gases. Nearly all conventional systems use commercial silicon point-contact diodes for detectors. As mentioned earlier these diodes are sufficiently noisy that they are the limiting factor in the sensitivity of most systems. Preamplifier systems can be readily built from either solid-state or vacuum tube components with noise figures an order of magnitude lower than the noise figures of the commercial diodes. The preamplifier is usually tuned to the modulation frequency with a bandpass of the order of 1 kHz and a gain around 10^4. Phase-sensitive detectors and display instruments are, of course, readily available in large variety and their selection becomes a matter of individual preference. The generator used for Stark-effect modulation typically consists of an oscillator operating in the region of 25 to 100 kHz coupled to a wave shaping and amplifying circuit. The output is usually a zero-based square wave variable in amplitude from 0 to 2000 V and capable of driving a capacitive load of 10 to 1000 pF. At the present time, we know of no commercially available instrument that meets these requirements, but several good circuit designs are in the literature.[48-51]

Absolute frequency measurements have traditionally been made in two steps. A cavity wavemeter is used first to give an approximate value good to ± 10 MHz or so. A more accurate value is then obtained by direct comparison with a harmonic from a low-frequency crystal-stabilized oscillator. The wavemeter value for the main source frequency establishes the correct harmonic of the crystal oscillator, and the beat note between the crystal harmonic and the source is amplified and detected by a radio receiver. The frequency of the receiver is then adjusted so that the beat note coincides with the spectroscopic signal being measured. The absolute accuracy of such measurements depends on three contributing factors: (1) the known resonant frequency of the reference crystal, (2) the calibration of the radio receiver, and (3) the accuracy with which the beat note can be matched to the center or peak of the spectral feature. At frequencies above 10 GHz it is usually necessary to use an intermediate frequency oscillator (0.5–2 GHz) which is phase locked to the low-frequency crystal oscillator in order to reduce the harmonic number and obtain strong beat signals. Many variations and improvements on this basic measurement system are in common use in laboratories throughout the world.

One important modern variation utilizes a commercial variable frequency oscillator containing an external cavity stabilized klystron. Devices of this kind have been available for many years and offer good short-term stability (~ 1 part in 10^7) with high output power levels (~ 200 mW). They are generally tunable by a factor of two with center frequencies between 1 and 10 GHz. Commercial electronic counters can now count frequencies well beyond 10 GHz using transfer oscillator techniques with an absolute accuracy of the order of a few parts in 10^8. The combination of a high-frequency cavity stabilized VFO and a direct reading counter eliminates the need for a wavemeter and allows the radio receiver to be replaced by a simple fixed-frequency tuned amplifier. These systems are extremely convenient and are finding application wherever absolute frequency measurements in the few parts in 10^7 range will suffice.

To achieve greater accuracy in frequency measurements, a number of commercial sources are now producing frequency synthesizers which translate the stable frequency of a precision frequency standard to any selected

[48] L. C. Hedrick, *Rev. Sci. Instrum.* **20**, 781 (1949).
[49] H. W. de Wijn, *Rev. Sci. Instrum.* **32**, 735 (1961).
[50] C. O. Britt, *Rev. Sci. Instrum.* **38**, 1496 (1967).
[51] P. A. Baron and D. O. Harris, *Rev. Sci. Instrum.* **41**, 1363 (1970).

frequency from dc up to 1 GHz or more. This stable but variable frequency can then be used in phase-lock loops or directly in place of the VFO in the system described earlier.

Modern applications of high-frequency VFO and synthesizer systems have placed some rather severe design restrictions on the mixer which generates the harmonics of the standard frequency and compares them with the main signal frequency. This mixer mount design problem has been solved in a variety of ways in different laboratories. One mixer mount design that has been successfully used on VFO systems at the National Bureau of Standards is shown in Fig. 9. The VFO signal is fed directly to the pin of a commercially packaged 1N26 diode through the BNC connector at the bottom of the figure. An rf line stretcher or double stub tuner is usually inserted between the VFO and the mixer mount to optimize the power transmitted to the diode junction. The beat frequency resulting from the mixing of the VFO frequency with the main frequency in the waveguide is then obtained from the crystal case through the BNC connector on the side of the mount. Figure 9 illustrates a device intended for operation in the frequency range

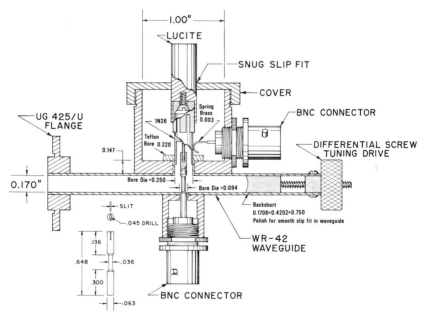

FIG. 9.   Microwave mixer mount for frequency-measurement applications using commercial cartridge diodes and a variable frequency oscillator (VFO) standard in the 2–12 GHz range. The dimensions given are for a device intended to operate in the 18–40 GHz region (1 in. = 2.54 cm).

18–26 GHz, but will produce adequate beat signals up to 40 GHz. The design concept is quite general and can be scaled to other waveguide sizes and commercial diode packages. A mixer of this type using RG-97 waveguide with 1N53 diodes is currently used for frequency measurements throughout the region from 40 to 130 GHz.

In 1965, a complete microwave spectrometer system was introduced to the commercial market by the Hewlett Packard Company. Named the MRR (for *M*olecular *R*otational *R*esonance) spectrometer, it was engineered as a complete package which provided broad-banded scanning capabilities over entire waveguide bands. During the next few years, the MRR spectrometer continued to evolve in sophistication. For a variety of reasons the manufacturer terminated production of the complete spectrometer and many of its components late in 1974. It is the authors' opinion that the MRR spectrometer and similar instruments that are still available from other manufacturers have made a lasting contribution to microwave spectroscopic technology. We will attempt to identify specific advances that these instruments have made in the state of the art and indicate some of the consequences that they have had on microwave spectroscopy. Although the MRR system is no longer available, it was one of the first on the market and is still the most widely used instrument of its type. The discussion below refers specifically to the last model produced.

The microwave source for the MRR spectrometer is a BWO phase locked to a digitally programmed reference oscillator. In a given band the microwave frequency can be dialed manually or set by computer. The frequency can be rapidly stepped maintaining the phase lock across any portion of the band. This kind of broad-banded source allows one to scan the entire band continuously or to change frequency rapidly for a specific investigation in any part of the band. More significantly, it allows the resolution of the spectrum to be controlled.

The absorption cell for this system is a waveguide septum-type Stark cell which has been carefully engineered to provide good microwave propagation characteristics. A broached waveguide design is used in which the septum and a 0.01 in. dielectric fit snugly in a groove in the waveguide wall. The amount of dielectric protruding into the waveguide is kept to a minimum. The septum itself is tapered linearly and asymmetrically at each end to minimize reflections. This design has produced an X-band absorption cell with microwave propagation characteristics comparable to other high-quality microwave components (i.e., low insertion loss, low VSWR, etc.) and these characteristics are maintained from 8 to 40 GHz. Because of manufacturing limitations, it has not been possible to produce this cell in lengths longer than 3 ft so it is customary to use two cells in series with parallel

Stark modulators. The broad-banded characteristics of this absorption cell allows the full scanning capabilities of the programmable BWO source to be realized. In addition to these source and absorption cell developments, numerous other engineering refinements were made primarily aimed at making the instrument portable and convenient to operate.

The impact of this new technology on microwave spectroscopy has been significant. The broad-banded rapid scans of spectra under low resolution have been demonstrated to be useful for determining certain structural parameters of large molecules.[52] An excellent collection of representative spectra from low-resolution experiments has been presented by Jones and Cook.[38] The well-characterized nature of the microwave fields in these instruments have made it possible conveniently and routinely to measure properties of microwave spectra, other than frequency, which have been difficult to measure in the past. For example, measuring the relative intensities of microwave lines is useful for both assignment of spectra and identification of vibrational states. It is also interesting to note that for the first time in the history of microwave spectroscopy a large number of laboratories (~ 24) have essentially identical instrumentation. For certain types of experiments it is possible, therefore, to eliminate instrumental effects as the source of discrepancies in the measurements made by different investigators. As a result, certain systematic errors in these measurements can be identified. The problem of absolute pressure calibration in line shape studies is the classic example of such a situation.

Thus far we have restricted our discussion to spectrometer systems that are in common use for general purpose spectroscopy. Microwave spectroscopy is a very versatile technique and a wide variety of special instruments and experimental methods have been developed to extend its capabilities. In the remainder of this section, we shall briefly mention a few specific examples of these efforts and refer the reader to the literature for a more detailed treatment.

Conventional microwave spectroscopic techniques require a sample pressure on the order of 0.01 Torr (1.33 Pa). Many molecules do not have a vapor pressure this high at room temperature but can be volatilized at higher temperatures. Still other species are thermodynamically stable only at elevated temperatures. One obvious approach is to heat the absorption cell until the required amount of sample is in the vapor phase. For temperatures up to 200°C or so, conventional absorption cell designs and materials will usually be adequate. At higher temperatures, however, some of the common cell construction materials begin to decompose and others are attacked by the vapors. The application of Stark modulation becomes increasingly

[52] W. E. Steinmetz, *J. Amer. Chem. Soc.* **96**, 685 (1974).

difficult because of the breakdown of insulating materials and because of thermionic emission from the electrodes. The design of a high-temperature cell, which has satisfactory microwave transmission properties and is chemically inert to the species being studied, presents a formidable problem in materials technology and construction techniques. Lovas and Lide[53] have extensively reviewed the literature of high-temperature microwave spectroscopy and describe techniques for studying the spectra of molecules at temperatures from 200 to above 2000°C.

Spectrometers using resonant cavity absorption cells have been built for a variety of special applications.[54-57] The most frequently used cavity design seems to be the semiconfocal Fabry–Perot system although other designs have been successfully used. With suitable coupling of the source radiation, a cavity absorption cell can in principle be made to resonate at any particular frequency over a wide frequency range. Because of this resonance the microwave transmission characteristics of these cells are inherently narrow banded and scanning in frequency becomes a twofold problem. It is necessary to vary the resonant frequency of the cavity and the frequency of the source oscillator simultaneously. Tracking the two frequencies smoothly over wide frequency ranges has proven to be difficult in practical instruments. Most cavity spectrometers have used source modulation, but both Stark and Zeeman systems have been built. The long effective path length that can be achieved in a small cavity cell with low surface-to-volume ratio offers a distinct advantage in sensitivity over conventional single-pass systems of comparable dimensions. Cavity absorption cells have proven to be particularly useful for studies of gases at high pressures[57] and are finding increasing applications in single-frequency monitoring.[55]

It has long been recognized that the millimeter and submillimeter regions of the spectrum offer a wealth of spectroscopic potential if conventional microwave techniques could be applied there. For many of the lighter molecules, the thermal population distribution highly favors those levels which produce transitions in the millimeter and submillimeter regions. This and other factors generally cause rotational absorption coefficients to increase rapidly with frequency, so the advantages to be gained can be significant.

Historically the experimental problems encountered at frequencies above 50 GHz were so formidable that only a small number of laboratories ventured

[53] F. J. Lovas and D. R. Lide, Jr., *Advan. High Temp. Chem.* **3**, 177–212 (1971). See also J. Hoeft, F. J. Lovas, E. Tiemann, and T. Törring, *Z. Angew. Phys.* **31**, 265 (1971); **31**, 337 (1971).

[54] M. Lichtenstein, J. J. Gallagher, and R. E. Cupp, *Rev. Sci. Instrum.* **34**, 843 (1963).

[55] L. W. Hrubesh, R. E. Anderson, and E. A. Rinehart, *Rev. Sci. Instrum.* **41**, 595 (1970).

[56] H. E. Radford, *Rev. Sci. Instrum.* **39**, 1687 (1968).

[57] U. Minglegrin, R. G. Gordon, L. Frenkel, and T. E. Sullivan, *J. Chem. Phys.* **57**, 2923 (1972).

beyond this limit. As indicated in the preceding sections, modern developments in sources, detectors, and absorption cells have pushed this limit beyond 200 GHz. This newly opened region from 50 to more than 200 GHz can now be worked by almost conventional techniques using fundamental power from commercial millimeter wave oscillators, packaged detectors, and parallel plate absorption cells with Stark modulation. The frontier that remains is in the submillimeter region which bridges the gap between the microwave and far infrared.

Until very recently, submillimeter spectroscopy has been carried out using the methods developed by Gordy and co-workers.[13] Power at submillimeter wave frequencies was obtained for these experiments from lower-frequency reflex klystrons using harmonic generation. Silicon point contact diodes in a run-in mount or cyrogenic bolometers have been used as detectors. Using a variety of absorption cells and modulation techniques, these methods have produced significant results for a large number of molecules. This approach, however, is not without drawbacks. The microwave power available at harmonic frequencies is low. Producing working frequency multipliers and detectors requires good mechanical and electrical design, proper semiconductor material, and considerable skill and patience on the part of the experimenter. Moreover, many of the experiments have been done in free space absorption cells without the benefit of Stark or Zeeman data to facilitate assignment. Searching a wide frequency range by these methods is tedious, and the interpretation of spectra is complicated by the fact that transitions produced by several harmonically related microwave frequencies are observed simultaneously.

An entirely different approach to submillimeter spectroscopy has been developed by Krupnov and co-workers.[15] A block diagram of their spectrometer is shown in Fig. 10. Microwave power is obtained from submillimeter backward-wave oscillators which have been described in the references given below.[58] The electronic tuning range of this family of oscillators is given as $\pm 20\%$ near 215 GHz and $\pm 10-12\%$ near 1.07 THz, which compares favorably with lower frequency BWO's. The corresponding midband output power is given as 30 and 0.6 mW, respectively. No data on source noise or stability are reported. At frequencies below 500 GHz the BWO can be phase locked with a continuous tuning range 3–6 GHz. Above 500 GHz the BWO's are used without frequency stabilization. A free space absorption cell and acoustical detector are designed as an integral unit, and two cells can be placed in series to record simultaneously an unknown and a reference

[58] M. B. Golant et al., Prib. Tekh. Eksp. No. 4, 136 (1965); M. B. Golant et al., Prib. Tekh. Eksp. No. 3, 231 (1969); Instrum. Exp. Tech. (USSR) (English Transl.) No. 4, 887 (1965); No. 3, 801 (1969).

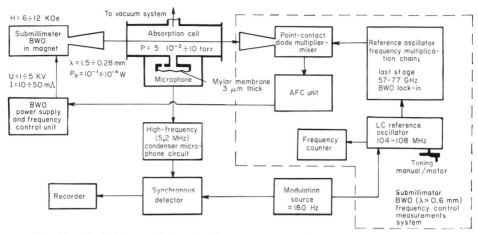

FIG. 10.   Block diagram of the submillimeter wave scanning spectrometer (RAD) used by Krupnov and his collaborators.

spectrum. Source modulation at 180 Hz and synchronous detection are employed.

Frequency measurements are made by two methods. When the BWO is phase locked, the reference oscillator at the base of the frequency multiplication chain is counted directly. When the BWO is not stabilized, frequency measurements appear to be made relative to a known reference spectrum, such as that of SO_2.

Spectra taken with acoustic detection are generally made at higher sample pressures (0.05–10 Torr or 6.7–1333.2 Pa) and higher source power levels (0.1–100 mW) than generally employed in conventional spectrometers. The ability to vary both these parameters as well as the BWO sweep rate over a wide range enables the resolution of this spectrometer to be varied by at least five orders of magnitude. In addition, a wide range of tradeoffs between sensitivity and resolution are possible. Although, the lines recorded with this instrument are broad compared to measurements made at low milliTorr pressures with conventional spectrometers, under favorable conditions the relative accuracy of the frequency measurements possible with both methods is comparable.

Figure 11 is a moderate resolution scan of the spectrum of SO_2 near 1.06 THz taken at room temperature. The derivative presentation is due to the source modulation. The structure of the Q-branch series $J_{11,J-10} \leftarrow J_{10,J-9}$ is shown from $J = 48$ to the bandhead near $J = 25$. This is the highest frequency spectrum obtained to date by scanning microwave techniques. To place these results in perspective, it is instructive to compare

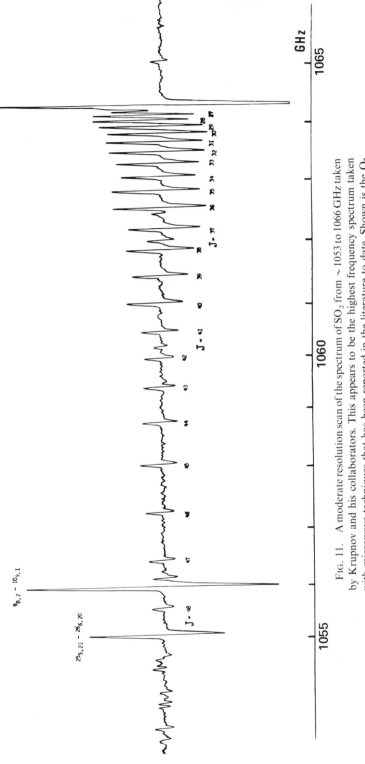

FIG. 11. A moderate resolution scan of the spectrum of SO_2 from ~ 1053 to 1066 GHz taken by Krupnov and his collaborators. This appears to be the highest frequency spectrum taken with microwave techniques that has been reported in the literature to date. Shown is the Q-branch series $J_{10,J-9} \rightarrow J_{11,J-10}$.

Fig. 11 to the SO_2 spectra obtained in the same frequency range by Fourier transform far-infrared techniques.[59] This spectrometer is also capable of high sensitivity. One figure in Krupnov and Burenein[15] shows the $J = 16 \leftarrow 15$ line of the $(01^{1d}0)$ vibrational state of $^{14}N^{15}N^{16}O$ in natural abundance, with a signal-to-noise ratio approaching 10:1 at 402,754.6 MHz. This spectrum was also taken at room temperature.

The high source power levels and sample pressures possible with acoustic detection should be advantageous in the study of molecules with small dipole moments or rotation-induced spectra.[60] The absence of Stark, Zeeman, or double-resonance data to facilitate the assignment of new spectra may be a serious disadvantage of this detection method. The broadbanded scanning capabilities of this instrument make it feasible, however, to observe the patterns characteristic of different spectra and to use assignment techniques similiar to those employed in the infrared.

4.3.6. Applications

As indicated in the introduction of this chapter, the challenge that faces modern microwave spectroscopists is to apply their technology and experimental expertise to other areas of science. The discussions in the preceeding sections should serve to define the existing technology and to indicate where the significant advances in instrumentation are likely to come in the future. Throughout that discussion we have attempted to indicate the particular instrumental configurations that naturally lend themselves to new applications. It is not the intention of this chapter to spell out the range of potential new applications for microwave spectroscopic techniques in detail. It seems appropriate, however, to make a few concluding remarks that will lead the reader into some of these new areas.

Chemical analysis, in the traditional sense, using microwave spectroscopic techniques has been dealt with in detail by other authors.[38] Microwave techniques can also be used to detect and monitor short-lived molecules that result from gas phase reactions, discharges, or pyrolysis. Several short-lived molecules with open-shell ground state electronic configurations have been studied by these methods and are reviewed by Carrington.[8] Many other short-lived molecules with closed shells, such as BF,[61] SF_2,[62] and CH_2NH,[63] have also been detected. The investigation of molecules produced by pyrolytic cracking of larger organic molecules is a new area of chemical

[59] J. W. Fleming, *IEEE Trans. Microwave Theory Tech.* **MTT-22**, 1023 (1974).
[60] J. K. G. Watson, *J. Mol. Spectrosc.* **40**, 536 (1971).
[61] F. J. Lovas and D. R. Johnson, *J. Chem. Phys.* **55**, 41 (1971).
[62] D. R. Johnson and F. X. Powell, *Science* **164**, 950 (1969).
[63] D. R. Johnson and F. J. Lovas, *Chem. Phys. Lett.* **15**, 65 (1972).

analysis for which microwave techniques appear particularly well suited.[64]

Both the specificity and sensitivity of microwave methods can offer significant advantages in applications where specific molecules must be monitored. Continuous monitoring with flow-through absorption cells is both possible and desirable for many of these applications. The monitoring of atmospheric pollutants such as SO_2, NH_3, and H_2CO at part per million levels has already been demonstrated.[65, 66] The continuous flow through capability may well have far-reaching applications for on-line production monitoring in chemical plants.

The field of molecular radio astronomy, which has made rapid advances during the past few years, is another area that is heavily dependent on laboratory microwave spectroscopy for its data base. More than 38 molecules and molecular ions have now been detected in the gas clouds of the interstellar medium. The spectra obtained from the telescopes are rapidly becoming as complex as those encountered in the laboratory. The detection of molecules such as CH_3CHO,[67] CH_3NH_2,[68] and CH_3OCH_3[69] has prompted further laboratory studies because their exceedingly complex spectra were not well understood. At least two molecules, H_2CS[70] and CH_2NH,[63] were produced and studied in the laboratory specifically with their anticipated telescope detection in mind. Several others, notably CN,[71] CH,[72] HCC,[73] HNN^+,[74] and HCO^+,[75, 76] were first identified from the telescope data alone. The demand for new laboratory data in support of these effects is great. Molecular radio astronomy has been nicely summarized through mid-1974 by Zuckerman and Palmer.[77]

Experiments which simultaneously employ sources of microwave and laser radiation also offer a number of exciting possibilities. Double-resonance experiments, similar to those described earlier, can be performed with

[64] F. J. Lovas, F. O. Clark, and E. Tiemann, *J. Chem. Phys.* **62**, 1925 (1975).

[65] L. W. Hrubesh, *Radio Sci.* **8**, 167 (1973).

[66] H. Uehara and Y. Ijuuin, *Chem. Phys. Lett.* **28**, 597 (1974).

[67] C. A. Gottlieb, *in* "Molecules in the Galactic Environment" (M. A. Gordon and L. E. Snyder, eds.), p. 181. Wiley (Interscience), New York, 1973.

[68] N. Kaifu, M. Morimoto, K. Nagane, K. Akabane, T. Iguchi, and K. Takagi, *Astrophys. J.* **191**, L135 (1974).

[69] L. E. Snyder *et al.*, *Astrophys. J.* **191**, L79 (1974).

[70] D. R. Johnson and F. X. Powell, *Science* **169**, 679 (1970).

[71] K. B. Jefferts, A. A. Penzias, and R. W. Wilson, *Astrophys. J.* **161**, L87 (1970).

[72] B. E. Turner and B. Zuckerman, *Astrophys. J.* **187**, L59 (1974).

[73] K. D. Tucker, M. L. Kutner, and P. Thaddeus, *Astrophys. J.* **193**, L115 (1974).

[74] S. Green, J. A. Montgomery, Jr., and P. Thaddeus, *Astrophys. J.* **193**, L89 (1974).

[75] W. Klemperer, *Nature (London)* **227**, 1230 (1970).

[76] B. L. Ulich, J. M. Hollis, L. E. Snyder, F. J. Lovas, and D. Buhl, *Astrophys. J.* (to be published).

[77] B. Zuckerman and P. Palmer, *Annu. Rev. Astron. Astrophys.* **12**, 279 (1974).

microwave and laser frequencies. Two-photon experiments are also possible in which neither the microwave frequency nor the laser frequency is resonant with the molecule, but their sum or difference frequency is.[78] These are just a few of the areas where microwave applications are blossoming. Most certainly the future holds even more exciting possibilities.

ACKNOWLEDGMENTS

The authors wish to thank Dr. Frank Lovas and Dr. Jon Hougen for their comments and careful review of this manuscript. We are also grateful to our many colleagues who over the past several years have discussed these topics with us and contributed to the ideas presented in this work.

[78] T. Oka, in "Laser Spectroscopy" (R. G. Brewer and A. Mooradian, eds.). Plenum Press, New York, 1974.

4.4. Radio-Frequency Region *

4.4.1. Introduction

4.4.1.1. Introduction. Radio-frequency spectroscopy is conducted in absorption rather than in emission; the material is exposed to coherent electromagnetic radiation, the absorption and refraction of which are measured, often simultaneously. The range of frequency is, broadly speaking, from 10^{-5} to 10^{11} Hz, but the upper bound merges into the far-infrared region of the spectrum. There are within the radio-frequency region some characteristic emissions and absorptions arising from transitions between quantum states of atoms and molecules, but it is the intention of this section to concentrate instead on the absorptions arising from molecular motion in solids and liquids. Because these are very often nonresonant effects, in which thermal motion plays a large part, they give rise to absorption bands more than a frequency decade in breadth, which would be regarded as continua by most spectroscopists. From the shape of these continua information is obtained about the characteristic molecular motions. This subject has been called "dielectric relaxation spectroscopy." Related information can be derived from the inelastic scattering of higher-frequency radiation, involving frequency shift.

The refraction and absorption of radio-frequency radiation by molecules arises principally from its interaction with their electrical moments, which may be permanent or induced by the radiation itself, or by other applied fields. Magnitudes of electrical moments are derived from the interaction of the materials or molecules with what are essentially static electric fields; the derivations are an important part of dielectric theory, which we shall discuss principally in the time-varying or frequency-varying domain. The most important electrical moments involved are the permanent dipole moment, if the molecule is polar, and the scalar electrical polarizability.

4.4.1.2. The Electrical and Optical Properties of Materials. The permittivity or dielectric constant ε of a material is defined as the ratio of electrical displacement $\mathbf{D}$ to applied electric field $\mathbf{E}$

$$\mathbf{D} = \varepsilon\mathbf{E}. \tag{4.4.1}$$

When this equation is written in electrostatic units, ε is a dimensionless quantity, and may be related directly to the optical constants of the material.

---

*Chapter 4.4 is by J. B. Hasted.

If the equation is written in SI units, then ε must be multiplied by the free space permittivity $\varepsilon_0 = (36\pi \times 10^9)^{-1} = 8.85 \times 10^{-12} \text{ F}^{-1}$:

$$\mathbf{D} = \varepsilon\varepsilon_0\mathbf{E}. \tag{4.4.2}$$

The discussion of dielectric theory in its relation to molecular motion is normally carried out in electrostatic units; and where it is necessary to introduce SI units for the discussion of conduction, mention will be made of the change.

When a time-varying electric field $E = E_0 \exp j\omega t$ is applied to a material, the electrical displacement is in general a complex quantity, the real part representing its magnitude and the imaginary part its phase. The permittivity, or dielectric constant, is also complex:

$$\hat{\varepsilon} = \varepsilon' - j\varepsilon'' \tag{4.4.3}$$

with ε' and ε'' real. The choice of negative sign is arbitrary, but is becoming increasingly common, and therefore will be adopted here; ε'' is termed the dielectric loss, and ε' the dielectric constant or (real) permittivity.

The complex refractive index of the material is written in a similar form

$$\hat{n} = n - j\kappa, \tag{4.4.4}$$

where n is the (real) refractive index and κ the extinction coefficient. The relations between dielectric and optical properties are

$$\varepsilon' = n^2 - \kappa^2, \tag{4.4.5}$$

$$\varepsilon'' = 2n\kappa. \tag{4.4.6}$$

The extinction coefficient κ is defined by

$$I = I_0 \exp(-2\pi\kappa l/\lambda), \tag{4.4.7}$$

where I_0 is the intensity of electromagnetic radiation, of wavelength λ, impinging on the material, and I the intensity after passing through a distance l in the material.

The power absorption coefficient α which is more familiar in spectroscopy is, however, not defined in terms of the number of wavelengths through the material, but in terms of distance only:

$$I = I_0 \exp(-\alpha l). \tag{4.4.8}$$

The power absorption coefficient, which is proportional to line strength and oscillator strength, is thus very different from dielectric loss, being related as

$$\kappa = \alpha/4\pi\bar{\nu}, \qquad \varepsilon'' = n\alpha/2\pi\nu, \tag{4.4.9}$$

where $\bar{\nu}$ cm^{-1} is the wavenumber of the radiation.

When one takes into account the frequency variation of both dielectric and optical properties, one finds that what are regarded as lossless dielectrics may also be highly absorbing optically. Furthermore an absorption band will not maximize at the same frequency in α as in ε''.

4.4.2. Experimental Techniques of Radio-Frequency Spectroscopy

4.4.2.1. Introduction.   4.4.2.1.1. CAPACITANCE CELLS. We have seen that in dielectric relaxation spectroscopy two measurements are of central importance, that of the real permittivity ε' and that of the imaginary part of the permittivity or dielectric loss ε''. Fortunately most methods of measurement are capable of yielding both parts so that we shall discuss each method in the context both of ε' and ε''. There is a great variety of techniques, since many investigators have preferred to face up to their own problems rather than make use of available, but perhaps little-known, designs.

The basic method of measurement consists of measuring the complex impedance of an air-filled capacitor cell, and remeasuring it with the cell filled with material. The ratio of the capacitances C is then a measure of ε':

$$\varepsilon' = C(\text{filled})/C(\text{vacuum}), \tag{4.4.10}$$

and the loss is given by

$$\varepsilon'' = [\omega R(\text{filled})C(\text{vacuum})]^{-1} \tag{4.4.11}$$

when the resistance $R(\text{vacuum})$ of the empty cell is infinite. The pulsatance $\omega = 2\pi f$ at frequency f.

A comparison between the dielectric properties of two different materials, for example, those of a solvent and those of a solution, is made by evaluating the complex impedances of the cell filled with the two materials. Although it is often possible to calculate the capacitance of the empty cell from electrostatic theory, this is not usually necessary, since $C(\text{vacuum})$ can be measured. For Eq. (4.4.10) to hold, however, the substitution of material for air or vacuum in a cell must be electrostatically complete; the fringing field must have the same configuration in both filled and empty cells. One way of achieving this is by ensuring that the field continues to be uniform outside the capacitance-measuring region, with the material or the air continuing uniformly also. Thus the capacitance electrodes must be surrounded by guard electrodes, which are also filled with material or are empty as appropriate. Only one of these guard electrodes need be insulated from its neighboring capacitance electrode; but the other guard electrode must be maintained always at the same potential as its neighboring measurement electrode. Such cells are known as "three-terminal" cells, and require special circuitry.

The measurement of capacitance requires the use of an accurate standard of capacitance or empty capacitance cell whose design enables an absolute calculation of capacitance to be made from electrostatic theory.

The principal problems of capacitance cell design can be summarized as follows:

(1) stray capacitance calibration problems;
(2) maintenance of adequate electrical contact between electrodes and material;
(3) physicochemical problems associated with filling and emptying.
(4) temperature and pressure control and measurement; and
(5) electrode polarization.

1.   Nearly all cells are constructed with either planar or concentric cylindrically curved electrodes; typical cells are illustrated in Figs. 1 and 2. The simplest type of two-terminal cylindrical cell is shown in Fig. 1a. In the

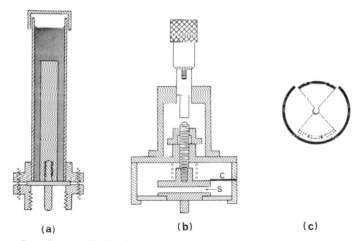

(a)   (b)   (c)

FIG. 1.   Capacitance cells. Insulating material is shown shaded, metal is black or cross-hatched. (a) Two-terminal coaxial cell shown filled with (shaded) liquid, and constructed so as to be connected directly to a concentric socket. (b) Two-terminal parallel plate cell with adjustable distance between plates: S—specimen, C—spring contact. (c) Cross section through variable standard rotary action capacitance, shown in position of minimum capacitance. Broken line shows position of maximum capacitance.

cylindrical electrode configuration it is possible to eliminate stray capacitance by use of a cylindrical electrode of variable length. A variable standard capacitance cylindrical cell (Fig. 1c) has also been produced using rotary motion.[1] The inner and outer cylinders are cut away, so that a 90° rotation

[1] A. M. Thompson, *Proc. Inst. Elec. Eng.* **106B**, 307 (1959).

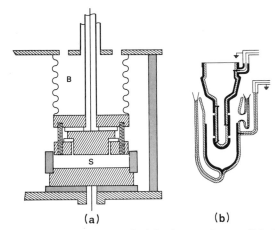

(a)   **(b)**

Fig. 2.   Capacitance cells. (a) Three-terminal fixed-separation parallel plate cell with electrodes sprung against solid specimen, to allow for thermal expansion: S—specimen of ice, which
is frozen in situ. B—temperature bath. Insulating material shaded; Metal cross-hatched.
(b) Glass cylindrical three-terminal cell for liquids built onto taper joints. Electrodes (black) are
made of fired silver paste or platinizing solution, and are connected through the glass (cross-
hatched) by means of platinum wire.

varies the capacitance from its maximum value down to its nominal zero
value.

In the case of planar electrodes, however, one of the two plates can be
adjusted in distance from the other by means of a micrometer, as in Fig. 1b.[2]
Planar electrode cells, two and three terminal, are suitable both for solids
and for liquids.

In two-terminal cells the mechanical variation of electrode area is necessary in order to eliminate stray capacitance, which may be different when
the cell is empty and when it is full. The linear relationship between capacitance and electrode area for ideal systems enables stray capacitance to be
eliminated when, with different areas exposed, two measurements are made
full and two empty. In three-terminal cells, i.e., cells with guard electrodes,
the most significant part of the stray capacitance, namely that which arises
from fringing field, is increased on filling by factor ε'. It is possible to calibrate
the cell using a material of accurately known permittivity, and in this way
avoid the unknown fringing field.

Among the most suitable liquids for calibration purposes are those
presented in the accompanying tabulation.[3]

[2] M. G. Broadhurst and A. J. Bur, *J. Res. Nat. Bur. Std.* **69C**, 165 (1965).
[3] F. I. Mopsik, *J. Chem. Phys.* **50**, 2559 (1969).

Liquid	$T\,(°C)$	ε'
Carbon tetrachloride	0	2.2786
	50	2.1789
Carbon disulfide	−50	2.8363
	25	2.6344
Isopentane	−50	1.9512
	25	1.8275
Toluene	−50	2.5724
	25	2.3807

2. For liquids and gases there are no serious problems of electrode contact other than those of electrode polarization; but a rigid well-machined electrode will normally be in contact with a well-machined solid dielectric specimen only at a small number of points. It would be possible to improve the contact by filling the space with paste of (approximately) the same complex permittivity. A simpler alternative has been to use foil electrodes, metallizing paint, colloidal graphite, evaporated metal films, or similar conducting coatings. These often prove satisfactory but of course the machining of the solid specimen must be good. It is possible to make measurements with an air space deliberately left on each side of the dielectric specimen, between its surfaces and the electrodes.[4, 5]

The contact problem is intensified when measurements are required over a large temperature range, when thermal expansion or contraction can be expected. It can only be solved by arranging for one electrode to be forced against the dielectric specimen by spring contact. A three-terminal cell used in studies of ice[6] is shown in Fig. 2a. The contact is made by freezing the electrode assembly when immersed in water; it is possible that the freezing technique could be more widely applied.

3. A pure liquid dielectric must occupy the entire volume between the electrodes, and must not attack them chemically. The avoidance of skin effect losses has encouraged many workers to gold-plate electrodes; the chemical inactivity of gold offers an additional advantage. Teflon insulation and sealing is also chemically inert, but the need for flushing of the cell and for rapid handling of liquid samples has resulted in many cells being constructed from silica or glass. A glass three-terminal cell[7] is shown in Fig. 2b.

Another problem encountered with liquid dielectrics is the removal of bubbles. Usually these originate in the evolution of dissolved gas, which is

[4] A. C. Lynch, *Proc. Inst. Elec. Eng.* **104B**, 359 (1957).
[5] M. G. Broadhurst and A. J. Bur, *J. Res. Nat. Bur. Std.* **69C**, 165 (1965).
[6] R. P. Auty and R. H. Cole, *J. Chem. Phys.* **20**, 1309 (1952).
[7] G. A. Vidulich and R. L. Kay, *Rev. Sci. Instrum.* **37**, 1662 (1966).

most likely when there is a local temperature change. It is necessary to remove as much dissolved gas as possible by previous boiling.

4. While temperature control is relatively easy with the aid of thermostat water pumped through jackets, pressure control can present difficult problems. For high-pressure measurements adjustable electrodes must be avoided, so that three-terminal cells are necessary. Pressure variation can cause difficulties with electrode contact and with cell loading.

4.4.2.1.2. ELECTRODE POLARIZATION. When an electrolytic or biological sample whose conductance is higher than 10^{-8} ohm^{-1} cm^{-1} is in contact with electrodes, a potential difference can develop at the surface as a result of charge migration. This "electrode polarization" effect produces a polarization impedance Z_p, which is in series with the sample impedance and which must be eliminated in measurements. Increasing the electrode area, for example, by constructing electrodes of platinized platinum, increases the polarization capacitance C_p, which in the absence of polarization is infinite. Use of a cell in which the sample thickness is variable while the electrode area is unchanged can overcome the polarization problem, but with biological specimens this is not always possible.

A four-terminal cell can be used to minimize electrode polarization with a circuit as shown in Fig. 3. Two electrodes are used to apply the emf, while two large area pickup electrodes are used to measure the potential difference across the sample. The pickup electrodes P are as distant as possible from the application electrodes E, and are connected to the high-impedance amplifier A_1. The output is combined with the output of the identical amplifier A_2 which amplifies the potential developed across an impedance standard.

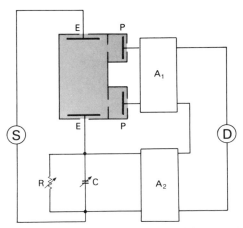

FIG. 3. Four-terminal bridge circuit for specimens subject to electrode polarization: S—source; D—detector; A—amplifiers; P—pickup electrodes; E—electrodes.

Electrode polarization is due to the transport of ions to the electrode surface, where they are often unable to recombine, so that they form a local space charge at the interface of specimen and electrode. This space-charge layer can greatly influence the capacitance measurement, and its discharge will affect the dielectric loss or conductance.

Proton semiconductors, such as ice, and also electrolytic solutions, are particularly prone to polarization effects, and the lowest frequency relaxation in ice[8] is probably due to this space charge. It is possible to make allowance for polarization effects by ascribing an impedance pattern to the electrode layer and including lumped circuit elements. It has been found[9] that below 5 kHz frequency the real permittivity of pure liquid formic acid varies as ω^{-2}, while the conductivity is frequency independent. This might be accounted for by representing the electrode impedance as a pure capacitance C_{el} in series with the dielectric, so that

$$\varepsilon'(\text{apparent}) = \varepsilon' + (G^2/\omega^2 C_{el} C_0), \qquad (4.4.12)$$

where G is the conductance of the specimen and C_0 the capacitance of the empty dielectric cell.

More usually in solids the electrode impedance Z_{el} is complex, and one can make an empirical fitting to the relation

$$Z_{el} = Z_0 (i\omega)^{-n}. \qquad (4.4.13)$$

The value of n must be found by measurement at sufficiently low frequencies for ε' to make a negligible contribution, and

$$\frac{G(\text{apparent})}{C_0} + j\omega\varepsilon'(\text{apparent}) = \frac{(G/C_0) + j\omega\varepsilon'}{1 + \hat{Z}_{el}\{(G/C_0) + j\omega\varepsilon'\}} \qquad (4.4.14)$$

In the case when $G \gg \varepsilon'\omega C_0$ and $|\hat{Z}_{el}|G \ll 1$, one can use

$$G(\text{apparent})/C_0 = (G/C_0) - (Z \cos \tfrac{1}{2}n\pi)\omega^{-n}(G^2/C_0), \qquad (4.4.15)$$

$$\varepsilon'(\text{apparent}) = \varepsilon' + (Z_0 \sin \tfrac{1}{2}n\pi)\omega^{-(n+1)}(G^2/C_0). \qquad (4.4.16)$$

Trial computations must be made to test whether the data can be fitted to these expressions.

In making such a correction to measurements it is assumed that there is no intrinsic dielectric dispersion in the low-frequency region. Only in cases where such an assumption is permissible can information be extracted from otherwise confusing data.

[8] A. von Hippel, D. B. Knoll, and W. B. Westphal, *J. Chem. Phys.* **54**, 134, 150 (1971); A. von Hippel, *J. Chem. Phys.* **54**, 145 (1971).

[9] C. F. Johnson and R. H. Cole, *J. Amer. Chem. Soc.* **73**, 4536 (1951).

4.4.2.2. Bridge Measurements.[10] 4.4.2.2.1. INTRODUCTION. The four-arm alternating current Wheatstone-type bridge (Fig. 4a) is under balance when

$$Z_1/Z_2 = Z_3/Z_4 \qquad (4.4.17)$$

provided that both real and imaginary parts of the balance equation are satisfied by the adjustment of two separate variables. The impedance of a lossy capacitance C having series resistance R is

$$Z = (1/j\omega C) + R \qquad (4.4.18)$$

and the phase angle ϕ is $-(\pi/2 - \delta)$ with "loss angle" tangent

$$\tan \delta = \omega CR = \varepsilon''/\varepsilon' \qquad (4.4.19)$$

for a capacitor filled with dielectric of permittivity $\hat{\varepsilon}$. When the bridge is in balance the phase angles in the arms satisfy the equation

$$\phi_1 = \phi_2 + \phi_3 - \phi_4. \qquad (4.4.20)$$

When the capacitance cell being tested contains a very low loss material, small errors in standard components can lead to large errors in loss measurement, since δ is very small compared with $(\pi/2) - \delta$, which is the measured phase angle.

An important error inherent in bridge measurements arises from capacitative coupling between the leads at either side of a bridge element; this adds a capacitance which cannot be eliminated, but it can be made constant by enclosing each element within a screen connected to one end of the element. The bridge has then the form of Fig. 4b. The screening capacitances of course modify the impedances of the bridge elements, so that substitution methods become necessary.

In the substitution technique the unknown capacitance is placed in series or parallel with a standard capacitance after the bridge has been balanced; the bridge is then rebalanced. A variable standard capacitance is most satisfactory, but even with this the range of unknown capacitance that can be measured in this way is somewhat limited. Errors can arise from neglect of the series inductance possessed by the dielectric cell.

The most frequently used four-arm bridge for capacitance measurement is that due to Schering,[11] which is shown in Fig. 4c. The balance equations are

$$C_1R_1 = C_3R_3, \qquad C_1R_2 = C_2R_3, \qquad \tan \delta = \omega C_1 R_1. \qquad (4.4.21)$$

where R_3 is a fixed resistor; balance is obtained by adjustment of C_3 and either C_2 or R_2.

[10] R. G. Bennett and J. H. Calderwood, in "Complex Permittivity" (B. K. P. Scaife, ed.), Chapter 3. English Univ. Press, London, 1971.
[11] H. Schering, Z. Inst. 40, 124 (1920).

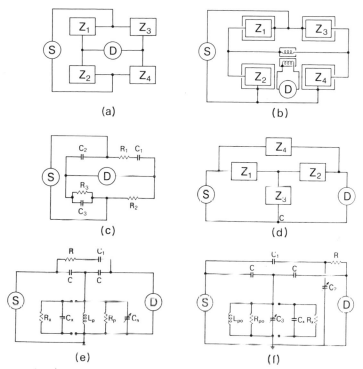

Fɪɢ. 4. Bridge circuits: S—ac source; D—detector; subscript x—unknown circuit element; subscript s—standard circuit element. (a) Four-arm ac Wheatstone bridge; (b) four-arm bridge with screened elements; (c) Schering bridge; (d) bridged-T circuit; (e) substitution bridged-T circuit; (f) substitution twin-T circuit.

4.4.2.2.2. THE BRIDGED-T CIRCUIT. The bridged-T circuit,[12] shown in Fig. 4d, has the advantage that source and detector are joined to a common terminal C, which is earthed. The condition for balance is

$$Z_1Z_2 + Z_1Z_3 + Z_2Z_3 + Z_3Z_4 = 0. \qquad (4.4.22)$$

Unfortunately at least one bridge arm must be inductive for balance to be achieved, and it is best that a fixed inductor be used, so that the substitution technique is necessary, as in Fig. 4e. The balance conditions for this bridge, without substitution, are as follows:

$$RR_p = 1/\omega^2C^2 \qquad \text{and} \qquad (1/\omega L) - (\omega C^2/C_1) - \omega C_s - 2\omega C = 0.$$

The test capacitor C_x and loss R_x are connected in parallel with L_p after initial balance, and a new balance obtained by adjustment of C_s and R. The

[12] W. N. Tuttle, *Proc. I.R.E.* **28**, 23 (1940).

change in C_s will be equal to C_x, and the value of R_x is related to the change ΔR in R:

$$1/R_x = \omega^2 C^2 \, \Delta R. \tag{4.4.23}$$

Twin-T versions of this system have been developed[13–15] as in Fig. 4f. Initial balance is by adjustment of C_2 and C_3. After substitution a fresh balance of these capacitances yields the unknown C_x, R_x from the equations

$$C_x = \Delta C_3, \tag{4.4.24}$$

$$1/R_x = \Delta C_2 \, RC^2 \omega^2 / C_1. \tag{4.4.25}$$

4.4.2.2.3. THREE-TERMINAL CELL BRIDGES AND THE WAGNER EARTH.[†] The equivalent circuit of a three-terminal impedance connected as one arm of a four-arm bridge is illustrated in Fig. 5a, where Z_x is the impedance to be measured, while Z_{x1} and Z_{x2} are the impedances between the electrodes and the guard ring. If E is connected to A, then Z_{x1} is in parallel with the source, where it has no effect, and Z_{x2} is in parallel with Z_2, which is made as low as possible to reduce the error. A circuit of this type is shown in Fig. 5b; it has the balance conditions

$$C_3 R_3 = C_1 R_1, \qquad C_2 R_3 = C_1 R_2 \tag{4.4.26}$$

and after substitution

$$C_x = \Delta C_3, \qquad 1/R_x = (C_1/R_3) \, \Delta(1/C_2). \tag{4.4.27}$$

There are, however, still errors arising from the admittances of the guard rings; a complete elimination is only possible with the aid of the Wagner earth (Fig. 5c). Two further bridge arms Z_5 and Z_6 are used to rebalance the bridge with the switch in position E, after it has already been balanced with the switch at B. Iteration of this procedure is necessary until the bridge is balanced with the switch in either position, when points B, D, and E are at the same potential, so that the current in Y_2 is zero and Z_x is effectively not shunted. Bridged-T networks can also be used for three-terminal impedance measurements.[16]

A method[17, 18] of avoiding the Wagner earth is shown in Fig. 5d. The high-gain amplifier A drives V_2 to ground, so that the applied voltage is

[13] D. E. Sinclair, *Proc. I.R.E.* **28**, 310 (1940).

[14] D. Woods, *J. Inst. Elec. Eng.* **104**, Part C, 506 (1957).

[15] K. Posel, *Trans. S. African Inst. Elec. Eng.* **48**, 243, (1958); *Proc. I.E.E.* **110**, 126 (1963).

[16] R. G. Bennett, *J. Sci. Instrum.* **37**, 195 (1960).

[17] S. Roberts, Rep. No. 66-C-333, General Electric Res. and Develop. Center, Schenectady, New York, 1966.

[18] H. P. Hall and R. G. Fulks, *Elec. Eng. N.Y.* **81**, 368 (1962).

[†] "Earth" in the U.K. is synonymous with "ground" in the U.S.A.

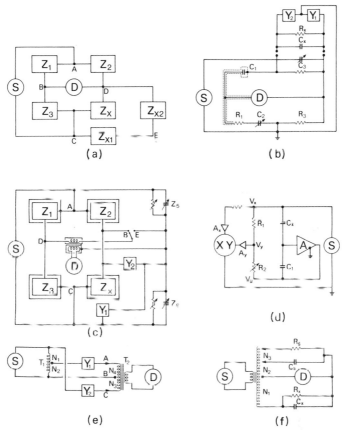

FIG. 5. Three-terminal cell bridges and inductively coupled bridges: S—ac source; D—detector; subscript x—unknown circuit element; subscript s—standard circuit element; A—amplifiers. (a) Equivalent circuit of three-terminal impedance in four-arm bridge; (b) three-terminal impedance substituted in four-arm bridge; (c) four-arm bridge with Wagner earth; (d) XY-plotter as detector; (e) inductively coupled bridge; (f) tapped inductance ratio-arm bridge.

nearly equal to the voltage across the cell. Unit gain amplifiers are placed at each input of the XY-recorder in order to isolate it from the bridge. Then

$$V_y \left(1 + \frac{R_2}{R_1}\right) = \frac{V_x R_2}{R_1} - \frac{q}{C_1}, \tag{4.4.28}$$

where q is the charge on the cell dielectric, which is taken to be lossless. The circuit is used as a bridge in the range $0.01-10$ Hz by varying R_2 or C_1 until V_y is zero, when

$$C_x = R_2 C_1 / R_1. \tag{4.4.29}$$

4.4.2.2.4. INDUCTIVELY COUPLED BRIDGES. An inductively coupled bridge,[19-21] is shown in Fig. 5e, where N_i represent the number of windings on transformers T_i.

$$Y_{1,2} = C_{1,2} + j/R_{1,2}$$

are the admittances to be compared. Under balance,

$$Y_1/Y_2 = N_1N_3/N_2N_4. \qquad (4.4.30)$$

Source and detector may be earthed at one point, and since at balance all of the transformer is then earthed, interwinding capacitances are unimportant. Leakage resistance and reactance in the transformer windings do, however, introduce errors. Three-terminal capacitances can be measured at Y_2 with guard ring connected to B.

The Cole and Gross bridge[22] differs from the above in that source and detector are interchanged, and one transformer winding is omitted. Conductance is balanced with the aid of the conductance standard LR. Commercial versions of this bridge are available,[23-27] in the range 50 Hz–5 MHz.

In the Thompson bridge[28] the equivalent of a standard variable conductance is obtained by a quadrature amplifier with unit gain connected in series with a standard variable capacitance.

In the Lynch bridge[29] the measurement is performed by substitution in which the test capacitor is connected in parallel with a variable standard capacitor and also with a variable conductance network. Different networks have been designed for different problems. An interesting variable conductance network, incorporating gauged capacitances and a resonant inductor (different for each frequency) was reported by Weir and Dryden.[30]

Another type of inductively coupled bridge makes use of tapped inductance ratio arms.[31-34] A simplified circuit is shown in Fig. 5f, where R_s and C_s

[19] A. D. Blumlein, British Patent No. 323037 (1928).

[20] H. L. Kirke, *J. Inst. Elec. Eng.* **92**, Part III, 2 (1945).

[21] H. A. M. Clark and P. B. Vanderlyn, *Proc. Inst. Elec. Eng.* **96**, Part II, 365 (1949).

[22] R. H. Cole and P. M. Gross, *Rev. Sci. Instrum.* **20**, 252 (1949).

[23] A. G. Mungall and D. Morris, *Rev. Sci. Instrum.* **34**, 839 (1963).

[24] J. F. Hersh, *Gen. Radio Exp.* **36**, 3 (1962).

[25] Type 1615 A. Capacitance Bridge, Gen. Radio Co., West Concord, Massachusetts, 1964.

[26] R. Calvert, Wayne Kerr Monograph No. 1, Wayne Kerr, New Malden, Surrey, U.K.

[27] B. Rogal, *Proc. Inst. Elec. Eng.* **107B**, 427 (1960).

[28] A. M. Thompson, *J. Inst. Elec. Eng.* **102**, Part B, 704 (1956).

[29] A. C. Lynch, *J. Inst. Elec. Eng.* **104**, Part B, 363 (1957); *Proc. Inst. Elec. Eng.* **112**, 426 (1965).

[30] K. G. Weir and J. S. Dryden, *J. Sci. Instrum.* **40**, 318 (1963).

[31] J. J. Hill and A. P. Miller, *Proc. Inst. Elec. Eng.* **109**, Part B, 157 (1962).

[32] J. F. Hersh, *Gen. Radio. Exp.* **36**, No. 8 (1962).

[33] W. H. P. Leslie, *J. Inst. Elec. Eng.* **108**, Part B, 539 (1961).

[34] B. Hague, "Alternating Current Bridge Methods." Pitman, London, 1959.

are standards, and balance is achieved by varying the tapping points on the upper half of the secondary winding. At balance

$$R_s = R_x N_3/N_1, \qquad C_x = C_s N_2/N_1. \tag{4.4.31}$$

4.4.2.3. Resonance Methods of Measuring Impedance. In the so-called "Q-meter" a voltage or current is injected into a resonant circuit having inductance L and capacitance C, and the voltage across the capacitor is measured when the circuit is turned to resonance. In the resistance injection circuit (Fig. 6a) the voltmeter V_2 measures the voltage across test capacitor C_x. At resonance V_2 is a maximum:

$$\omega L_p = 1/\omega C_x \tag{4.4.32}$$

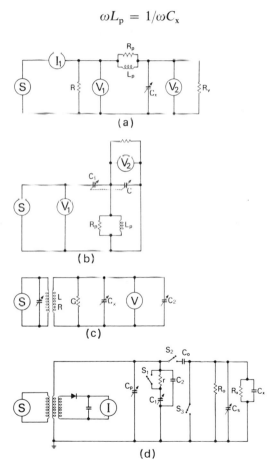

Fig. 6. Resonance methods of measurement: S—variable frequency source; V—voltmeter; I—current meter; subscript x—unknown circuit element; subscript s—standard circuit element. (a) Resistance injection Q-meter; (b) capacitor injection Q-meter; (c) Hartshorn–Ward circuit; (d) double-resonance circuit for lossy liquid measurements.

and

$$\left|\frac{V_2}{V_1}\right| = \left[\omega L_p \left(\frac{1}{R_p} + \frac{1}{R_v}\right)\right]^{-1}, \tag{4.4.33}$$

which is the selectivity Q of the coil loaded by the resistance R_v.

In the capacitor injection circuit (Fig. 6b)[35] resonance is found when

$$\omega L_p = 1/(C + C_1), \tag{4.4.34}$$

and then

$$\left|\frac{V_2}{V_1}\right| = \left(\frac{C_1}{C + C_1}\right) \cdot \frac{R_p}{\omega L_p} = \frac{C_1 Q}{C + C_1}. \tag{4.4.35}$$

C_1 is ganged to C in such a way that $C/(C + C_1)$ is constant, and V_2 is then proportional to selectivity Q.

Injection into the resonant circuit is also possible via a tapped inductance and a mutual inductance. In the circuit of Hartshorn and Ward[36] (Fig. 6c) a variable capacitor cell C_1 is used for the test material. C_2 is a small variable capacitor with the aid of which the width of the resonance curve is determined. As before

$$\omega L = 1/(C_x + C_2) \tag{4.4.36}$$

at resonance. The capacitance C_2 is varied by an amount δC until the voltage is reduced by a factor of $\sqrt{2}$:

$$|V|_{\delta C} = (1/\sqrt{2})|V|_{res}. \tag{4.4.37}$$

The total capacitance change between the plates is $\Delta C = 2\delta C$, and the conductance

$$G = \omega \, \Delta C/2(q - 1)^{1/2}, \tag{4.4.38}$$

where

$$q = |V|^2_{res}/|V|^2_{\delta C}. \tag{4.4.39}$$

If the values of ΔC with and without the dielectric are ΔC_1 and ΔC_0, then the conductance G_x of the sample is

$$G_x = (\Delta C_1 - \Delta C_0)/2(q - 1)^2, \tag{4.4.40}$$

and for the dielectric

$$\tan \delta = G_x/\omega C_x. \tag{4.4.41}$$

Further studies of resonance methods are available.[10, 37—39]

[35] A. J. Biggs, and J. E. Houldin, *Proc. Inst. Elec. Eng.* **96**, Part III, 295 (1949).

[36] L. Hartshorn and W. H. Ward, *J. Inst. Elec. Eng.* **79**, 597 (1936).

[37] I. T. Barrie, *Proc. Inst. Elec. Eng.* **112**, 408 (1965).

[38] R. G. Bennett, *J. Sci. Instrum.* **39**, 417 (1962).

[39] L. Hartshorn, "Radio Frequency Measurements." Chapman & Hall, London, 1947.

For high-conductance or high-loss liquids the resonance technique must be modified if accuracy is not to be sacrificed owing to the great breadth of the resonance curve. The resonant properties of fixed geometry resonant circuits, possessing both inductance and capacitance, such as those illustrated in Fig. 7, can be measured both in air and in the liquid.[40, 41] When placed near a variable-frequency oscillator, they resonate at the appropriate frequency and cause a depression in the anode current of the oscillator. For resonant frequencies f_0 empty and f immersed,

$$\varepsilon' = f_0^2/f^2, \tag{4.4.42}$$

but only if the conductivity of the liquid is small.

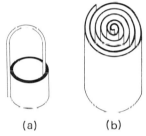

(a)   (b)

FIG. 7.   Fixed geometry resonant circuits [J. Wyman, *Phys. Rev.* **35**, 623 (1930); A. S. Brown, P. M. Levin, and E. W. Abrahamson, *J. Chem. Phys.* **19**, 1226 (1951)]: (a) cylindrical type, 5 × 7 cm, natural period 10^{-8} sec; (b) spiral type, diameter 1.7 cm, natural period 10^{-9} sec.

It is also possible to measure the current flowing in a resonant circuit by means of a thermocouple; when this current is maximized at resonance, the frequency so obtained is, for high-conductance liquids, different from that obtained using the oscillator anode current diagnostic. The latter gives only 0.2% accuracy in ε' for liquids with conductivity less than 30 times that of water, and the former gives this accuracy for liquids with conductivities less than 100 times. For low-loss liquids such as benzene, an accuracy of 0.01% has been claimed.

For very lossy liquids a sharp primary resonant circuit can be coupled to a broad secondary resonant circuit containing the dielectric cell,[42−44] as in Fig. 6d. When the switch S_1 is open and S_2 and S_3 are closed, the isolated primary circuit is tuned to resonance, by maximizing I. Now S_3 is opened with the dielectric cell C_x empty, and C_s is adjusted until I is again a minimum; this adjustment sets the secondary circuit at resonance. Now R_0 is

[40] J. Wyman, *Phys. Rev.* **35**, 623 (1930).
[41] A. S. Brown, P. M. Levin, and E. W. Abrahamson, *J. Chem. Phys.* **19**, 1226 (1951).
[42] B. Ichijo, *J. Appl. Phys.* **24**, 307 (1953).
[43] J. C. Anderson, "Dielectrics." Chapman & Hall, London, 1964.
[44] B. Ichijo and T. Arai, *Rev. Sci. Instrum.* **32**, 122 (1961).

found by adjusting C_p, with all switches closed, until I is a maximum. Then S_3 is opened and C_s adjusted until I is a minimum, and the values of C_s and I are recorded. Again S_3 is closed, S_1 opened and C_1 varied until I returns to its previous value. Thus R_0 is given by

$$R_0 = C_1{}^2 r/C_p{}^2. \tag{4.4.43}$$

When the cell is filled the procedure is repeated, thus determining the new C_x and the parallel resistance R_r, R_x being obtained from

$$1/R_r = (1/R_0) + (1/R_x). \tag{4.4.44}$$

4.4.2.4. Ultrahigh Frequency and Microwave Techniques.

4.4.2.4.1. GUIDED WAVE TRANSMISSION. The radio-frequency techniques of Section 4.4.2.3 become inaccurate when the wavelength is sufficiently short to be comparable with the dimensions of the apparatus. In this case voltage variations occur along the circuit wires and over the capacitance plates. The upper limit of frequency depends on cell dimensions and specimen permittivity, but is normally of the order of 100 MHz. From there to about 3 GHz the impedance must be distributed along the circuit, which is built of transmission line, usually concentric. Above about 3 GHz the skin effect losses at the center conductor make transmission lines impracticable, so that transmission down hollow waveguides becomes necessary. The impedances are still distributed, and many of the concepts and techniques are the same, so that the two frequency bands may be considered together. At the same time guided wave transmission has much in common with free-wave transmission, so that optical techniques such as interferometry are viable above about 10 GHz. The methods available include absorption, reflection, standing wave measurement, bridges, and cavity resonators.

A unit length of transmission line is considered to possess a series inductance L and resistance R with shunt capacitance C and conductance G. The voltage and current at point along a transmission line are given by

$$V = V_+ \exp(-\gamma z) + V_- \exp \gamma z, \tag{4.4.45}$$

$$IZ_0 = V_+ \exp(-\gamma z) - V_- \exp \gamma z, \tag{4.4.46}$$

where the characteristic impedance Z is given by

$$Z_0^2 = (R + j\omega L)/(G + j\omega C) \tag{4.4.47}$$

and the propagation coefficient γ by

$$\gamma = \alpha + j\beta = \{(R + j\omega L)(G + j\omega C)\}^{1/2}, \tag{4.4.48}$$

where α is the attenuation coefficient[†] and β the phase coefficient.

[†] This must be distinguished from α the power absorption coefficient, and from α the relaxation time spread parameter.

For a loss-free vacuum-filled transmission line, the parameters will be written with subscript zero:

$$\alpha_0 = 0, \qquad \beta_0 = 2\pi/\lambda_0, \qquad Z_0{}^2 = L/C, \qquad (4.4.49)$$

but when the line is filled with dielectric

$$\gamma = \gamma_0\sqrt{\hat{\varepsilon}} \qquad \text{and} \qquad Z = Z_0/\sqrt{\hat{\varepsilon}} \qquad (4.4.50)$$

so that

$$\varepsilon' = (\lambda_0/2\pi)^2(\beta^2 - \alpha^2), \qquad (4.4.51)$$

$$\varepsilon'' = (\lambda_0/2\pi)^2 2\alpha\beta, \qquad (4.4.52)$$

where λ_0 is the free-space wavelength. Measurement of both phase and amplitude are necessary in order to derive $\hat{\varepsilon}$.

Efficient propagation of electromagnetic waves down a hollow metal tube is possible only when the tube dimensions are sufficiently large; when this is not the case, only an exponentially attenuated "evanescent mode" will be found. The "cut-off" or limiting wavelength λ_c, which can be propagated without evanescence, is dependent on the waveguide dimensions and on the mode of propagation. The wavelength in the guide λ_g is related to λ_c by

$$1/\lambda_g{}^2 = (1/\lambda_0{}^2) - (1/\lambda_c{}^2), \qquad (4.4.53)$$

and the transverse components of the fields in a loss-free vacuum-filled guide are given by

$$E = E_+ \exp(-j\beta_0 z) + E_- \exp(j\beta_0 z), \qquad (4.4.54)$$

$$H = \{E_+ \exp(-j\beta_0 z) - E \exp(j\beta_0 z)\}/Z_0, \qquad (4.4.55)$$

where $\beta_0 = 2\pi/\lambda_g$ and z is the distance along the waveguide.

When a waveguide is completely filled with dielectric, the free-space wavelength λ_0 is replaced by $\lambda_0/\sqrt{\hat{\varepsilon}}$, so that

$$\gamma = 2\pi j[(\hat{\varepsilon}/\lambda_0{}^2) - (1/\lambda_c{}^2)]^{1/2} \qquad (4.4.56)$$

from which

$$\varepsilon' = (\lambda_0/2\pi)^2(\beta^2 - \alpha^2) + (\lambda_0/\lambda_c)^2, \qquad (4.4.57)$$

$$\varepsilon'' = 2\alpha\beta(\lambda_0/2\pi)^2, \qquad (4.4.58)$$

which is of similar form to Eqs. (4.4.51) and (4.4.52), so that, as before, both phase and amplitude are required for a knowledge of $\hat{\varepsilon}$.

4.4.2.4.2. BRIDGE MEASUREMENTS. For liquid specimens, phase and amplitude can be measured in either transmission line, waveguide or free wave by means of bridge circuits[45-47] such as those illustrated in Figs. 10a and

[45] T. J. Buchanan, *Proc. Inst. Elec. Eng.* **99**, Part III, (1952).
[46] T. J. Buchanan and E. H. Grant, *Brit. J. Appl. Phys.* **6**, 64 (1955).
[47] F. Hufnagel and G. Klages, *Z. Angew Phys.* **12**, 202 (1960).

10c. In principle these are more closely related to a Mach–Zehnder inter-ferometer than to a radio-frequency bridge. One part of the divided signal passes through a phase shifter and an attenuator, both calibrated, while the other part passes through a variable length cell; by means of a probe tra-versable along the wavepath, the signal $V(z)$ is measured in phase and amplitude. The phase change between two signals is measured by "backing off" each signal in turn to zero, using the phase shifter. At a distance z from a short circuit the voltage in the dielectric is given by

$$V = V_+\{\exp \gamma z - \exp(-\gamma z)\}$$
$$= 2V_+\{\sinh \alpha z \cos \beta z + i \cosh \alpha z \sin \beta z\}. \tag{4.4.59}$$

Its magnitude is

$$|V| = \sqrt{2}|V_+|\{\cosh 2\alpha z - \cos 2\beta z\}^{1/2}, \tag{4.4.60}$$

and its phase θ is given by

$$\tan \theta = (\tan \beta z)/(\tanh \alpha z). \tag{4.4.61}$$

For a representative dielectric with $\alpha/\beta = 0.1$, the calculated[48] variation of phase and amplitude with distance βz in front of a short circuit is shown in Fig. 8. It is, however, not trivial to obtain explicit values for α and β in terms of V and θ from Eqs. (4.4.59)–(4.4.61); the simplest procedures are as follows[45, 46]:

1.   When measurements of V are made at distances ξ and 2ξ from the short circuit,
$$\cosh 2\alpha\xi = \tfrac{1}{4}R^2 + \{(1 - \tfrac{1}{4}R^2)^2 + R^2 \sin^2 \eta\}^{1/2}, \tag{4.4.62}$$

$$\cos 2\beta\xi = \tfrac{1}{4}R^2 - \{(1 + \tfrac{1}{4}R^2)^2 + R^2 \sin^2 \eta\}^{1/2}, \tag{4.4.63}$$

where R is the ratio of the magnitudes of the induced voltages and η the phase difference.
2.   The phase angles θ_1 and θ_2 at two equal amplitude points, which are a distance Δz from each side of the voltage minimum z_0, are related to α as

$$\tanh \alpha z_0 = 2 \tan(\beta \Delta z)/(\theta_1 - \theta_2). \tag{4.4.64}$$

3.   If z_1 and z_2 are values of z at which amplitudes are equal, then

$$\sin \alpha z_1 \simeq \cosh 2\alpha z_2. \tag{4.4.65}$$

In modern bridge experiments the response curve is recorded and curve fitted to an expression containing the electric properties of the liquid as

[48] R. G. Bennett, *J. Sci. Instrum.* **37**, 195 (1960).

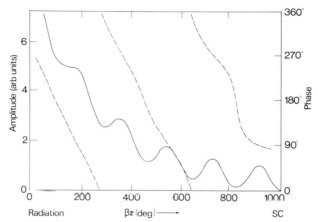

FIG. 8.   Calculated variation of phase (broken line) and amplitude (full line) along short-circuited (SC) dielectric-filled line or guide with $\alpha/\beta = 0.1$. [R. G. Bennett and J. H. Calderwood, in "Complex Permittivity" (B. K. P. Scaife, ed.), Chapter 3. English Univ. Press, London, 1971.]

adjustable parameters. Bridge elements that serve to divide or merge signals include the directive feed, the magic T, and the hybrid T or "rat-race." High-frequency bridges that use only a single multiarm junction are analogs of low frequency bridges. One type of admittance meter is commercially available[49] in a concentric line version suitable for study of liquids in the range 250–1000 MHz[50]; it is shown schematically in Fig. 9, and consists of a T-junction of four arms, containing the generator, the dielectric cell, and standards of conductance G_0 and susceptance X_s (imaginary part of admittance). Variable-length transmission line is used to adjust the distance between cell and T-junction. The arms are fitted with miniature rotatable pickup loops, the signals from which are balanced at the detector. The signal amplitudes are proportional to the cosines of the angles ϕ between the loop planes and the line axes, and these angles are indicated on a scale. A null point is obtained with susceptance zero and empty cell with plunger $\lambda_0/4$ from the window. This is achieved by adjustment of the variable line length until the empty cell is located an integral number of half wavelengths from the T-junction. The cell is now filled and a null point reestablished by adjustment of the conductance arm loop and the cell plunger to liquid thickness d. The conductance arm loop angle is calibrated in conductance G and the (real) admittance Y of the dielectric cell is calculated from

$$Y = (\varepsilon_0/\mu_0)^{1/2}[G/(G_0 \cos \phi)]. \qquad (4.4.66)$$

[49] General Radio, Type 1602-AB.
[50] F. I. Mopsik, Thesis, Brown Univ., Providence, Rhode Island (1964).

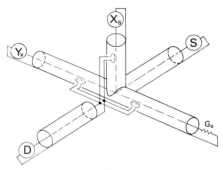

FIG. 9.   Schematic diagram of concentric line admittance meter: S—source; D—detector; subscript x—unknown circuit element; subscript s—standard circuit element.

Then ε' and ε'' are calculated in terms of the variables $a = 2\alpha d$ and $b = 2\beta d$, using

$$\varepsilon' = (\lambda_0/4\pi d)^2(b^2 - a^2), \tag{4.4.67}$$

$$\varepsilon'' = (\lambda_0/4\pi d)^2 2ab, \tag{4.4.68}$$

with

$$\gamma = \left(\frac{\lambda_0}{4\pi d}\right)\left(\frac{b \sinh a - a \sin b}{\cosh a - \cos b}\right) \tag{4.4.69}$$

and

$$a \sinh a + b \sin b = 0. \tag{4.4.70}$$

This commercial bridge is essentially similar to versions developed by Cole and his collaborators,[51–53] which did not possess rotatable coupling loops. A similar waveguide bridge using a six-arm junction has also been reported.[54]

4.4.2.4.3. HIGH-LOSS LIQUIDS AND FREE-WAVE TECHNIQUES. High-loss dielectrics can be matched to the characteristic impedance of the waveguide, so that the standing waves are almost completely damped out. A standing wave will always arise when a sudden dielectric boundary occurs, but it is possible to minimize it in waveguides by the following techniques:

(1)   introduction of other reflecting elements in the form of adjustable matching stubs;
(2)   use of dielectric windows whose dimensions are such as to transform

[51] S. H. Glarum, Rev. Sci. Instrum. 29, 1016 (1958).
[52] F. I. Mopsik, Thesis, Brown Univ., Providence, Rhode Island (1964).
[53] F. I. Mopsik and R. H. Cole, J. Chem. Phys. 44, 1015 (1966).
[54] E. L. Ginzton, "Microwave Measurements." McGraw-Hill, New York, 1957.

the impedance of the empty waveguide to that of the dielectric-filled one; and

(3) use of wedge-shaped dielectric boundaries, which act by canceling the wave phases reflected from different parts of the wedge.

The minimization of the standing wave in front of the dielectric is monitored by means of a movable miniature probe known as a *standing wave indicator*. Such a wire probe projects into the broad side of a rectangular waveguide through a slot cut along its center. For a concentric line, the probe passes through a hole in the center conductor as in Fig. 10e, and the center conductor is made adjustable in length, by sliding within a fixed central tube.

The standing wave within the lossy liquid itself can be used in the derivation of the complex permittivity $\hat{\varepsilon}$. The liquid is terminated by a short circuit, and a computer fit to the recorded response data is made, using adjustable parameters, which include the complex permittivity and the impedance of the short circuit.

When the standing wave ratios (maximum–minimum signal) both before and behind the dielectric are sufficiently close to unity, there is very little standing wave in the dielectric itself. A lossy dielectric will then absorb the wave approximately exponentially,

$$V(z) = V_i \exp(-\alpha z), \qquad (4.4.71)$$

where V_i is the signal incident on the dielectric. For this type of measurement,[55] it is important that a pure single mode be propagated in the waveguide. The method is only valuable for lossy liquids, and can be regarded as a special case of the bridge technique; it is available for both waveguides and concentric lines, but the absorption measurement must be made over several standing wavelengths π/β, otherwise the exponential character of the absorption cannot be separated from any errors arising from superposed standing waves. Waveguide measurements on lossy liquids may be adapted to yield also the real dielectric constant by measuring the enhancement of absorption coefficient when the waveguide is only slightly larger than the cutoff dimension.[56] The two absorption coefficients κ_1 and κ_2 are appropriate to rectangular TE_{01} guides of widths a_1 and a_2, or cylindrical TE_{11} guides of radii r_1, and r_2. Then

$$\varepsilon' = \tfrac{1}{2}c^2[(k_1{}^2\kappa_1{}^2 - k_2{}^2\kappa_2{}^2)/(\kappa_1{}^2 - \kappa_2{}^2)] - (\kappa_1{}^2 + \kappa_2{}^2), \quad (4.4.72)$$

$$\varepsilon'' = 2\kappa_1[\varepsilon' - (k_1{}^2c^2/\omega^2) + \kappa_1{}^2]^{1/2} \qquad (4.4.73)$$

[55] C. H. Collie, D. M. Ritson, and J. B. Hasted, *Trans. Faraday Soc.* **42A**, 129 (1946).
[56] C. H. Collie, J. B. Hasted, and D. M. Ritson, *Proc. Phys. Soc.* **60**, 145 (1948).

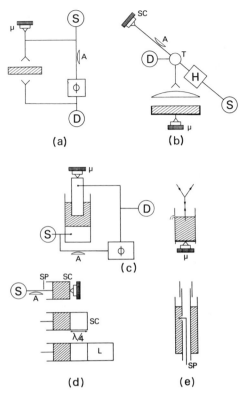

FIG. 10. Schematic diagrams of microwave techniques. Key: S—source; D—detector; A—variable attenuator; ϕ—phase changer; SC—fixed or movable short-circuit; SP—standing wave probe; L—load; μ—micrometer with short-circuit plunger; T—hybrid T-junction; Y—horn; dielectric specimens are shown cross-hatched. (a) Free-wave transmission interferometer for millimeter waves; (b) free-wave reflection interferometer for millimeter waves; (c) bridge circuit with specimen of variable thickness (mechanical problems of varying the thickness without using flexible waveguide have been solved by the use of the variable depth liquid reflection cell shown schematically on the right); (d) reflection standing-wave techniques in waveguide; (e) fixed position probe in liquid-filled coaxial line, terminated at top by slotted section of variable length.

with $k_{1,2} = \pi/a_{1,2}$ for rectangular TE_{01} guide and $k_{1,2} = 1.841/r_{1,2}$ for cylindrical TE_{11} guide.

The absorption of a free wave in a lossy liquid will also tend to be an exponential function of the thickness of the liquid, in the limit of large thickness, provided that the distance from the transmitter to the liquid is much larger than λ_0. This has been made the basis of a millimeter-wave technique for measurement of κ,[57, 58] and may be combined with free-wave

[57] J. A. Lane and J. A. Saxton, *Proc. Roy. Soc.* **A214**, 531 (1952).
[58] M. Yasumi, *Bull. Chem. Soc. Japan* **24**, 53, 60 (1951).

measurements of amplitude reflection coefficient R and Brewster angle[57, 59] $\theta_B = \arcsin n^{-1}$. Unfortunately R is not very sensitive to changes in refractive index n, being given by

$$R = (n^2 + \kappa^2 + 1 - 2n)/(n^2 + \kappa^2 + 1 + 2n). \qquad (4.4.74)$$

Free-wave bridge methods are also available.[60] Reflection coefficient measurement in waveguides has also been attempted, and combined with measurement of wavelength in a filled guide.[61-65] Fixed-length absorption measurement in guides has also been employed.[66, 67]

Free-wave techniques are most useful at wavelengths of less than 4 mm, since the fabrication of miniature dielectric cells, not to mention standing wave indicators and other waveguide components, becomes increasingly difficult as the wavelength becomes smaller. Although the early free-wave techniques mentioned previously are suitable for lossy liquids, low-loss liquids require interferometric techniques to be used. An interferometer[68, 69] essentially analogous to the Michelson is shown schematically in Fig. 10b. A plane wave signal produced by horn and lens is reflected from a metal surface fronted by a variable thickness of liquid. The incident wave is split into two equal parts at the hybrid ring T (or "rat-race"), which acts both as the splitter and recombiner of the signal. One part passes to the free-wave cell, and the other through a variable attenuator to a movable short circuit. Both parts, being recombined at the ring, pass to the detector.

The detector current is a complicated function of the liquid thickness d:

$$I \propto \frac{1 + \exp(-4\beta_0\kappa d) - 2\exp(-2\beta_0\kappa d)\cos 2\beta_0 nd}{[(1 + n)^2 + \kappa^2 + \{(1 - n)^2 + \kappa^2\}\exp(-4\beta_0\kappa d) - 2\exp(-2\beta_0\kappa d)\{(1 - n^2 - \kappa^2)\cos 2\beta_0 nd + 2\kappa \sin 2\beta_0 nd\}]}$$
$$(4.4.75)$$

where n and κ are the optical constants [Eq. (4.4.9)]. The fitting of this equation to a measured interferogram $I(d)$ requires a computer program, although some simplification is possible. For low-loss, small κ liquids, the

[59] T. E. Talpey, *Onde Elec.* **33**, 561 (1953).

[60] R. M. Redheffer and E. D. Winkler, *M.I.T. Radiat. Lab. Rep.* **483**, 15, 18, 145.

[61] W. M. Heston, E. J. Hennelly, and C. P. Smyth, *J. Amer. Chem. Soc.* **70**, 4093 (1948); H. L. Lacques and C. P. Smyth, *J. Am. Chem. Soc.* **70**, 4097 (1948); W. H. Surber, *J. Appl. Phys.* **19**, 514 (1948).

[62] G. E. Crouch, *J. Chem. Phys.* **16**, 364 (1948).

[63] H. C. Bolton, *Proc. Phys. Soc.* **61**, 294 (1949).

[64] J. Ph. Poley, *Appl. Sci. Res.* **B4**, 173 (1954).

[65] W. H. Surber, *J. Appl. Phys.* **19**, 514 (1948).

[66] J. G. Powles, *Trans. Faraday Soc.* **44**, 537 (1948).

[67] D. H. Whiffen and H. W. Thompson, *Trans. Faraday Soc.* **42a**, 114 (1946).

[68] W. E. Vaughan, W. S. Lovell, and C. P. Smyth, *J. Chem. Phys.* **36**, 535 (1962).

[69] S. K. Garg, H. Kilp, and C. P. Smyth, *J. Chem. Phys.* **43**, 2341 (1965).

maxima and minima are spaced in the ratio of wavelengths in the dielectric, and the slope of the plot of $n(I_{max} - I_{min})$ for extrema numbered N is of approximate slope $-2\pi\kappa/N$.

4.4.2.4.4. STANDING WAVE TECHNIQUES. A wide variety of standing wave methods in waveguide[58] and also in concentric line[63] are available for solid dielectrics whose dimensions are not variable and which cannot therefore be as accurately studied by phase and attenuation measurements. A signal proportional to the field is obtained at a miniature wire probe inserted through a longitudinal slot in the empty waveguide or concentric line, and traversable along its length. This operation can also be carried out in liquid-filled guides and concentric lines, as illustrated in Fig. 10e. Commercially available precision "standing wave indicators" enable measurements to be made of the input impedances of dielectric surfaces, which are related to their characteristic impedances and dielectric properties.

The standing wave is studied in front of a specimen of dielectric of length d (Fig. 10d). Let the front dielectric surface impedance be Z_f and the back dielectric surface impedance be Z_b; its characteristic impedance is, as before, Z, and that of the empty waveguide Z_0. From Eq. (4.4.59) it can be shown that

$$Z_f = Z(Z_b + Z \tanh \gamma d)/(Z + Z_b \tanh \gamma d), \tag{4.4.76}$$

where the propagation coefficient γ refers to the dielectric. In a loss-free waveguide or concentric line the ratio of the amplitudes of the incident and reflected waves is given by the reflection coefficient

$$R = |V_-|/|V_+| = |(Z_f - Z_0)/(Z_f + Z_0)|. \tag{4.4.77}$$

When the incident and reflected waves are in phase, at the maximum of the standing wave,

$$|V_{max}| = |V_+| + |V_-| = |V_+|(1 + R), \tag{4.4.78}$$

and at minimum

$$|V_{min}| = |V_+| - |V_-| = |V_+|(1 - R), \tag{4.4.79}$$

so that the voltage standing wave ratio

$$|V_{min}/V_{max}| = S = (1 - R)/(1 + R). \tag{4.4.80}$$

At a voltage minimum the magnitude of the current is

$$|I| = (|V_+| + |V_-|)/Z_0 = |V_+|(1 + R)/Z_0 \tag{4.4.81}$$

and the impedance

$$Z_{min} = |V_{min}/I_{min}| = Z_0[(1 - R)/(1 + R)] = SZ_0. \tag{4.4.82}$$

At a distance z in front of the front dielectric surface

$$Z_z = Z_0(Z_f + jZ_0 \tan \beta_0 z)/(Z_0 + jZ_f \tan \beta_0 z). \tag{4.4.83}$$

Therefore at z_0 where there is a voltage minimum

$$Z_f = Z_0(S - j \tan \beta_0 z_0)/(1 - jS \tan \beta_0 z_0). \qquad (4.4.84)$$

When the back surface of the dielectric is a short circuit, $Z_b = 0$ and

$$Z_f = Z \tanh \gamma d. \qquad (4.4.85)$$

Using

$$Z = Z_0(j\beta_0/\gamma) \quad \text{and} \quad \beta_0 = 2\pi/\lambda_g, \qquad (4.4.86)$$

one finds

$$\frac{\tanh \gamma d}{\gamma d} = \frac{S - j \tan(2\pi z_0/\lambda_g)}{(2j\pi d/\lambda_g)/[1 - jS \tan(2\pi z_0/\lambda_g)]}. \qquad (4.4.87)$$

This equation was derived by Roberts and von Hippel,[70] and charts were prepared[71] to enable solutions to be found. The measured quantities are S and z_0, the standing wave ratio and position of first minimum, when the dielectric specimen is short-circuited at its rear. In modern research computer solutions are normally used, but other related techniques have been worked out in which the computations are much simpler. One such method[72, 73] involves two measurements, one of S and z_0 with short circuit at the rear surface, after which the short circuit is withdrawn until the minimum again appears at its original position; the withdrawal distance is the second measurement.

Alternatively,[73, 74] S and z_0 are measured with rear surface short circuited and open-circuited; the open circuit is simulated by withdrawing the short circuit by a quarter of a guide wavelength (Fig. 10d) (since a real open circuit will radiate and thereby cease to be perfectly open). The front face impedances for open circuit $Z_{f\infty}$ and short circuit Z_{fs} are given by

$$Z_{fs} = Z \tanh \gamma d \qquad (4.4.88)$$

and

$$Z_{f\infty} = Z/\tanh \gamma d \qquad (4.4.89)$$

so that

$$Z = (Z_{fs}Z_{f\infty})^{1/2} = jZ_0\beta_0/\gamma \qquad (4.4.90)$$

from which γ is calculated.

When the dielectric is terminated by a short circuit of variable position, this termination can be regarded as a reactive impedance jX, which may be measured in the absence of dielectric specimen by determining the standing wave ratio and phase.

[70] D. D. Roberts and A. R. von Hippel, *J. Appl. Phys.* **17**, 610 (1946).

[71] A. R. von Hippel, "Dielectrics and Waves." Wiley, New York, 1954.

[72] R. G. Bennett and J. H. Calderwood, *Proc. Inst. Elec. Eng.* **112**, 416 (1965).

[73] E. Fatuzzo and P. R. Mason, *J. Sci. Instrum.* **41**, 694 (1964).

[74] A. A. Oliner and H. M. Altshuler, *J. Appl. Phys.* **26**, 214 (1955).

Using equations

$$Z_{f1,2} = (ZjX_{1,2} + Z \tanh \gamma d)/(Z + jZ_{1,2} \tanh \gamma d), \qquad (4.4.91)$$

where subscripts 1 and 2 represent the two different positions of the short circuit, one finds

$$Z^2 = [X_1 X_2 (\Delta Z_f/\Delta X) + jZ_{f1}Z_{f2}]/[j - (\Delta Z_f/\Delta X)], \qquad (4.4.92)$$

where

$$\Delta Z_f = Z_{f2} - Z_{f1} \qquad \text{and} \qquad \Delta X = X_2 - X_1. \qquad (4.4.93)$$

Where a measurement on a liquid is to be made, or where a solid specimen is available in more than one length, the calculation is also simplified. For lengths d, $2d$, yielding $Z_{f1,2}$,

$$\tanh^2 \gamma d = (2Z_{f1} - Z_{f2})/Z_{f2}, \qquad (4.4.94)$$

which is readily soluble. For variable d[75] a graph is plotted of the positions $y(n)$ of the minima, against d, with $y(0)$ taken as zero (n is an integer). At the intersection of the $y(d)$ graph with the straight line

$$y(n) = -d + (n\lambda_0/2) \qquad (4.4.95)$$

one has

$$dy/dd = -n^2\pi^2(d/\beta)^2 = -n^2\pi^2m^2. \qquad (4.4.96)$$

If Δy denotes the change in y when d is decreased sufficiently for the $(n + 1)$th minimum to fall where the nth one previously did, namely at the interface, then

$$\frac{\beta}{\beta_0} = \frac{1 - m^2\{1 - m^2[1 - (2\pi/3)(3n^2 + 3n + 1)]\}}{1 - \beta_0 \Delta y/\pi} \qquad (4.4.97)$$

from which β may be derived; α is determined similarly.

The literature contains descriptions of many more standing wave experiments,[76-84] in some of which it is necessary to terminate the guide in an artificial load L (Fig. 10d).

[75] E. Fatuzzo and P. R. Mason, *J. Appl. Phys.* **19**, 87 (1959).

[76] G. Williams, *J. Phys. Chem. Itaca* **63**, 534 (1959).

[77] M. G. Corfield, J. Horzelski, and A. H. Price, *Brit. J. Appl. Phys.* **12**, 680 (1961).

[78] D. A. A. S. N. Rao, *Brit. J. Appl. Phys.* **17**, 109 (1966).

[79] V. I. Little, *Proc. Phys. Soc.* **66B**, 175 (1953).

[80] W. H. Surber and G. E. Crouch, *J. Appl. Phys.* **19**, 1130 (1948).

[81] J. G. Powles, *Nature (London)* **162**, 614 (1948).

[82] H. Okabayashi, *Bull. Chem. Soc. Japan* **35**, 163 (1962).

[83] "Techniques of Microwave Measurements," Vol 11. M.I.T. Radiat. Lab. Ser., Boston Tech. Publ., Lexington, Massachusetts, 1964.

[84] W. B. Westphal, *in* "Dielectric Materials and Applications" (A. von Hippel, ed.). Wiley, New York, 1954.

4.4.2.4.5. CAVITY RESONATORS. Cavity resonators are essentially the extension of the radio-frequency resonance techniques to the concentric line and waveguide bands. The cavity resonator is also familiar at optical frequencies in the form of the Fabry–Perot etalon. Cavity resonators are uniquely suitable for studying very low-loss dielectrics for which the measurement of ε'' by bridge or standing wave techniques would be less accurate.

The resonator is a length of waveguide or concentric line terminated at each end by a short circuit, so that a resonant standing wave can be set up if the dimensions are suitable. When this is the case the electromagnetic energy is stored for a period of time, long compared with the wave period, but is lost by dielectric and wall resistive loss. The resonator is coupled loosely to a generator by means of one or more small holes or magnetic pickup loops, and a detector may be similarly coupled. With sufficiently loose coupling, high wall conductivity, and low dielectric loss, the detector power P at frequency f is given by

$$P = P_{max}/\{1 + 4Q^2(f - f_{max})^2/f_{max}^2\}, \qquad (4.4.98)$$

where P_{max} is the maximum power that is observed at the detector when the cavity is in resonance at the resonance frequency $f = f_{max}$. The selectivity Q of the cavity is given in terms of its capacitance C and conductance G by

$$Q = 2\pi C f_{max}/G, \qquad (4.4.99)$$

as at lower frequencies.

An empty cavity resonator with inside walls plated with silver or gold possesses a Q of order 10^4. As the frequency is varied, the detector signal, which is normally very small, passes through a narrow resonance peak at $f = f_{max}$. The shape of this peak is determined by Eq. (4.4.98) and its width is a measure of Q. In addition to fixed frequency resonators one can also build resonators of variable length for use with a fixed-frequency generator.

When a cavity resonator is completely filled with dielectric, the resonance frequency, which is a function of the cavity dimensions and propagation mode, changes proportionately to $\sqrt{\varepsilon'}$:

$$\varepsilon' = f_{0max}^2/f_{max}^2, \qquad (4.4.100)$$

where f_{0max} and f_{max} are the resonant frequencies, respectively, empty and full. From Eq. (4.4.99) it follows that the loss is given by

$$\varepsilon'' = (1/Q) - (1/Q_0), \qquad (4.4.101)$$

where Q_0 refers to the empty cavity and Q to the dielectric-filled cavity. Thus for low-loss materials the completely filled resonator is an instrument of great simplicity and precision. To determine Q, the frequency is varied across the resonance from one-half power to one-half power by a decrement Δf; then

$$Q = f_{max}/\Delta f. \tag{4.4.102}$$

In a tunable resonator the change of length may be considered as producing a change of capacitance ΔC, which is related to C_T by

$$1/Q = \Delta C/2C_T, \tag{4.4.103}$$

where C_T is the total capacitance in the equivalent circuit of the cavity.

With solid specimens, with lossy liquids, and with all measurements at wavelengths greater than about 10 cm, filling of the entire resonator is impracticable, and partially filled resonators become necessary. The most important of these are the following:

1.   Concentric circular line resonator (TEM mode) with disk dielectric specimen mounted in a gap in the center conductor.[85–88] When a solid specimen is in position in the resonator, no systematic turning in length is possible, so that a variable frequency generator must be used.

2.   Cylindrical TM waveguide resonator with axially mounted rod (or liquid-filled tube) specimen. This is also a nontunable resonator, since a change of length does not imply a proportionate change of resonant frequency.[89]

3.   Cylindrical TE waveguide resonator with disk specimen. This resonator is tunable in length by means of a plunger with proportionate change in resonant frequency.[89] The resonator can also be used with a rod (or liquid-filled tube) specimen, without loss of tunability.[90]

The analysis of partially filled cavity measurements is complicated and will not be discussed in detail. For the calculation of propagation coefficient it is necessary that the boundary conditions at the specimen surface be correctly satisfied. Resonators are not normally commercially available, but must be designed for the problem in hand. The same applies to specimen cells used in attenuation, bridge, and reflection measurements. Nearly all other microwave components are, however, commercially available in most of the bands. Most of the waveguide techniques described above are suitable for operating at a single frequency, or over a small tunable range, say, 1% of the frequency. This is not only because of the characteristics of the resonators

[85] C. N. Works, *J. Appl. Phys.* **18**, 605 (1947).

[86] C. N. Works, T. W. Dakin, and F. W. Boggs, *Trans. Amer. Inst. Elec. Eng.* **63**, 1092 (1944).

[87] T. W. Dakin and C. N. Works, *J. Appl. Phys.* **18**, 789 (1947).

[88] J. G. Powles, *Nature (London)* **162**, 614 (1948).

[89] F. Horner, T. A. Taylor, R. Dunsmuir, J. Lamb, and W. Jackson, *J. Inst. Elec. Eng.* **93**, Part III, 53 (1946); R. Dunsmuir and J. G. Powles, *Phil. Mag.* **37**, 747 (1946).

[90] C. H. Collie, J. B. Hasted, and D. M. Ritson, *Proc. Phys. Soc.* **60**, 71 (1948).

or cells, but because of those of the commercial waveguide components, generators, and detectors. Concentric line techniques can of course be used over a much wider frequency range (possibly 100% change). Nevertheless the fact that each dielectric relaxation process extends over more than a decade of frequency means that studies in the microwave region consume much time and money, being mostly conducted at a number of spot frequencies in the range of the process. More detailed discussions of microwave measurement techniques are available.[91]

4.4.2.5. Time Domain Techniques. 4.4.2.5.1. INTRODUCTION. Time domain techniques have certain advantages over the frequency domain techniques considered in previous sections. Once the data-handing equipment is set up, the complex permittivity is measured in one operation over what corresponds to four or five decades of frequency. The accuracy is at present not quite the equal of that obtainable with frequency domain techniques; however, there is one further advantage, which is that the time domain technique is particularly suitable in the subHertzian frequency range $(10^{-5}-1$ Hz). Within the time domain techniques there is a division between the slower response region ($<10^6$ Hz) and the faster response region (10^6-10^{10} Hz), which requires the use of transmission lines, as is the case in the frequency domain. In the faster response region, however, "time domain spectroscopy" has only recently become experimentally available, with the advent of pulses of picosecond rise time and of fast response storage oscilloscopes.

4.4.2.5.2. THE SLOW RESPONSE REGION. A voltage step is applied across a capacitance cell, and the current response $I(t)$ is recorded as a function of time. The Fourier transform of this response function is proportional to $\hat{\varepsilon}(\omega)$:

$$\hat{\varepsilon}(\omega) \propto \int_0^\infty I(t) \exp(-j\omega t)\,dt. \tag{4.4.104}$$

This relation holds as an equality for the application of a unit voltage step to a filled unit capacitor. Accurate transformation of a set of data requires the storage of at least 500 data points, followed by a fast Fourier transform computation. This can be avoided in the slow response region[5, 7, 92, 93] by use of the Hamon approximation,[94] which can be written as

$$\varepsilon''(f) = I(0.1f)/2fC_0, \tag{4.4.105}$$

[91] H. M. Barlow, and A. L. Cullen, "Microwave Measurements." Constable Press, London, 1966.

[92] W. Reddish, Soc. Chem. Ind. Monograph No. 5 (1959).

[93] G. Williams, Polymer 4, 27 (1963); G. Williams and D. C. Watts, Trans. Faraday Soc. 66, 80 (1970); G. Williams, D. C. Watts, S. B. Dev, and A. M. North, Trans. Faraday Soc. 67, 1323 (1971).

[94] B. V. Hamon, Proc. Inst. Elec. Eng. 99, 151 (1952).

where C_0 is the capacitance of the empty capacitor and I the current for unit voltage application.

A different approach[95] is as follows. When a voltage step function is applied to the capacitor in a circuit of the type of Fig. 11a, the charge q varies approximately with the logarithm of the time. Therefore the electrometric

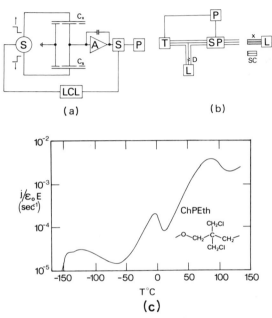

(a)   (b)

(c)

FIG. 11. (a) Schematic diagram of slow response time domain capacitance measurement: S in square—sampler and digital store; P—processer and graphical display; LCL—logarithmic clock; A—amplifier; S in circle—source. (b) Schematic diagram of fast response time domain reflectometry: T—tunnel diode; D—discontinuity for time marking; L—matched load; SP—sampler and digital store; P—processer and graphical display; x—sample; SC—short circuit. (c) Thermally stimulated depolarization spectrum, or thermogram, for polymer ChPEth [J. van Turnhout, *Advan. Stat. Elec.* 1, 56 (1971); Thesis, Univ. of Leyden (1972)].

sampling of the charge at equal intervals of log t provides a number of approximately equal samples, the most efficient minimal set for mathematical operations. The transforms are

$$\varepsilon'(\omega) = \sum_{p=-\infty}^{\infty} \Delta q(n)\, x(n), \qquad (4.4.106)$$

$$\varepsilon''(\omega) = \sum_{p=-\infty}^{\infty} \Delta q(n)\, y(n), \qquad (4.4.107)$$

[95] P. J. Hyde, *Proc. Inst. Elec. Eng.* **117**, 1891 (1970).

where $n = 2^p$ and $x(n)$ and $y(n)$ are coefficients. Only a small number of right-hand-side terms are used in practice.

Spectrometers have been constructed in the time ranges 10^3-10^{-3} and $10^{-1}-10^7$ sec, the former sampling from a single pulse, and the latter from a series of equal length pulses.

4.4.2.5.3. FAST RESPONSE REGION. For times faster than about 1 μsec, a voltage pulse must be propagated down a transmission line. When it passes from a region of characteristic impedance Z_0 into a dielectric-filled region of impedance Z, the complex reflection coefficient and transmission coefficient are, respectively,

$$R = (Z - Z_0)/(Z + Z_0) \tag{4.4.108}$$

and

$$T = 2Z/(Z + Z_0). \tag{4.4.109}$$

The ratio of the Fourier transforms of the incident and reflected or transmitted pulses gives the frequency response of R or T:

$$R(\omega) = \frac{F_r(\omega)}{F_i(\omega)} = \frac{\int_{-\infty}^{\infty} f_r(t) \exp(-j\omega t)\, dt}{\int_{-\infty}^{\infty} f_i(t) \exp(-j\omega t)\, dt}. \tag{4.4.110}$$

A suitable method of Fourier transforming step pulses is by means of the Samulon[96] modification of the Shannon[97] sampling theorem, expressed as

$$F(\omega) = \frac{\tau \exp(j\omega\tau/2)}{2j \sin(\omega\tau/2)} \sum_{n=-\infty}^{\infty} f(n\tau) - f(n\tau - \tau) \exp(-jn\omega\tau),$$

$$\tag{4.4.111}$$

where τ is the time interval between the sampled data points (not a relaxation time).

Fast pulse generators and sampling techniques, developed with network analysis as the end in view, are commercially available. A tunnel diode produces a pulse with a rise time on the order of 30 psec. It is used with precision concentric transmission line and sampling oscilloscope display in a circuit of the type shown in Fig. 11b. The pulses are repeated at regular intervals and sample signals are taken at a series of sampling times, different for each pulse transmission. A time reference signal is established for the sampling time base, by connecting via a junction a standard transmission line delay terminated by a reflecting discontinuity and load.

The dielectric properties of the specimen contained in the concentric line are related to the complex reflection coefficient by

$$\hat{\varepsilon} = [(1 - R)/(1 + R)]^2. \tag{4.4.112}$$

[96] H. A. Samulon, *Proc. I.R.E.* **39**, 175 (1951).
[97] C. Shannon, *Proc. I.R.E.* **37**, 10 (1949).

For this equation to be applied one must, however, be certain that the reflected pulse corresponds to reflection from the front dielectric interface, and includes no contribution from reflection at the rear interface, or any other discontinuity. Errors are introduced at the low-frequency end of the range by a reflection contribution at the rear end of the sampled reflection pulse; these can be avoided by making the cell of sufficient length.

A time domain transmission technique has been reported[98] in which the transmission pulses through two different lengths a and b of dielectric are compared:

$$T_a/T_b = \exp\{-j\omega(a - b)\sqrt{\hat{\varepsilon}}/c\}, \tag{4.4.113}$$

where c is the speed of light. When the ratio of transmission coefficients is expressed as $r \exp j\phi$,

$$\hat{\varepsilon} = [c^2/\omega^2(a - b)^2]\{\phi^2 - (\ln r)^2 - 2j\phi \ln r\}. \tag{4.4.114}$$

The method is insensitive to time reference errors, but the available frequency range is smaller than in the reflection method. A similar technique[99] makes use of the two signals R_1 and R_2 reflected, respectively, from the front dielectric surface and from a short circuit placed well behind the dielectric; the latter signal passes twice through the dielectric:

$$R_1 = (1 - \rho^2)z/(1 - \rho^2 z^2) \tag{4.4.115a}$$

and

$$R_2 = (1 - z^2)\rho/(1 - \rho^2 z^2), \tag{4.4.115b}$$

where

$$\rho = [(1/\hat{\varepsilon})^{1/2} - 1]/[(1/\hat{\varepsilon})^{1/2} + 1] \tag{4.4.116}$$

and

$$z = \exp(-jd\sqrt{\hat{\varepsilon}}/c) \tag{4.4.117}$$

from which $\hat{\varepsilon}$ can be calculated. Thin specimens can be used with this technique.

Errors in time domain spectroscopy can arise from computing, from timing errors, from noise, and from incorrect truncation in transformation.[100] Various methods of producing time reference points have been worked out.[98] The linearity of the sampling time base is of great importance, and is often ensured by digital methods. Noise errors can be reduced by signal averaging, and a more sophisticated three-point scanning technique[99] is available. In this technique repeated measurements are made at three points in time. At the first the slope of the waveform is zero, at the second it is finite; the third point is movable and incrementally scans the time window.

[98] H. W. Loeb, G. M. Young, P. A. Quickenden, and A. Suggett, *Ber Bunsengese. Phys. Chem.* **75**, 1155 (1971).

[99] A. M. Nicolson, and G. F. Ross, *IEEE Trans. Instrum. Measur.* **IM-90**, 377 (1970).

[100] A. M. Nicolson, *IEEE Trans. Instrum. Measur.* **IM-17**, 395 (1968).

Changes in the difference between the first two signals are used to generate correction voltages in the scanning circuit to return the second point, and thereby the time window, to its initial position. In this way the jitter in the sampling is greatly reduced.

It would be valuable to find a way of avoiding the Fourier transform in this type of experiment. It is well known that a dielectric responds to an applied step field by an exponential or similar response which can be written in the form

$$\Psi(t) = (\varepsilon_s - \varepsilon_\infty)\{1 - \exp(-t/\tau)\} \qquad (4.4.118)$$

for the Debye relaxation. The original time domain experiments of Fellner–Feldegg[101] made use of a formula

$$(\varepsilon_\infty - 1)\,\delta(t) + \Psi(t) = (2c/d)[R(t)/V_0], \qquad (4.4.119)$$

where $V_0(t)$ is the incident voltage pulse, $\delta(t)$ the Dirac function for an instantaneous response $(\varepsilon_\infty - 1)$, and $R(t)$ the reflected signal. It was shown[102] that this equation is inaccurate unless the sample thickness is very small. A correction has been worked out[103] for thicker samples:

$$(\varepsilon_\infty - 1) + \Psi(t) = \frac{2c}{d}\int_0^t dt'\,\frac{R(t')}{V_0} + \frac{2c}{d}\int_0^t dt'\left(\frac{R(t')}{V_0}\cdot\frac{R(t-t')}{V_0} + \frac{2R(t)}{V_0}\right)$$
$$(4.4.120)$$

This result requires only the numerical integration of $R(t)$ and its self-convolution to obtain $\Psi(t)$ from experimental data.

4.4.2.6. Thermally Stimulated Depolarization Spectra. A dielectric which takes more than a second to respond to an applied electric field will retain its polarization charge sufficiently long for its discharge to be observed continuously when it is short-circuited through an electrometer amplifier or picoammeter. A dielectric which retains its charge semipermanently is termed an *electret*[104–106]; under isothermal conditions the application of the charge takes just as long as does the discharge process.

In Section 4.4.3.7 the exponential temperature variation of relaxation time is discussed. An electret can be made by charging a dielectric at an elevated temperature where its relaxation time is short and then cooling it in the electric field to a temperature sufficiently low for the relaxation time to be conveniently long; the charge is retained and the field produced by it can

[101] H. Fellner-Feldegg and E. F. Barnett, *J. Phys. Chem.* **74**, 1962 (1970).
[102] M. J. C. van Gemert, *J. Phys. Chem.* **75**, 1323 (1971).
[103] R. H. Cole, *J. Phys. Chem* **78**, 1440 (1974).
[104] B. Gross, "Charge Storage in Solid Dielectrics," Elsevier, Amsterdam, 1964.
[105] H. J. Wintle, *J. Appl. Phys.* **42**, 4724 (1971).
[106] J. van Turnhout, *Advan. Stat. Elec.* **1**, 56 (1971); Thesis, Univ. of Leyden (1972).

be used, provided that the temperature remains sufficiently low. With some practical dielectrics, charges can be retained for many years at room temperature.

When a dielectric specimen is polarized by the application of a field, cooled to a sufficiently low temperature, and allowed to warm up slowly at a measured rate $T(t)$ while in contact with an electrometer, then the time variation of the current density $j(t)$ will represent a relaxation spectrum, known as the *thermally stimulated depolarization spectrum* (TSD spectrum or thermogram).

The expression relating the relaxation time temperature variation $\tau(T)$ to the TSD spectrum or current thermogram is

$$j(t) = -\tau^{-1}(T)E(\varepsilon_s - \varepsilon_\infty)[1 - \exp\{-\tau^{-1}(T_f)t_f - \tau_r^{-1}\xi(T_f, T_d)\}]$$
$$\times \exp\{-\tau^{-1}(T_d)(t_d - t_s)\} \exp\{-\tau_r^{-1}\xi(T, T_d)\},$$

where

$$\xi(T, T_d) = dt/dT \int_{T_d}^{T} \tau^{-1}(T)/\tau_r^{-1} \, dT,$$

T_d is the low temperature from which the depolarization process commences at time t_d, T_f the high temperature at which the charging in field E is carried out and terminated at time t_f, τ_r the natural time of movement of the dipoles in the absence of intermolecular restrictions, i.e., the relaxation time at infinite temperature, and t_s the time at which the specimen is stored at the low temperature T_d.

It is readily shown that the current maximum $dj/dT = 0$ occurs when $d\tau/dt = -1$, i.e., at a temperature which uniquely defines the magnitude rate of change of τ with temperature, i.e., the activation energy. The spectrum therefore displays activation energy peaks; the technique is particularly valuable for polymers, a typical spectrum being shown in Fig. 11c.

4.4.3. The Physical Basis of Dielectric Relaxation Spectra

4.4.3.1. Introduction: Electric Moments of Molecules. The origin of dielectric loss, and its counterpart, permittivity or dielectric constant, lies in the molecular electrical structure of matter which is distorted or polarized by the application of an electric field. This distortion or polarization is not instantaneous, but follows the time variation of the applied field with a certain phase lag, which causes an absorption of energy from the field. The phase lag is related to the molecular interactions, which govern the relaxation of the electrical structure.

Molecules, while maintaining an overall electrical neutrality, possess a distribution of electrical charge over their structure. Not only can this be distorted by electric fields, but also, orientation of the molecule is possible when electric fields are applied.

Electrical charge distribution can be described in terms of electrical multipole moments, of which the lowest order in a neutral molecule is a dipole moment; one positive and one negative electrical charge $\pm q$ are separated by distance x; the vector dipole moment

$$\boldsymbol{\mu} = q/\mathbf{x}. \tag{4.4.121}$$

Molecules possessing a permanent dipole moment are known as *polar molecules*, and μ will usually be of the order of several units of 10^{-18} esu, which are called *debye units* (D). Any chemical bond between two different atoms may be regarded as possessing a permanent dipole moment, but there are many molecules such as CCl_4 whose symmetry prevents the molecule from being polar, in the sense defined above, even though the individual C—Cl bonds do have dipole moments. CCl_4 and many similar nonpolar molecules possess higher-order multipole moments, but we shall not consider these at present.

In addition to permanent dipole moments, molecules possess dipole and higher-order multipole moments which are induced by the action of an applied electric field, and are absent when there is no such field. The induced dipole moment is proportional to the field strength $\mathbf{E}$, the constant of proportionality being known as the polarizability α:

$$\boldsymbol{\mu}_{ind} = \alpha \mathbf{E}. \tag{4.4.122}$$

Since $\boldsymbol{\mu}_{ind}$ is also a vector quantity, α is scalar. The polarizability is anisotropic, however, so that it actually has tensor properties. When the applied field is time varying, then the induced dipole will also be time varying, but with a phase lag arising from the failure of the induced molecular polarization to follow instantaneously the applied field. This phase lag becomes important at much higher frequencies than the phase lag for orientation of polar molecules. The molecular polarizability may be considered to be made up of two parts, that due to displacement of the electrons, and that due to displacement of the nuclei. The phase lag of the former becomes appreciable only at frequencies higher than those in the visible region of the spectrum, so that the electronic polarization contributes to the visible refractive index, but the atomic polarization arising from displacement of the nuclei lags behind the field when its frequency is comparable to the frequencies of molecular vibrations.

Thus we can distinguish three regions of phase lag, or "relaxation," corresponding in ascending order of frequency to molecular permanent dipole orientation, to nuclear motion, and to electronic motion. In a time-invariant or static electric field, all these types of polarization contribute to the static permittivity ε_s.

The polarization has the effect of reducing the field inside the dielectric by a factor ε_s. When a true charge $+\xi$ esu/unit area is applied to a metal plate of area A, and a charge $-\xi$ esu/unit area to a parallel plate separated by a distance d, the field between them is

$$E_v = 4\pi\xi \qquad \text{(when unfilled)} \qquad (4.4.123)$$

and

$$E = 4\pi\xi/\varepsilon_s \qquad \text{(filled with dielectric).} \qquad (4.4.124)$$

This drop in field strength might have been attributed to a reduction of the surface charge density by the amount

$$P = \xi[1 - (1/\varepsilon_s)]. \qquad (4.4.125)$$

Thus the presence of the dielectric is equivalent to charging the two dielectric surfaces with charges of opposite sign to those causing the field. The surface density of these opposite sign charges is known as the *polarization P*; it is defined as the total charge that passes through a unit area parallel to the plates when a dielectric is inserted. The surface charges $\pm PA$ give rise to a total dipole moment M of the dielectric:

$$M = PAd = P\mathscr{V}, \qquad (4.4.126)$$

where $\mathscr{V}$ is its volume. In a continuous model the polarization can therefore be regarded as the dipole moment per unit volume. Eliminating ξ, one finds

$$\varepsilon_s = 1 + (4\pi P/E), \qquad (4.4.127)$$

and since, neglecting end effects, the total true charge

$$Q = \xi A = CE/d, \qquad (4.4.128)$$

the capacitance

$$C = A\varepsilon/4\pi d. \qquad (4.4.129)$$

It is customary to use the electrostatic system of units in dielectric theory. The dielectric constant or permittivity ε is a dimensionless quantity, a multiplier to the permittivity of a vacuum $\varepsilon_0 = 1$. In the rationalized mks, or SI system, however, the vacuum permittivity $\varepsilon_0 = (36 \times 10^9 \pi)^{-1} = 8.85 \times 10^{-12}$ F/m^{-1}.

The calculation from ε_s of the microscopic polarization or molecular dipole moment is not simply a matter of dividing by molecular volume, since the actual "inner" field experienced by the molecule is not in general the same as the applied macroscopic field. The earliest method of calculating the inner field F was by considering a microscopic spherical region surrounding the molecule, but large compared with it. Within this sphere the interaction between the molecular dipoles can be calculated; for example, Lorentz showed that for a cubic lattice of polarizable atoms the dipoles inside the sphere produce zero field. In this case the total inner field therefore arises from the external contribution, which is the electric field inside the spherical region due to all sources other than the polarization within it. Electrostatic calculation shows this to be equal to $E + (4\pi P/3)$, so that using Eq. (4.4.127),

$$F = [(\varepsilon_s + 2)/3]E. \tag{4.4.130}$$

For a cubic lattice of atoms of polarizability α and molar volume V,

$$P = N_0 \alpha F/V, \tag{4.4.131}$$

where N_0 is the Avogadro constant. Therefore

$$(\varepsilon_s - 1)/(\varepsilon_s + 2) = N_0 \alpha/3V \tag{4.4.132}$$

which is the Clausius–Mosotti formula for the permittivity of a nonpolar system.

4.4.3.2. Permittivities in Static and Time-Varying Fields. The calculation of the polarization of a system of N rigid polar molecules was first performed by Debye,[29] who started from the expression

$$P = N\mu\langle\cos\theta\rangle, \tag{4.4.133}$$

where θ is the angle between the permanent dipole and the applied field. From the assumption that the orientations of the moments are distributed about that of the applied field in accordance with Boltzmann's law, there follows the expression

$$\langle\cos\theta\rangle = \coth(\mu F/kT) - (kT/\mu F), \tag{4.4.134}$$

which is known as the *Langevin function* $L(\mu F/kT)$. It approximates to $\mu F/kT$ when this is much smaller than unity.

Under these conditions of applied fields very much smaller than 10^7 V/cm,

$$P = N_0\mu^2 F/3VkT. \tag{4.4.135}$$

As with Eq. (4.4.132), there follows Debye's equation for the static dielectric constant of a polar and polarizable material:

$$(\varepsilon_s - 1)/(\varepsilon_s + 2) = (4\pi N_0/3V)[\alpha + (\mu^2/3kT)]. \tag{4.4.136}$$

The inadequacy of the Lorentz inner field results, however, in a failure of the Debye equation to account for the static dielectric constants of dense fluids in terms of their molecular dipole moments.

In Onsager's[107] treatment the molecule is represented as a point dipole of moment m within a spherical molecule-sized cavity of radius

$$a = (3V/4\pi N_0)^{1/3}. \tag{4.4.137}$$

[107] L. Onsager, *J. Amer. Chem. Soc.* **58**, 1486 (1936).

Within this cavity the inner field $\mathbf{F} = \mathbf{G} + \mathbf{R}$ is composed of

(i) the cavity field

$$\mathbf{G} = [3\varepsilon_s/(2\varepsilon_s + 1)]\mathbf{E}, \qquad (4.4.138)$$

which would be produced in the empty cavity by the applied field, and

(ii) the reaction field

$$\mathbf{R} = 2(\varepsilon_s - 1)\mathbf{m}/(2\varepsilon_s + 1)a^3. \qquad (4.4.139)$$

The molecular dipole moment is composed of a permanent and an induced component

$$\mathbf{m} = \boldsymbol{\mu} + \alpha\mathbf{F}. \qquad (4.4.140)$$

As with the Debye calculation the average $\langle \cos \theta \rangle$ is calculated using Boltzmann statistics. The final Onsager expression is

$$(\varepsilon_s - n^2)(2\varepsilon_s + n^2)/\varepsilon_s(n^2 + 2)^2 = 4\pi N_0 \mu^2/9VkT, \qquad (4.4.141)$$

where n is the refractive index. This expression is superior to that of Debye in predicting static dielectric constants of dense fluids. Nevertheless, both expressions assume an inner field which increases without limit for increasing E, thereby incorrectly predicting universal ferroelectricity in sufficiently high fields. Also, both theories are inadequate in that no account is taken of the local forces between neighboring dipoles, except insofar as they contribute to Boltzmann statistics. The permittivity is sensitive to these forces, and its magnitude and frequency variation can be used to assess the adequacy of structural models.

Kirkwood[108] first took these forces into account by considering a spherical specimen containing molecular dipoles. The field inside such a specimen is smaller than that outside by a factor of $3/(\varepsilon_s + 2)$. Within this specimen it is supposed that, in the absence of applied field, it is possible to calculate the dipole moment $\boldsymbol{\mu}^*$ of the specimen when one central dipole $\boldsymbol{\mu}$ is held in a fixed orientation. This calculation is structural; it depends purely on the directional binding forces that exist between groups of molecules in "associated" liquids. Several such calculations of $\boldsymbol{\mu}^*$ exist.[109, 110] The dipole moment of the spherical specimen is, however, enhanced by polarization induced by the central dipole itself, yielding a total dipole moment

$$\overline{\mathbf{M}} = \int \mathbf{P} \cdot dV + \boldsymbol{\mu}^*, \qquad (4.4.142)$$

[108] J. G. Kirkwood, *J. Chem. Phys.* **4**, 592 (1936).

[109] G. Oster and J. G. Kirkwood, *J. Chem. Phys.* **11**, 175 (1943).

[110] G. H. Haggis, J. B. Hasted, and T. J. Buchanan, *J. Chem. Phys.* **20**, 1452 (1952); J. A. Pople, *Proc. Roy. Soc.* **A221**, 498 (1954).

where the integral is taken over the whole specimen except the spherical part itself. Calculation of $\mathbf{P}$ in terms of a radial expansion of potential yields

$$\bar{\mathbf{M}} = 9\varepsilon_s \boldsymbol{\mu}^* / (\varepsilon_s + 2)(2\varepsilon_s + 1). \tag{4.4.143}$$

The permittivity is related to $\bar{\mathbf{M}}$ by

$$\varepsilon_s - 1 = 4\pi\bar{\mathbf{M}}/VE. \tag{4.4.144}$$

Substitution yields the Kirkwood equation

$$(\varepsilon_s - 1)(2\varepsilon_s + 1)/3\varepsilon_s = (4\pi N_0/3VkT)\boldsymbol{\mu} \cdot \boldsymbol{\mu}^*, \tag{4.4.145}$$

which is often written with a scalar correlation parameter g defined by

$$g\boldsymbol{\mu} = \boldsymbol{\mu} \cdot \boldsymbol{\mu}^*. \tag{4.4.146}$$

In the calculation of $\boldsymbol{\mu}^*$ the radius of the spherical specimen must be sufficiently large for the material outside it to be regarded as electrically homogeneous from the point of view of the dipole.

In the Kirkwood model an unpolarizable point-dipole molecule was assumed, but incorrect account was taken of the cavity field acting on the polarizable molecule. The resulting expression did not, as it should, reduce to the Onsager equation (4.4.141) for the case of uncorrelated molecules for which $g = 1$. Other criticisms were also leveled at it, but if the total local field F is taken to be acting on the polarizable molecule, then a different expression is obtained:

$$(\varepsilon_s - n^2)(2\varepsilon_s + n^2)/\varepsilon_s(n^2 + 2)^2 = 4\pi N_0 g\mu^2/9kTV. \tag{4.4.147}$$

The free molecule dipole moment μ must be employed in this equation, which is known as the *Kirkwood–Fröhlich equation*, and was originally derived by Fröhlich using a different technique.[111, 112]

For time-varying electric fields the variation of polarization with time leads to relaxation spectra. The assumption was made by Debye[113] that when a step function field is applied to the dielectric, the orientation polarization P_2 approaches exponentially with time t to its maximum value:

$$P_2 = (P - P_1)\{1 - \exp(-t/\tau)\}. \tag{4.4.148}$$

The characteristic time τ of exponential rise, or decay, is known as the *macroscopic relaxation time*. The total polarization P also includes the distortion polarization P_1:

$$P(t) = P_1(t) + P_2(t), \tag{4.4.149}$$

[111] R. H. Cole, *J. Chem. Phys.* **27**, 33 (1957).
[112] H. Fröhlich, "Theory of Dielectrics." Oxford Univ. Press, London and New York, 1949.
[113] P. Debye, "Polar Molecules." Chemical Catalog Co., New York, 1929.

and P_2 increases at a rate proportional to its departure from equilibrium

$$dP_2/dt = (P - P_1 - P_2)/\tau, \tag{4.4.150}$$

which is the solution for step application of field, and as

$$P_2 = (P - P_1) \exp(-t/\tau) \tag{4.4.151}$$

for step removal of field.

The static permittivity is related to the total polarization by Eq. (4.4.127):

$$E(\varepsilon_s - 1) = 4\pi P, \tag{4.4.152}$$

and the refractive index n to the distortion polarization

$$E(n^2 - 1) = 4\pi P_1. \tag{4.4.153}$$

For alternating field $E = E_0 \exp j\omega t$ of pulsatance $\omega = 2\pi f$, where f represents frequency, one finds

$$\frac{dP_2}{dt} = \left(\frac{\varepsilon_s - n^2}{4\pi}\right) E_0 \exp j\omega t - \frac{P_2}{\tau}, \tag{4.4.154}$$

the solution of which is

$$P_2 = (\varepsilon_s - n^2)E/4\pi(1 - j\omega\tau). \tag{4.4.155}$$

The form of this equation implies that the orientation polarization lags the phase of the field. The permittivity is given by

$$\hat{\varepsilon} = n^2 + [(\varepsilon_s - n^2)/(1 + j\omega\tau). \tag{4.4.156}$$

Writing $n^2 = \varepsilon_\infty$ and rationalizing, one obtains

$$\varepsilon' = \varepsilon_\infty + [(\varepsilon_s - \varepsilon_\infty)/(1 + \omega^2\tau^2)], \qquad \varepsilon'' = \omega\tau(\varepsilon_s - \varepsilon_\infty)/(1 + \omega^2\tau^2). \tag{4.4.157}$$

These are known as the *Debye equations*[†]; their behavior is illustrated by typical data on a fully logarithmic scale in Fig. 12. The real permittivity falls with increasing frequency from ε_s to ε_∞. The dielectric loss ε'' reaches a maximum value equal to

$$\varepsilon''_{max} = (\varepsilon_s - \varepsilon_\infty)/2 \tag{4.4.158}$$

when $\omega\tau = 1$, and is equal to one-half of this value for $\omega\tau = 0.27$ or 3.73. The great breadth of this "absorption line" is due to its origin in an exponential decay or relaxation rather than a resonance process.

The relaxation wavelength ("Sprungwellenlange") λ_s is defined as

$$\lambda_s = 2\pi c\tau, \tag{4.4.159}$$

[†] They are also referred to as the "Debye–Drude" or the "Debye–Pellat equations."

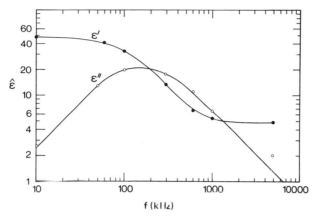

FIG. 12. Full line shows frequency variation of complex permittivity calculated according to the Debye equations. Data points (open circles ε'', filled circles ε') are for 2-propanol at 177.5 K [F. X. Hassion and R. H. Cole, *J. Chem. Phys.* **23**, 1756 (1955)].

and its inverse $\bar{\nu}_s$ is the relaxation wavenumber; the relaxation frequency

$$f_s = \tfrac{1}{2}\pi\tau \tag{4.4.160}$$

and the relaxation pulsatance $\omega_s = \tau^{-1}$.

The Debye equations can be represented by a "Cole–Cole diagram"[113a] in the complex $\varepsilon''(\varepsilon')$ plane as a semicircle stretching from $\varepsilon' = \varepsilon_s$, $\varepsilon'' = 0$ in the low-frequency limit to $\varepsilon' = \varepsilon_\infty$, $\varepsilon'' = 0$ in the high-frequency limit, as in Fig. 13a.

Denoting the chords to each data point from ε_s, 0 and ε_∞, 0, respectively, by v and u, the plot of $\log(v/u)$ against $\log \omega$ should be linear, of slope 45°; since $v/u = \omega\tau$, the intersection of this plot with $\omega = 1$ provides τ. Another analysis technique is by means of the relations

$$\varepsilon' = \varepsilon_s - \tau(\omega\varepsilon''), \tag{4.4.161}$$

$$\varepsilon' = \varepsilon_\infty + (1/\tau)(\varepsilon'/\omega). \tag{4.4.162}$$

Frequently a material exhibits a relaxation process in which the high-frequency limit of the real permittivity is not equal to n^2, but to some rather larger value which has been called the "infinite frequency" dielectric constant ε_∞. One or more further relaxations at higher frequencies contribute to the fall from ε_∞ to n^2.

4.4.3.3. Dielectric Response and Molecular Correlation.

We have seen that measurements of the frequency variation of the complex permittivity lead to a knowledge of the response of the polarization to time

[113a] K. S. Cole and R. H. Cole, *J. Chem. Phys.* **9**, 341 (1949).

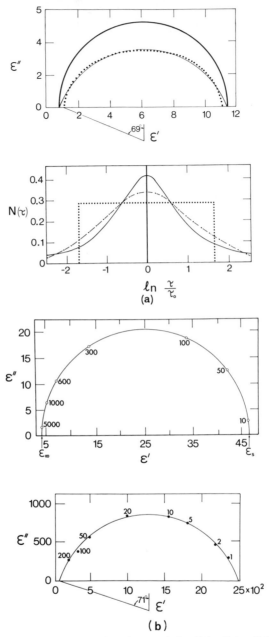

FIG. 13. (a) Distributions of relaxation times and the Cole–Cole $\varepsilon''(\varepsilon')$ diagrams corresponding to them. The heavier solid line represents single relaxation time; the lighter solid line, the Cole–Cole distribution with $\alpha = 0.23$; the dotted line, the rectangular distribution. Gaussian distribution is shown as the dot–dash line, but the $\varepsilon''(\varepsilon')$ dependence is almost indistinguishable from that of the Cole–Cole distribution and is not shown. (b) Cole–Cole diagrams for 2-propanol data of Fig. 12 and for colloidal suspension (30% by volume) of 0.188-μm diameter spherical polystyrene particles in water at 23°C (below) [H. P. Schwan, G. Schwarz, J. Maczak, and H. Pauly, *J. Phys. Chem. Ithaca* **66**, 2626 (1962)]. Frequencies are denoted in kilohertz.

variation of the electric field. The transformation from the time to the frequency domain is effected by equations of the type

$$E(t) = \int_{-\infty}^{\infty} E(\omega) \exp(j\omega t) \, d\omega. \tag{4.4.163}$$

The Fourier transform of the first moment of the macroscopic polarization function $P(t)$ yields the complex permittivity:

$$(\hat{\varepsilon} - \varepsilon_{\infty})/(\varepsilon_s - \varepsilon_{\infty}) = -\int_0^{\infty} P(t) \exp(-j\omega t) \, dt. \tag{4.4.164}$$

For an exponential macroscopic polarization function this equation leads to the Debye equations (4.4.156) and (4.4.157).

The purpose of radio-frequency spectroscopy is to throw light not on the macroscopic polarization function but on the molecular dipole response function in a system of coupled polarizable polar molecules. The dipole response function often contains a large measure of exponential behavior, but the question must be answered about the characteristic time of the dipole response, the so-called microscopic relaxation time τ': Is it the same as the macroscopic dielectric relaxation time τ of Eqs. (4.4.156) and (4.4.157)?

In Debye's original theory the relation between τ' and τ is

$$\tau' = [(\varepsilon_{\infty} + 2)/(\varepsilon_s + 2)]\tau. \tag{4.4.165}$$

The Onsager model was applied to the relaxation of the molecular dipole by Collie et al.,[114] and later by other workers; the relaxation equations tend to the same form as the Debye equations when $\varepsilon_s \gg 1$, and under these conditions $\tau' \to \tau$. The same result has since been obtained by other methods. Other treatments by Powles[115] and Glarum[116] lead, however, to a factor of approximately $\frac{3}{2}$:

$$\tau' = [(2\varepsilon_s + \varepsilon_{\infty})/3\varepsilon_s]\tau. \tag{4.4.166}$$

O'Dwyer and Sack[117] obtained a different factor:

$$\tau' = \{3\varepsilon_s\varepsilon_{\infty}(2\varepsilon_s + \varepsilon_{\infty})/(2\varepsilon_s^2 + 6\varepsilon_s^2\varepsilon_{\infty} + \varepsilon_{\infty}^2)\}\tau. \tag{4.4.167}$$

The problem is that of the treatment of the time-dependent behavior of the reaction field $\mathbf{R}$ [Eq. (4.4.139)] which arises in the Onsager cavity from the polarization induced by the dipole in its surroundings. The reaction field does not follow the phase of either the applied field or the dipole

[114] C. H. Collie, J. B. Hasted, and D. M. Ritson, *Proc. Phys. Soc.* **60**, 145 (1948).
[115] J. G. Powles, *J. Chem. Phys.* **21**, 633 (1953).
[116] S. H. Glarum, *J. Chem. Phys.* **33**, 1371 (1960).
[117] J. J. O'Dwyer and E. Hastings, *Prog. Dielec.* **7**, 1 (1967).

orientation. Fatuzzo and Mason[118] derived the relation

$$\mathscr{L}[-\dot{P}(t)] = \left(\frac{\hat{\varepsilon} - 1}{\varepsilon_s - 1}\right)\frac{\varepsilon_s}{\hat{\varepsilon}}\left(\frac{2\hat{\varepsilon} + 1}{2\varepsilon_s + 1}\right), \tag{4.4.168}$$

where $\mathscr{L}$ represents the Laplace transform operator. This relation does not lead to the Debye equations but to a more complicated frequency dependence of permittivity. A slight bulge is predicated on the high-frequency side of the Cole–Cole diagram. Since there are nearly always complications on the high-frequency side, this bulge has not appeared in any data, but its presence cannot be ruled out. It is predicted that $\tau = \tau'$.

A different treatment of the lag in the reaction field was given by Glarum,[116] leading to

$$\mathscr{L}[-\dot{P}(t)] = [(\hat{\varepsilon} - 1)/(\varepsilon_s - 1)][3\varepsilon_s/(\hat{\varepsilon} - 2\varepsilon_s)]. \tag{4.4.169}$$

This produces the Debye equations with

$$\tau = [3\varepsilon_s/(2\varepsilon_s + 1)]\tau'. \tag{4.4.170}$$

If it were shown experimentally that the Debye equations held exactly, then the Fatuzzo–Mason equation (4.4.168) would reduce to

$$P(t) = \frac{1}{2\varepsilon_s + 1}\left\{2\varepsilon_s \exp\left(\frac{-t}{\tau}\right) + \exp\left(\frac{-\varepsilon_s t}{\tau}\right)\right\}. \tag{4.4.171}$$

A dual form such as this is intrinsically unlikely. The difference between the two treatments arises from different assumptions about the states on the inside and outside of the cavity.

The modern approach to relaxation spectroscopy is not to derive the permittivity frequency characteristic from an assumed dipole response function, but rather to derive the latter from a measured permittivity frequency characteristic. Treatment by Nee and Zwanzig,[119] Klug et al.[120] and others using the Onsager model and the methods of Kubo nonequilibrium statistical mechanics,[121] lead essentially to the following formalism: The transformation of applied field from frequency to time domain is

$$\mathbf{E}(\omega) = \int_{-\infty}^{\infty} \mathbf{E}(t) \exp j\omega t \, dt. \tag{4.4.172}$$

[118] E. Fatuzzo and P. R. Mason, Proc. Phys. Soc. 90, 741 (1967).

[119] T. W. Nee and R. W. Zwanzig, J. Chem. Phys. 52, 6453 (1970).

[120] D. D. Klug, D. E. Kranbuehl, and W. E. Vaughan, J. Chem. Phys. 50, 3904 (1969).

[121] R. Kubo, J. Phys. Soc. Japan 9, 935 (1954); 12, 570 (1957); Rep. Progr. Phys. 29, 295 (1966).

The field inside the Onsager cavity is

$$\mathbf{G}(\omega) = \{3\hat{\varepsilon}(\omega)/(\hat{\varepsilon}(\omega) + \varepsilon_\infty)\}\mathbf{E}(\omega) \qquad (4.4.173)$$

in the frequency domain; it transforms into the time domain as

$$\mathbf{G}(\omega) = \int_{-\infty}^{\infty} \mathbf{G}(\omega) \exp(-j\omega t) \, d\omega/2\pi. \qquad (4.4.174)$$

For a dipole $\boldsymbol{\mu}$ placed within an Onsager cavity the interaction energy of the moment with the field is given by the perturbation Hamiltonian

$$H' = -\boldsymbol{\mu}\mathbf{G}(t). \qquad (4.4.175)$$

Within the cavity the field produces a mean polarization which contains a term due to the time-averaged value $\langle \boldsymbol{\mu}(t) \rangle$ of the dipole moment taken in the presence of the time-varying field. This is calculated by means of Kubo nonequilibrium statistical mechanics to be

$$\langle \boldsymbol{\mu}(t) \rangle = (1/3kT) \int_0^t \langle \boldsymbol{\mu}(0)\dot{\boldsymbol{\mu}}(s) \rangle \mathbf{G}(t-s) \, ds \qquad (4.4.176)$$

to terms linear in $\mathbf{E}$.

The permittivity is related to the average moment in the presence of the field by

$$\{(\hat{\varepsilon}(\omega) - \varepsilon_\infty)/(4\pi N_0/V)\}\mathbf{E}(\omega) = \langle \boldsymbol{\mu}(\omega);\mathbf{E} \rangle. \qquad (4.4.177)$$

Application of the convolution theorem leads to

$$\langle \boldsymbol{\mu}(\omega);\mathbf{E} \rangle = (1/3kT) \int_0^\infty \langle \boldsymbol{\mu}(0)\dot{\boldsymbol{\mu}}(t) \rangle \mathbf{G}(\omega) \exp(j\omega t) \, dt \qquad (4.4.178)$$

A normalized autocorrelation function of dipole moment is defined as

$$\gamma(t) = \langle \boldsymbol{\mu}(t)\boldsymbol{\mu}(0) \rangle / \langle \boldsymbol{\mu}(0)\boldsymbol{\mu}(0) \rangle. \qquad (4.4.179)$$

The permittivity frequency characteristic is obtained in the form

$$\frac{\{\hat{\varepsilon}(\omega) - \varepsilon_\infty\}\{2\hat{\varepsilon}(\omega) + \varepsilon_\infty\}\varepsilon_s}{(\varepsilon_s - \varepsilon_\infty)\hat{\varepsilon}(\omega)(2\varepsilon_s + \varepsilon_\infty)} = \int_0^\infty -\frac{d\gamma}{dt} \exp(j\omega t) \, dt. \qquad (4.4.180)$$

Whatever the form of the real and imaginary parts of the permittivity, it can be shown[40] that mutual transformation relations hold between them:

$$\varepsilon'(f) - \varepsilon_\infty = (2/\pi) \int_0^\infty f'\varepsilon''(f) \, df'/(f'^2 - f^2), \qquad (4.4.181)$$

$$\varepsilon''(f) = (2/\pi) \int_0^\infty (\varepsilon'(f) - \varepsilon_\infty)/(f'^2 - f^2) \, df'. \qquad (4.4.182)$$

These are known as the *Kramers–Kronig relations*; they have sometimes been used for the calculation of $\varepsilon'(f)$ from $\varepsilon''(f)$ or vice versa; but great caution must be exercised, since the entire frequency range and not merely a part of it must be taken into account in the integral.

4.4.3.4. Analysis of Multiple Relaxation Processes. The problem of resolving a set of $\hat{\varepsilon}(\omega)$ experimental data into two or more overlapping absorptions has been solved in a variety of ways. The sum of two Debye equations in the form

$$\hat{\varepsilon} = \frac{\varepsilon_s - \varepsilon_{\infty 1}}{1 + j\omega\tau_1} + \frac{\varepsilon_{\infty 1} - \varepsilon_{\infty 2}}{1 + j\omega\tau_2} \tag{4.4.183}$$

may be expressed in terms of reduced parameters as[121a]

$$
\begin{aligned}
Y &= \varepsilon''/(\varepsilon_s - \varepsilon_{\infty 2}), & Z &= (\varepsilon' - \varepsilon_{\infty 2})/(\varepsilon_s - \varepsilon_{\infty 2}), \\
Y_1 &= \omega\tau_1/(1 + \omega^2\tau_1^2), & Z_1 &= 1/(1 + \omega^2\tau_1^2), \\
Y_2 &= \omega\tau_2/(1 + \omega^2\tau_2^2), & Z_2 &= 1/(1 + \omega^2\tau_2^2).
\end{aligned}
\tag{4.4.184}
$$

The problem is then one of finding the best values of the "spectral parameters"

$$C_1 = (\varepsilon_s - \varepsilon_{\infty 1})/(\varepsilon_s - \varepsilon_{\infty 2}), \qquad C_2 = (\varepsilon_{\infty 1} - \varepsilon_{\infty 2})/(\varepsilon_s - \varepsilon_{\infty 2}) \tag{4.4.185}$$

that fit the data in the form

$$Y = C_1 Y_1 + C_2 Y_2, \qquad Z = C_1 Z_1 + C_2 Z_2. \tag{4.4.186}$$

The data points are plotted in the form $Y(Z)$ and a semicircle is drawn on the Z-axis between $(0, 0)$ and $(0, 1)$. The chord of the circle which is drawn through (Y, Z) from (Y_1, Z_1) to (Y_2, Z_2) must be divided in the proportion

$$(Y_1, Z_1 \text{ to } Y, Z)/(Y, Z \text{ to } Y_2, Z_2) = C_2/C_1 \tag{4.4.187}$$

for each data point. Trial chords are drawn in order to maintain a constant proportion for each data point. In certain cases there is independent evidence for assuming a certain value of this proportion, but often one must proceed without such assistance.

A more powerful computational technique has been proposed,[122] making use of the variables

$$
\begin{aligned}
x_\infty &= [(\varepsilon' - \varepsilon_\infty)/(\varepsilon_s - \varepsilon_\infty)]f^2 = x_0 f^2, \\
y_\infty &= \varepsilon''f/(\varepsilon_s - \varepsilon_\infty) = y_0 f^2, \\
z_\infty &= (\varepsilon_s - \varepsilon')/(\varepsilon_s - \varepsilon_\infty) = z_0 f^2.
\end{aligned}
\tag{4.4.188}
$$

[121a] K. Bergmann, D. M. Roberti, and C. P. Smyth, J. Phys. Chem. Ithaca 64, 665 (1960).
[122] A. M. Bottreau, J. Moreau, J. M. Laurent, and C. Marzat, J. Chem. Phys. 62, 360 (1975).

These are fitted to polynomials in f:

$$x_\infty = A_2 - \frac{A_4}{f^2} + \frac{A_6}{f^4} + \cdots, \qquad x_0 = A_0 - f^2 A_{-2} + f^4 A_{-4} + \cdots,$$

$$y_\infty = A_1 - \frac{A_3}{f^2} + \frac{A_5}{f^4} + \cdots, \qquad y_0 = A_{-1} - f^2 A_{-3} + f^4 A_{-5} + \cdots,$$

$$z_\infty = A_0 - \frac{A_2}{f^2} + \frac{A_4}{f^4} + \cdots, \qquad z_0 = A_{-2} - f^2 A_{-4} + f^4 A_{-6} + \cdots,$$
(4.4.189)

thus determining parameters A_h, which can usually be found for $4 \geqslant h \geqslant -4$; also $A_0 = 1$. For up to four separate relaxation frequencies f_i, $i = 1, 2, 3, 4$, one can write five linear equations

$$A_4 = A_3 X_1 - A_2 X_2 + A_1 X_3 - A_0 X_4,$$
$$A_3 = A_2 X_1 - A_1 X_2 + A_0 X_3 - A_{-1} X_4,$$
$$A_2 = A_1 X_1 - A_0 X_2 - A_{-1} X_3 - A_{-2} X_4,$$
$$A_1 = A_0 X_1 - A_{-1} X_2 - A_{-2} X_3 - A_{-3} X_4,$$
$$A_0 = A_{-1} X_1 - A_{-2} X_2 - A_{-3} X_3 - A_{-4} X_4,$$
(4.4.190)

from which the X_h parameters are derived. These are related to the relaxation frequencies f_i as

$$X_1 = \sum_i f_i,$$

$$X_2 = f_1 f_2 + f_1 f_3 + f_1 f_4 + f_2 f_3 + f_2 f_4 + f_3 f_4,$$

$$X_3 = f_1 f_2 f_3 + f_2 f_3 f_4 + f_3 f_4 f_1 + f_4 f_1 f_2,$$

$$X_4 = \prod_i f_i.$$
(4.4.191)

From the calculated X_h four relaxation frequencies f_i can be derived by solving an equation of the fourth degree. Moreover the parameters A_h are related to the spectral parameters C_i by

$$A_h = C_i f_i^h$$
(4.4.192)

and once the f_i and A_h are known, the spectral parameters may be determined. Further optimization of these parameters by least squares fitting is usually necessary.

4.4.3.5. Distributions of Relaxation Times. Many molecular species exhibit relaxation spectra that are broader than Debye equation spectra, and that are considered more likely to arise from a large number of relaxation times than from two or three only. A variety of distributed relaxation time equations have been used to interpret such spectra as follows:

1. The Cole–Cole equation[123] is written as

$$\hat{\varepsilon} = (\varepsilon_s - \varepsilon_\infty)/[1 + (j\omega\tau)^{1-\alpha}], \qquad (4.4.193)$$

which rationalizes to

$$\varepsilon' = \varepsilon_\infty + \frac{(\varepsilon_s - \varepsilon_\infty)\{1 + (\omega\tau)^{1-\alpha}\sin(\alpha\pi/2)\}}{1 + 2(\omega\tau)^{1-\alpha}\sin(\alpha\pi/2) + (\omega\tau)^{2(1-\alpha)}}, \qquad (4.4.194)$$

$$\varepsilon'' = \frac{(\varepsilon_s - \varepsilon_\infty)(\omega\tau)^{1-\alpha}\cos(\alpha\pi/2)}{1 + 2(\omega\tau)^{1-\alpha}\sin(\alpha\pi/2) + (\omega\tau)^{2(1-\alpha)}}. \qquad (4.4.195)$$

The Cole–Cole equation corresponds to a symmetrical distribution of relaxation times of width α. The $\varepsilon''(\varepsilon')$ resembles the Debye function in having the form of a semicircle, but its center lies below the horizontal $\varepsilon'' = 0$ axis, on the line drawn from $(0, \varepsilon_\infty)$ and making an angle $\alpha\pi/2$ with the horizontal axis.

The relaxation time distribution that gives rise to the Cole–Cole equation is

$$f(\tau) = \frac{\sin \alpha\pi}{2\pi \cosh\{(1 - \alpha)\ln(\tau/\tau_0)\} - \cos \alpha\pi}. \qquad (4.4.196)$$

The principal relaxation time at the center is denoted by τ_0. The distribution function with $\alpha = 0.23$ is illustrated[124] in Fig. 13a, and is compared with a Gaussian distribution of similar width, and with a rectangular relaxation time distribution function; Fig. 13a also shows the $(\varepsilon, \varepsilon')$ arcs appropriate to the Cole–Cole and the rectangular distributions and the Debye equation. The precise form of a symmetrical distribution function is difficult to separate from the dielectric data; what can be determined is its width and the fact that it is symmetrical. Experimental data for a colloid solution are contrasted with single relaxation time data in Fig. 13b. The distribution of the frequency parameter along the semicircular arc is dependent only on $f(\omega\tau)$. The data can be analyzed by means of chords u and v described in the previous section. The quantity $\log u/v$ plotted against $\log v$ is a straight line of slope $1 - \alpha$. Without this analysis the existence of a circular arc with depressed center cannot be taken as establishing the Cole–Cole relationship.

Least squares fitting of sets of data to the Cole–Cole equation is often carried out, but the question must also be answered: Is the Debye or the Cole–Cole equation (or perhaps another set of equations) the most accurate

[123] K. S. Cole and R. H. Cole, *J. Chem. Phys.* **9**, 341 (1949).
[124] H. P. Schwan, *Advan. Biol. Med. Phys.* **5**, 147 (1957).

fit to the data set? Since the Cole–Cole equation contains one more adjustable parameter than the Debye equation, it would be expected to reduce the fitting errors, even though it is not necessarily a more correct equation. In order to test whether the data actually fit one equation better than the other, complicated techniques of nonlinear regression analysis must be employed.[125]

A modified Cole–Cole equation has been used in the analysis of some ionic solution data:

$$\hat{\varepsilon} = (\varepsilon_s - \varepsilon_\infty)/[1 + j^{1-\alpha}(\omega\tau)^{1-\beta}]. \tag{4.4.197}$$

2. The Cole–Davidson[126] equation is written as

$$\hat{\varepsilon} - \varepsilon_\infty = (\varepsilon_s - \varepsilon_\infty)/(1 + j\omega\tau)^\alpha.$$

It represents an asymmetrical distribution of relaxation times which gives rise to a skewed arc $\varepsilon''(\varepsilon')$ diagram, an example of which is illustrated in reduced form in Fig. 14.

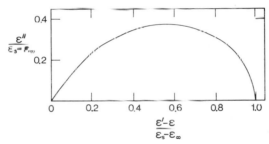

FIG. 14. Skewed arc representation [F. I. Mopsik and R. H. Cole, *J. Chem. Phys.* **44**, 1015 (1965)], using reduced parameters, of relaxation of liquid 1-octyl iodide ($CH_2I(CH_2)_6CH_3$). The curve was traced through data points covering the temperature range -40 to $+40°C$.

The relaxation time distribution for the skewed arc is

$$f(\tau) = \begin{cases} \dfrac{\sin \alpha\pi}{\alpha\pi} \left(\dfrac{\tau}{\tau_0 - \tau} \right)^\alpha & \text{for} \quad 0 \leqslant \tau \leqslant \tau_0 \\ 0 & \text{for} \quad \tau > \tau_0. \end{cases} \tag{4.4.198}$$

This type of distribution will be discussed further in Section 4.4.3.6.

A graphical analysis of data can be made by plotting $\tan\{\alpha^{-1} \arctan \varepsilon''/(\varepsilon' - \varepsilon_\infty)\}$ against $\log \omega$. Where the function is unity, $\tau = 1/\omega$.

[125] P. R. Mason, J. B. Hasted, and L. Moore, *Advan. Mol. Relax. Proc.* **6**, 217 (1974).
[126] D. W. Davidson and R. H. Cole, *J. Chem. Phys.* **18**, 1417 (1951).

3. The Fuoss–Kirkwood analysis[127] concerns the molecular as opposed to the macroscopic relaxation time. In terms of the variable $z = \omega/\omega_{max}$ the Debye equation leads to

$$\varepsilon'' = 2\varepsilon''_{max}/(z - z^{-1}), \qquad (4.4.199)$$

and hence to

$$\varepsilon'' = \varepsilon''_{max} \operatorname{sech}(\ln \omega/\omega_{max}), \qquad (4.4.200)$$

where ω_{max} is the radial frequency corresponding to the maximum value of ε''. In the Fuoss–Kirkwood analysis, however,

$$\varepsilon'' = \varepsilon''_{max} \operatorname{sech}(\beta \ln \omega/\omega_{max}), \qquad (4.4.201)$$

where β is an empirical parameter between 0 and 1. The relation between β and the Cole–Cole parameter α is

$$\beta\sqrt{2} = (1 - \alpha)/\cos \tfrac{1}{4}(1 - \alpha)\pi. \qquad (4.4.202)$$

4. The Fröhlich distribution[128] is derived in terms of a reorientation process which takes place by a jump over a potential barrier. There is assumed to be a uniform distribution of barrier heights between $E = E_0$ and $E = E_0 + E_1$. A relaxation time is in general related to the barrier height by equations of the type

$$\tau'_0 = A \exp(E_0/kT) \qquad (4.4.203)$$

and

$$\tau' = A \exp(E_0 + E)/kT = \tau_0' \exp(E/kT). \qquad (4.4.204)$$

In this model the upper limit of relaxation times is

$$\tau_1' = \tau_0' \exp(E_1/kT) \qquad (4.4.205)$$

and the distribution of times is given by

$$f(\tau') = \begin{cases} kT/E_1 & \text{for} \quad \tau_0' < \tau' < \tau_1' \\ 0 & \text{otherwise.} \end{cases} \qquad (4.4.206)$$

Thus the logarithmic distribution function of τ' is rectangular.

The importance of this treatment is that, provided E_1 is temperature independent, as is likely, it predicts a distribution parameter that becomes narrower as the temperature increases; this behavior is qualitatively that which is observed in practice.

An important difference between a Fröhlich distribution and distributions such as the Cole–Cole one is that the former is only of finite width, and the $\varepsilon''(\varepsilon')$ plot from it intersects the ε'-axis vertically at either end. It has been

[127] R. M. Fuoss and J. G. Kirkwood, *J. Amer. Chem. Soc.* **63**, 385 (1941).
[128] H. Fröhlich, "Theory of Dielectrics." Oxford Univ. Press, London and New York, 1949.

shown[129] that the Fröhlich distribution arc is approximately that of a semi-ellipse. A similar finite distribution has been examined[130] in which

$$f(\tau) = 1/A\tau, \qquad \tau_1 < \tau < \tau_2. \qquad (4.4.207)$$

The more general form

$$f(\tau) = 1/A\tau^n \qquad (4.4.208)$$

represents a right-hand skewed arc when $n < 1$.

4.4.3.6. Defect Diffusion and Free-Volume Models. A model was proposed by Glarum[131] which leads to an $\varepsilon''(\varepsilon')$ function approximately the same as the Cole–Davidson skewed arc (Fig. 14). There are assumed to be lattice defects in the liquid, in the presence of which the dipole can adapt its orientation almost instantaneously to the applied field. In the absence of such a defect, a dipole can still reorient, with relaxation time τ'. The overall dipole correlation function $\gamma(t)$ is then given by a product of the two separate correlation functions:

$$\gamma(t) = \exp(-t/\tau')\{1 - p(t)\}, \qquad (4.4.209)$$

where $p(t)$ is the probability that the arrival of a defect has caused relaxation by time t.

The motion of the defects is assumed to be governed by diffusion, with a characteristic diffusion time τ_d, which is the average time for a defect to diffuse over the mean separation of defects. The general form of the dispersion function is

$$\frac{\hat{\varepsilon}(j\omega) - \varepsilon'}{\varepsilon_s - \varepsilon'} = \frac{1}{y} + \frac{y - 1}{y\{1 + (a_0 y)^{1/2}\}}, \qquad (4.4.210)$$

where $y = 1 + j\omega\tau$ and $a_0 = l_0/D\tau$, and where D is the defect diffusion coefficient and l_0 the mean defect separation.

Three special cases may be discerned:

1. When $\tau' \ll \tau_d$, the relaxation due to the arrival of a defect is rare, and a Debye relaxation is predicted.

2. When $\tau' = \tau_d$, a skewed arc is predicted similar to a Cole–Davidson arc with parameter $\alpha = 0.5$, but the arc approaches the ε'-axis at an angle of $\pi/4$ rather than $\alpha\pi/2$, which is the angle of approach for the Cole–Davidson equation.

[129] F. Bergmann, Thesis, Univ. of Freiburg-im-Breslau (1957).

[130] K. Higasi, Dielectric Relaxation and Molecular Structure. Res. Inst. Appl. Elec., Sapporo, Japan, 1961.

[131] S. H. Glarum, *J. Chem. Phys.* **33**, 369 (1960).

3.   When $\tau' \gg \tau_d$, the diffusion of defects is the dominant process, and a symmetrical spread of relaxation time is predicted with Cole–Cole parameter $\alpha = 0.5$.

Thus there is a theoretical basis for the possibility of a relaxation function whose form changes with temperature.

It was proposed by Anderson and Ullman[132] that reorientation is affected by another time-dependent parameter, the free volume v, which fluctuates about a time-averaged value $\langle v \rangle$. The time variation of $x = v - \langle v \rangle$ is assumed to be controlled by a random walk among the elements of free-volume space. Diffusion and viscous flow are also considered to be related to the local free volume available to the molecule.[133] The reorientation rate is a monotonically increasing function of x, which is taken as

$$f(x) \propto \begin{cases} \exp\{R(x/X)^a\} & \text{for} \quad x \geqslant 0 \\ \exp\{-R(|x|/X)^a\} & \text{for} \quad x < 0, \end{cases} \qquad (4.4.211)$$

where R and a are constants and $1 \leqslant a \leqslant 2$.

When the rate of free-volume fluctuation is slow compared with the rate of reorientation, a symmetrical Cole–Cole arc is predicted; its distribution parameter α increases with increasing R, that is, with increasingly strong dependence of relaxation rate on free volume. The molecular environment remains almost constant during the reorientation process, but different molecules possess different environments.

When the rate of change of v is much greater than the reorientation rate, then each molecule reorients in effectively the same environment, and a single relaxation time is predicted.

In the intermediate case a skewed-arc plot is predicted. For molecules in a momentarily large free volume the reorientation rate will be high and a distribution of relaxation times is expected, but for molecules in a small free volume the reorientation rate is low and a single relaxation time is appropriate. The high-frequency behavior is therefore spread, while the low-frequency behavior is unique, giving in effect a left-skewed arc.

Examples[134–136] of skewed-arc behavior in nonhydroxyl (i.e., non-hydrogen-bond-forming) molecules include: isobutyl chloride and bromide, isoamyl bromide, n-octyl iodide, n-propyl nitrite, tolyl xylyl sulfone, citral; but all these examples are at temperatures below $-50°C$. At laboratory temperatures there are also polychlorinated vinyl derivatives, polyvinyl

[132] J. E. Anderson and R. Ullman, *J. Chem. Phys.* **47**, 2178 (1967).
[133] M. H. Cohen and D. Turnbull, *J. Chem. Phys.* **31**, 1164 (1959).
[134] D. J. Denney, *J. Chem. Phys.* **27**, 259 (1957).
[135] A. H. Glarum, *J. Chem. Phys.* **33**, 639 (1960).
[136] F. I. Mopsik and R. H. Cole, *J. Chem. Phys.* **44**, 1015 (1965).

acetate, 1,2-dibromopentane, and 2-methyl-1,2-dichloropropane. In addition eight dihydroxylic alcohols, as well as glycerol, demonstrate skewed-arc behavior.

The skewed-arc data for 1-octyl iodide are shown in Fig. 14, in reduced form. Nine temperatures between -40 and $+40°C$ are represented on this arc; but the precise Cole–Davidson form is not demonstrated, since the maximum value of $\varepsilon''/(\varepsilon_s - \varepsilon_\infty)$ is 0.34, while that corresponding to the best value of α which would be derived from the high-frequency asymptotic slope is 0.373. In these data, however, the value of α is temperature invariant. The low-frequency asymptotic slope is close to $\pi/2$, whereas that appropriate to the Cole–Cole arc would be $(1 - \alpha)\pi/2$. The skewed-arc pattern has also been reported[137] in a simple molecular solid, crystalline hydrogen iodide. A defect theory would seem to be appropriate in such a case.

4.4.3.7. Activation Energies of Relaxation. Since the reorientation process might be regarded as a molecular collision, the empirical Arrhenius equation would predict a simple logarithmic dependence of log $1/\tau$ against T^{-1}.

Following Eyring[138] one can treat the relaxation process as analogous to a chemical rate process. The temperature variation of the microscopic relaxation time will then be approximately exponential, according to Eyring's equation

$$\tau = (h/kT) \exp(-\Delta S^*/R) \exp(\Delta H^*/RT), \qquad (4.4.212)$$

where ΔS^* is the molar entropy of activation and ΔH^* the molar enthalpy of activation for the relaxation process. Provided that ΔS^* and ΔH^* are temperature independent, a graph of $\log_{10}(\tau T)^{-1}$ against T^{-1} should be a straight line of negative slope, from which ΔH^* can be calculated. No great significance attaches to ΔS^* calculated in this way, but the order of magnitude of ΔH^* can give some clue to the molecular energy involved in the relaxation process. For example, the finding $\Delta H^* \simeq 4.5$ kcal/mole for the principal relaxation of liquid water[56] can indicate that the process involves the surmounting of a barrier equivalent in energy to the breakage of an OH $\cdots$ O hydrogen bond.

A different temperature dependence of relaxation time was predicted by Bauer,[139] with microscopic relaxation time proportional to $\delta(2\pi I/RT)^{1/2}$ where δ is the width of the potential barrier surmounted in the dipole reorientation process and I the moment of inertia of the polar molecule.

[137] N. E. Hill, W. E. Vaughan, A. H. Price, and M. Davies, "Dielectric Properties and Molecular Behaviour," pp. 360, 386. Van Nostrand-Reinhold, Princeton, New Jersey, 1969.

[138] S. Glasstone, K. J. Laidler, and H. Eyring, "The Theory of Rate Processes." McGraw-Hill, New York, 1941.

[139] E. Bauer, Can. Phys. 20, 1 (1944).

The Bauer equation is

$$\tau = \delta(2\pi I/RT)^{1/2} \exp(-\Delta S^*/R) \exp(\Delta H^*/RT), \qquad (4.4.213)$$

and the additional factor has the effect of making the entropy factor ΔS^* positive in certain cases where in application the Eyring equation to the data would make it negative. A negative value would imply that for dipole orientations with small energies of activation, the transition state involves an increase in the local molecular order, which would seem unlikely. Analysis of data for molecules of large moment of inertia accentuates this situation.

If one writes

$$\tau = ST^n \exp(\Delta H^*/RT) \qquad (4.4.214)$$

with S and n constants, then it is seen that

$$n\,(\text{Arrhenius}) = 0, \qquad n\,(\text{Eyring}) = -1, \qquad n\,(\text{Bauer}) = -\tfrac{1}{2}.$$

It is unclear from the experimental data of Whiffen and Thompson[140] for aromatic liquids over a temperature range -70 to $+80°C$ which temperature variation is to be preferred; but it is at any rate clear that the simple Arrhenius temperature dependence is inadequate to explain the data.

Usually the value of ΔS^* is only a few entropy units, so that the Eyring equation is an acceptable form. Polyhydroxylic alcohols show rather larger values (~ 10 e.u.). The Arrhenius activation energy for aliphatic alcohols[141] is temperature variant below about 200 K, but above this temperature it is temperature invariant. An analysis of this variation has been expressed in the form

$$\log \tau = a + [b/(T - T_0)] \qquad (4.4.215)$$

with a, b, and T_0 constants; this is a form identical with the Antoine equation for vapor pressures. The value of b is found to be common to most alcohols,[142] so that the energy barrier would appear to tend to a common value, defined by the network of hydroxyl groups. Solid alcohols are discussed in Section 4.4.4.4. In the case of water some temperature variation of ΔH^* has been found as high as $80°C$, and structural interpretations have been given.[143,144]

[140] D. H. Whiffen and H. W. Thompson, *Discuss Faraday Soc.* **42A**, 129 (1946).

[141] F. X. Hassion and R. H. Cole, *J. Chem. Phys.* **23**, 1756 (1955); R. H. Cole and D. W. Davidson, *J. Chem. Phys.* **20**, 1389 (1952); W. Dannhauser and R. H. Cole, *J. Chem. Phys.* **23**, 1762 (1955); C. P. Smyth and S. K. Garg. *J. Phys. Chem Ithaca* **69**, 1294 (1965); M. Bruma, Thesis, Sorbonne, Paris (1952); R. Arnoult, A. Lebrun, and C. Boullet, *Arch. Sci. Geneve* **9**, 44 (1956); C. Brot, Thesis, Sorbonne, Paris (1956); R. Arnoult, A. Lebrun, C. Moriamez, M. Moriamez, and R. Wemelle. *Arch. Sci. Geneve* **10**, 48 (1957); D. W. Davidson and J. Wheeler, *J. Chem. Phys.* **30**, 1357 (1959).

[142] D. W. Davidson, *Can. J. Chem.* **39**, 571 (1961).

[143] C. H. Collie, J. B. Hasted, and D. M. Ritson, *Proc. Phys. Soc.* **60**, 145 (1948).

[144] G. H. Haggis, J. B. Hasted, and T. J. Buchanan, *J. Chem. Phys.* **20**, 1452 (1952).

Other liquids such as isoalkyl halides show ΔH^* temperature dependence, and these also show skewed-arc $\varepsilon''(\varepsilon')$ plots. Since it has often been proposed that structural effects such as cybotactic groups (dynamic, statistically ordered, molecular aggregates[145]) are likely to exist in most liquids at temperatures close to freezing point, relaxation spectroscopy might be expected to show ΔH^* temperature dependence most markedly in this region.

4.4.3.8. Resonance Processes in the Submillimeter Region. The reorientation processes just discussed are dominant in the radio-frequency spectra of polar liquids and many solids. The processes causing infrared absorption bands in liquids and solids are, however, basically resonance processes, and the possibility that these extend into the millimeter-wave region must not be overlooked. Fröhlich[146] and Poley[147] proposed the existence of resonance absorptions and dispersions in polar liquids due to oscillations of the dipole under displacement from the temporary equilibrium position to which it is elastically bound by intermolecular forces. The resonance angular frequency ω_0 is given by [148]

$$(\varepsilon_g - n^2)/(\varepsilon_\infty - n^2) = I\omega_0^2/2kT, \qquad (4.4.216)$$

where I is the appropriate molecular moment of inertia. Typically the difference between ε_∞ for the principal Debye relaxation and n^2, the square of the refractive index (allowing for its enhancement by other resonance processes in the infrared), is about 0.2, which leads to values of ω_0 in the region $3-6 \times 10^{12}$ Hz; thus the absorption bands for this process of "hindered rotation," "libration," or "Poley absorption" occur in the submillimeter region in the wavenumber range $\bar{v} = 10-100$ cm^{-1}. Poley found that for substituted benzenes $\varepsilon_\infty - n^2$ was approximately proportional to μ^2, the square of the dipole moment, so that for molecules with large dipole moments such as nitrobenzene, for which $\mu^2 = 17.6$ D^2, $\varepsilon_\infty - n^2$ is as large as 2.5. Also for hydrogen-bond-forming molecules such as water and aliphatic alcohols, $\varepsilon_\infty - n^2$ is large, but μ^2 values are only about 3, so that additional processes probably contribute (Section 4.4.4.3).

In the limit $\omega\tau \gg 1$, where τ is the principal relaxation time, the contributions to ε' and ε'' of the resonant libration process may simply be added to the relaxation contributions in the form

$$\Delta\varepsilon' = (\varepsilon_\infty - n^2)\left\{\frac{1 - (\omega/\omega_0)^2}{[1 - (\omega/\omega_0)^2]^2 + (r\omega/\omega_0)^2} - 1\right\}, \qquad (4.4.217)$$

$$\Delta\varepsilon'' = \frac{(\varepsilon_\infty - n^2)r\omega/\omega_0}{[1 - (\omega/\omega_0)^2]^2 + (r\omega/\omega_0)^2}, \qquad (4.4.218)$$

[145] C. A. Benz and G. W. Stewart. *Phys. Rev.* **46**, 703 (1934).
[146] H. Fröhlich, "Theory of Dielectrics." Oxford Univ. Press, London and New York, 1949.
[147] J. Ph. Poley, *J. Appl. Sci. Res.* **B4**, 337 (1955).
[148] N. E. Hill, *Proc. Phys. Soc.* **82**, 723 (1963).

where r is a damping coefficient. These equations have the usual form for a damped resonance absorption, but when $\omega\tau \sim 1$, the resonance and relaxation contributions are not purely additive, and a more complicated set of equations is applicable.

For a narrow resonance line the frequencies of maximum absorption coefficient α and of maximum loss ε'' are the same, but for a broadened line, owing to their interdependence

$$\alpha = 2\pi\varepsilon''\bar{v}/n \qquad (4.4.219)$$

this is no longer the case, and it can be shown[149] that the maximum in ε'' occurs at a frequency lower than that for the maximum in α by a factor which tends to $1/r$ as r becomes large. Most observed lines have halfwidths in α of order 50 cm^{-1}, but this is only about half the width of a single Debye relaxation.

In order to study the form of these resonant libration absorptions it is necessary to subtract the contribution arising from the principal Debye relaxation; but a simple subtraction may be an overestimate, owing to the neglect in the derivation of the Debye equations of the molecular inertia term $I\,(d^2\theta/dt^2)$, which should properly be added to the viscous or frictional resistance term $\zeta\,d\theta/dt$ in the calculation of the time variation of θ when the molecule is subject to an electric force $-\mu F \sin \theta$:

$$-\mu F \sin \theta = \zeta(d\theta/dt) + I(d^2\theta/dt^2). \qquad (4.4.220)$$

Rocard[150] and Powles[151] independently calculated the value of this contribution, and arrived at almost identical results. The contribution is determined by $\frac{1}{2}IkT\tau^2$, and for aromatic molecules it is negligible at wavenumbers lower than 30 cm^{-1}, although rising to $\sim 10\%$ at 100 cm^{-1}. It is sometimes found that the tail of the calculated contribution rises above the experimental $\alpha(\bar{v})$, which indicates that the molecular inertia theory is an overestimate.

When the principal relaxation contribution has been subtracted from the submillimeter-band absorption function, there remains an absorption line of the form of Fig. 15 which may be attributed to resonant libration. It will be seen from the expression for $\Delta\varepsilon'$ in Eq. (4.4.217) that a minimum in the $\varepsilon'(\omega)$ function is predicted. Usually, however, only the absorption is measured and not the refractive index, so that no check can yet be made on this feature of resonance processes. An isolated resonance line will in the $\varepsilon''(\varepsilon')$ plane appear as a curlecue or as a full circle ◯, by contrast with the appearance of a relaxation process, which is that of a semicircle ◠.

[149] N. E. Hill, *Chem. Phys. Lett.* **2**, 5 (1968).
[150] Y. Rocard, *J. Phys. Radium.* **4**, 247 (1933).
[151] J. G. Powles, *Trans. Faraday Soc.* **44**, 802 (1948).

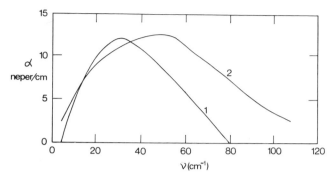

FIG. 15. Absorption bands [M. Davies, G. W. F. Pardoe, J. E. Chamberlain, and H. A. Gebbie, *Trans. Faraday Soc.* **64**, 847 (1968)] obtained from analysis of data for (1) liquid 1-1-1-trichloroethane (methyl chloroform) at 20°C and (2) liquid chlorobenzene at 20°C.

The form of the measured absorption lines can be interpreted with the aid of models of the libration process.[152] If a potential well is taken to have depth V and semiangular aperture ξ the harmonic libration frequency can be written[153]

$$\omega_0 = (\pi/\xi)(V/2I)^{1/2}. \tag{4.4.221}$$

When there are a number of available wells the mean time τ_r of residence in a well is related to the relative populations of wells and barriers and hence to the duration time of a jump τ_j by

$$\tau_r/\tau_j = \{1 - \exp(-V/kT)\}/\exp(-V(kT). \tag{4.4.222}$$

An observed librational absorption is fitted to chosen value of V and ξ and a collisional perturbation parameter τ_i, giving the total width

$$1/\tau = \tfrac{1}{4}\omega_0 + (\pi/\tau_i), \tag{4.4.223}$$

where τ_i is taken as an adjustable parameter.

In the "itinerant oscillator" model[153] the librator is taken to be within a temporary cage of Z nearest neighbors which by its inertial effect and lack of rigidity cause damping (r) of the librator. The diffusional reorientation time τ_d of the cage is related to a resistance ζ acting on the cage molecules:

$$\tau_d = \zeta/2kT \tag{4.4.224}$$

and ω_0 is obtained from Eq. (4.4.221) as in the first model; ζ and r are taken as adjustable parameters, but the aperture is calculated from the lattice structure, or estimated from Z.

[152] I. Larkin, *Trans. Faraday Soc.* **69**, 1278 (1973); **70**, 1444, 1457, (1974).

[153] B. Lassier and C. Brot, *Chem. Phys. Lett.* **1**, 581 (1968); N. E. Hill, *J. Phys. C* **4**, 2322 (1971).

Throughout this chapter it has been assumed that nonpolar liquids exhibit no dielectric dispersion, provided that they contain no appreciable polar impurities. Nonpolar liquids, however, exhibit some absorption in the microwave region,[154] and it is found to be approximately proportional to frequency, but insensitive to temperature. This absorption was supposed to arise from the action of temporary collision-induced dipoles with relaxation time corresponding to a wavelength of about 1 mm.

Intermolecular collisions distort chemical bonds by a small amount, e.g., 6° for C—C bond in CCl_4, thereby producing sufficient temporary dipole moment (~ 0.1 D) to account for the absorption band. This distortion requires very little energy (~ 0.3 kcal/mole), which explains the small temperature dependence.

The absorptivity α for benzene[155] rises linearly with frequency to a maximum of 5 Np/cm at $\bar{v} = 70$ cm^{-1}, and then falls off linearly with wavelength. This absorption does not correspond to any of the vibrational frequencies in the solid, and its width is rather greater than that of a Debye absorption. Carbon tetrachloride[156] shows an absorption maximizing at 50 cm^{-1} to 2.0 Np/cm, and carbon disulfide maximizing at 70 cm^{-1} to 3.5 Np/cm.

4.4.3.9. Dispersion of Conductivity. Relaxation spectroscopy, discussed in previous sections in terms of dipole orientation, is applied both to liquids and solids; although there are many insulators and semiconductors such as ice which exhibit Debye behavior, there is also a large group of solids, including ionic crystals, many polymers and other amorphous materials, whose behavior is quite different. A frequency variation of dielectric loss is apparent, but there is sometimes no maximum apparent in the function $\varepsilon''(\omega)$, and even if there is, it is of the order of 100 times as broad as the Debye peak. Very often an almost constant loss is found over several decades of frequency.

It is customary to describe these materials in terms of conductivity rather than permittivity; the theories and mechanisms of conduction are those of charge transport rather than dipole orientation. Nevertheless the equivalence of conduction and dielectric loss must again be stressed. If the direct current conductivity is denoted by σ_0,

$$\sigma(\omega) = \sigma_0 + \sigma'(\omega), \tag{4.4.225}$$

and the frequency-dependent part of the conductivity

$$\sigma' = \varepsilon_0 \omega \varepsilon''(\omega). \tag{4.4.226}$$

[154] D. H. Whiffen, *Trans. Faraday Soc.* **46**, 124 (1950).

[155] M. Davies, G. W. F. Pardoe, J. E. Chamberlain, and H. A. Gebbie, *Chem. Phys. Lett.* **2**, 411 (1968).

[156] G. Chantry, H. A. Gebbie, B. Lassier, and G. Wyllie, *Nature (London)* **214**, 163 (1967).

In the low-frequency limit of a Debye relaxation, $\varepsilon'' \propto \omega$, so that $\sigma' \propto \omega^2$. In the high-frequency limit $\varepsilon \propto \omega^{-1}$, so that σ' is constant. More typically for insulating solids ε'' falls with increasing ω, according to a power less than unity (see also Section 4.4.4.6). The "hopping" of carriers is responsible for this type of frequency-dependent conduction, but its discussion is beyond the scope of this chapter.

In some polymers, such as polyethylene terephthalate, polymethyl methacrylate and polydion carbonate, there is a low-frequency rise of $\varepsilon''(\omega)$. There is a broad maximum in $\varepsilon''(\omega)$, easily distinguishable from the Debye process maximum, since its full width at half maximum is about three decades instead of about one. An empirical equation[157] can be fitted to this type of behavior as follows:

$$1/\varepsilon'' = (\omega/\omega_2)^{-m} + (\omega/\omega_1)^{(1-n)}, \tag{4.4.227}$$

where $m < 1$ and $n < 1$. The constants ω_1 and ω_2 both vary with temperature, giving activation enthalpies. The corresponding behavior of ε' is a dependence on $\omega^{(n-1)}$; using the Kramers–Kronig relations (4.4.181) and (4.4.182), one can show that the behavior

$$\varepsilon'' = Aj\omega^{(n-1)} \tag{4.4.228}$$

yields

$$\varepsilon' = A \tan(n\pi/2) \cdot \omega^{n-1} \tag{4.4.229}$$

Therefore

$$\varepsilon''/\varepsilon' = \cot(n\pi/2). \tag{4.4.230}$$

In the complex $\varepsilon''\varepsilon'$-plane, the high-frequency side of the dispersion arc is a straight line sloping down to the $\varepsilon'' = 0$ axis, as in the Cole–Davidson equation.

Another instance of conductivity frequency variation, is the Debye–Falkenhagen relaxation in ionic conductivity of electrolytes.[158]

The theory of strong electrolytes, which applies to extreme dilutions ($< 10^{-2}$ M), relates the conductivity to the retarding effect produced on the transport ions by the surrounding atmosphere of ions of opposite sign. This atmosphere does not consist purely ions of opposite charge, but on balance there is in the neighborhood of each ion an excess of them. In applied alternating electric fields the atmosphere follows the oscillatory motion of the transport ions provided they do not oscillate at too high a frequency. The critical frequency at which the atmosphere ceases to follow the transport ion is related to the radial "plasma oscillation frequency"

$$\omega_p = (4\pi N_\pm e_\pm^2/m_\pm \varepsilon)^{1/2}, \tag{4.4.231}$$

[157] A. K. Jonscher, *Nature* (*London*) **253**, 717 (1975).
[158] H. Falkenhagen, "Eledtrolytes." Oxford Univ. Press, London and New York, 1934.

where $N_\pm$ is the ion density, $e_\pm$ the charge, $m_\pm$ the mass, and ε the permittivity of the medium (neglecting any effects on the permittivity which arise from the oscillating charges). This frequency also plays a critical role in the frequency variation $\hat{\varepsilon}(\omega)$ of the permittivity of the ionized gas or plasma according to the equation

$$\hat{\varepsilon} = 1 - \{\omega_p{}^2/\omega(\omega - j\omega_c)\}, \tag{4.4.232}$$

where ω_c is the radial collision frequency.

The ionic atmosphere in aqueous solution has the effect of depressing the mobility of the ions at frequencies below the relaxation frequency; at and above this frequency the static molar conductivity ($\Lambda_s = 1000\sigma/M$ at molarity M) rises to the value Λ_∞ which it would possess at infinite dilution.

The theory of the relaxation process leads to an expression for the relaxation time

$$\tau = (8.85 \times 10^{-11}\varepsilon)/\nu_i z_i M \Lambda_\infty{}^\pm, \tag{4.4.233}$$

where z_i is the ionic valency, and ν_i the number of ions formed by dissociation of the electrolyte molecule. For monovalent aqueous electrolytes, the relaxation time is about $10^{-10}/M$ sec. The expressions for the magnitude of the dispersion $\Lambda_\infty - \Lambda_s$ are complicated, but in solutions sufficiently dilute for the Debye–Hückel theory of strong electrolytes to be applicable, the dispersion is found to conform to the Debye–Falkenhagen theory. In more concentrated solutions some anomalies are found.

4.4.4. Molecular Structure and Dielectric Relaxation

4.4.4.1. Relaxation Times, Viscosities, and Shape Factors. Debye[159] proposed that the resistance to molecular dipole reorientation might be regarded as viscous. When the molecule is treated as a sphere rotating in a uniform viscous medium, it is found that the microscopic relaxation time

$$\tau = 4\pi\eta a^3/kT = 3\eta V/kT, \tag{4.4.234}$$

where V is the molar volume, η the viscosity, and a the molecular radius. Although the relaxation times so calculated are overestimated by a factor between 5 and 10, the prediction that τ and η/T will show that the same temperature dependence is well borne out by experimental data.

An attempt[160] to take into account the discontinuous nature of the molecular shell surrounding the polar molecule leads to a reduction of the predicted microscopic relaxation time, which then agrees much better with experimental values than does the value predicted by Eq. (4.4.234). This equation does not hold for dilute solutions of polar (B) in nonpolar (A) liquids. It is the AA molecular collisions which predominantly determine

[159] P. Debye, "Polar Molecules." Chemical Catalog Co., New York, 1929.
[160] A. Gierer and K. Wirtz, Z. Naturforsch 8a, 532 (1953).

the viscosity of the solution, while the BA collisions predominantly determine the dielectric relaxation of B molecules. It can be shown[161] that the solution relaxation time

$$\tau_s = x_A\tau_{AB} + x_B\tau_B, \tag{4.4.235}$$

where x_A and x_B are molar fractions, τ_{AB} the relaxation time in the limit of infinite dilution, and τ_B the relaxation time of the pure liquid B. It is the viscous force appropriate to η_{AB} which determines τ_s. The arguments can be extended to mixtures where both A and B are polar; two relaxation times are predicted

$$\tau_{sA} = x_B\tau_{AB} + x_A\tau_A, \tag{4.4.236}$$

$$\tau_{sB} = x_A\tau_{AB} + x_B\tau_B. \tag{4.4.237}$$

The experimental data do not, however, conform well to these relations.[162] A variety of geometrical factors are important in understanding the reorientation of molecules in solution.

Attempts have been made to assess the relationship between relaxation time and viscosity by comparing in the following manner the ratio of the two Arrhenius activation energies:

$$\eta = A \exp(\Delta H_\eta/RT), \tag{4.4.238}$$

$$\tau = B \exp(\Delta H_\tau/RT), \tag{4.4.239}$$

$$\Delta H_\tau = C \Delta H_\eta, \tag{4.4.240}$$

$$\tau = A^{-x}B\eta^x. \tag{4.4.241}$$

It has been proposed that a suitable choice of x might enable predictions to be made about τ on the basis of known η, but unfortunately[163] a purely exponential temperature dependence of η is rare in polar liquids. Furthermore τ/η usually varies strongly with temperature.

The Debye model of a spherical molecule in a viscous medium leading to Eq. (4.4.234) has been extended[164, 165] to ellipsoidal molecules of semiaxes a, b, c; three relaxation times are to be expected:

$$\tau_1 = (4\pi\eta/kT)abcs_i, \tag{4.4.242}$$

where s_i is a form factor depending on the ellipticities, b/a, c/b, a/c; numerical values of these factors s_a, s_b, and s_c have been calculated.[166]

[161] N. E. Hill, *in* "Dielectric Properties and Molecular Behaviour" (N. E. Hill, W. E. Vaughan, A. H. Price, and M. Davies, eds.), Chapter 1. Van Nostrand-Reinhold, Princeton, New Jersey, 1969.

[162] R. J. Meakins, *Trans. Faraday Soc.* **54**, 1160 (1958).

[163] S. Mallikartjun and N. E. Hill, *Trans. Faraday Soc.* **61**, 1389 (1965).

[164] F. Perrin, *J. Phys. Radium* **5**, 497 (1934).

[165] E. Fischer, *Phys. Z.* **60**, 645 (1939).

[166] A. Budo, E. Fischer, and S. Miyamoto, *Phys. Z.* **40**, 337 (1939).

For most molecules these relaxation times are likely to be fairly similar in magnitude, and in pure liquids widely separated relaxation processes are probably not to be attributed to this cause. Often, however, there are apparent small spreads of relaxation times, which are analytically very difficult to distinguish from two or three closely spaced relaxation times; it may be that this effect is attributable to the molecular shape factors.

The treatment is of greater value for large rod-shaped molecules in solution in nonpolar liquids whose molecules are sufficiently small to be regarded as a continuous viscous fluid. Studies[166, 167] of the relaxation of different quinones in the same solvent, benzene, have been used for investigation of shape factor effects. Although the observed relaxation times, based on the bulk liquid viscosity, are about five times too large, the correct sequence of relaxation time values is given by the theory.

4.4.4.2. Relaxation of Flexible Polar Molecules. One feature of the relaxation spectra of flexible organic polar molecules such as n-alkyl halides[168] is the marked increase in the Cole–Cole spread parameter α with increasing chain length. As the chain increases in length, there is an increasing variety of configurations which can be involved in dipole reorientation. By contrast many carboxylic acid esters show relaxation times that are not strongly dependent on chain length; this could indicate that it is only the polar segment of the molecule which reorients, the hydrocarbon chain having relatively little influence on the process.

When, in addition to rotation of an entire polar molecule, a polar segment is able to rotate, different relaxation times corresponding to each process may be found; numerous investigations of such cases have been made.[169]

The different relaxation times discussed in Section 4.4.4.1 for rigid ellipsoidal molecules are difficult to observe unless the ellipsoids have widely different semiaxes; but the separate relaxation times arising from rotation of part of a molecule are much more readily discerned.[170–172] For example,

[167] D. A. Pitt, and C. P. Smyth, *J. Amer. Chem. Soc.* **80**, 1061 (1958).

[168] R. W. Rampolla, R. C. Miller, and C. P. Smyth, *J. Chem. Phys.* **30**, 566 (1959).

[169] C. P. Smyth and D. M. Roberti, *J. Amer. Chem. Soc.* **82**, 2106 (1960); C. P. Smyth, K. Bergmann, and D. M. Roberti, *J. Phys. Chem. Ithaca* **64**, 665 (1960); C. P. Smyth and K. Higasi, *J. Amer. Chem. Soc.* **82**, 4759 (1960); C. P. Smyth, W. P. Purcell, and K. Fish, *J. Amer. Chem. Soc.* **82**, 6299 (1960); C. P. Smyth and W. P. Purcell, *J. Amer. Chem. Soc.* **83**, 1060, 1063 (1961); C. P. Smyth and E. L. Grubb, *J. Amer. Chem. Soc.* **83**, 4873, 4879 (1961); C. P. Smyth and F. K. Fong, *J. Amer. Chem. Soc.* **85**, 585, 1565 (1963); **40**, 2404 (1964); C. P. Smyth and A. A. Antony, *J. Amer. Chem. Soc.* **86**, 150 (1964); C. P. Smyth, W. E. Vaughan, S. B. Roeder, and T. Provder, *J. Chem. Phys.* **39**, 701 (1963).

[170] F. K. Fong and C. P. Smyth, *J. Phys. Chem. Ithaca* **67**, 226 (1963).

[171] C. P. Smyth and E. L. Grubb, *J. Amer. Chem. Soc.* **83**, 4873, 4879 (1961).

[172] G. Klages and A. Zentek, *Z. Naturforsch* **16a**, 1016 (1961); F. F. Hanna and G. Klages, *Z. Elektrochem.* **65**, 620 (1961).

partial rotation is readily observed in benzyl chloride ($C_6H_5 \cdot CH_2Cl$); but when adjacent methyl groups are introduced into the molecule, the CH_2Cl rotation is prevented.

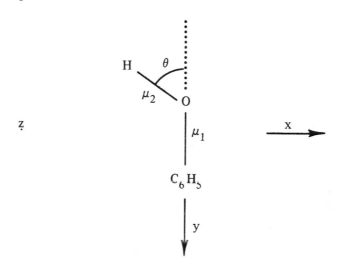

Consider[137] the phenol molecule C_6H_5OH for which the dipole moment can be regarded as composed of two components μ_1 and μ_2. Thus the total moment directed along the y-axis is $\mu_1 + \mu_2 \cos \theta$. The polarization associated with this component decays because of rotation about the x-axis with first-order rate coefficient

$$k_x = 1/\tau_x. \tag{4.4.243}$$

A similar decay occurs because of rotation about the z-axis:

$$k_z = 1/\tau_z. \tag{4.4.244}$$

The overall decay is first order with rate coefficient

$$k_1 = 1/\tau_1 = k_x + k_z. \tag{4.4.245}$$

This resultant relaxation time will be observed for the z-axis component, and the component along the x-axis will contribute

$$k_2 = 1/\tau_2 = k_z + k_y, \tag{4.4.246}$$

and that along the y-axis

$$k_3 = 1/\tau_3 = k_x + k_y. \tag{4.4.247}$$

Due to the different moments of inertia and shape factors, k_y can differ strongly from k_x and k_z, which are very similar to each other. Therefore

$k_2 \simeq k_3$. Two and only two relaxation times are to be expected, and

$$\tau_2 = 1/(n + 1)k_2, \qquad (4.4.248)$$

where $k_y/k_z = n$.

Unless $n > 3$, OH rotation about the y-axis will hardly be detectable, and for really good separation of relaxation times one requires $n > 10$. A good example of a molecule satisfying this requirement is 2,4,6-tri-*tert*-butyl phenol.[173] The two relaxation times are separated by a factor of about 20, and their relative intensities are in the expected ratio of the squares of the dipole moments of the rigid molecule and of the rotating part. Substitution of bromine in the molecule leads to changes which are consistent with the model. More frequently the two relaxations are much closer together and must be separated analytically.[169, 170, 172]

4.4.4.3. Relaxation Spectra of Hydrogen-Bonded Liquids. Hydrogen-bonded liquids such as water and alcohols are characterized by considerable structural correlation between the molecular dipoles. The Kirkwood correlation parameters g can be larger than 2, and the static permittivities are as high as 80 for water and 20–30 for aliphatic alcohols. Their temperature variation provides information about the hydrogen-bond structure of the liquid, which is taken to be a three-dimensional network in water[174] and a set of folded chains in alcohols.[175] It is possible to treat the dipole orientation process as involving the breakage of hydrogen bonds, or alternatively the bending of the bonds[176] or even the formation of multimer ring and chain species.[177]

The magnitudes of the activation energies of the principal dielectric relaxations indicate that breakage of one or more hydrogen bonds takes place in dipole reorientation. In water the Arrhenius energy is $\simeq 4.5$ kcal/mole, and is not strongly temperature dependent. In aliphatic alcohols at low temperatures the energy of both principal and high-frequency relaxations are strongly temperature dependent; they are of the same order as that of water at room temperature, but twice as large at low temperatures, so that two bonds may require breaking.

The oxygen orbitals in H_2O are hybridized sp^3, tetrahedrally arranged; in ice each oxygen is surrounded by two hydrogens and two bonds from hydrogens of neighboring H_2O molecules, so that each oxygen forms four hydrogen bonds. In liquid water and alcohols there is incomplete bonding,

[173] M. Davies and R. J. Meakins, *J. Chem. Phys.* **26**, 1584 (1957).

[174] G. H. Haggis, J. B. Hasted, and T. J. Buchanan, *J. Chem. Phys.* **20**, 1452 (1952).

[175] W. Dannhauser and R. H. Cole, *J. Chem. Phys.* **23**, 1762 (1955).

[176] J. Lennard-Jones and J. A. Pople, *Proc. Roy. Soc. A* **205**, 155 (1951); J. A. Pople, *Proc. Roy. Soc. A* **221**, 498 (1954).

[177] J. del Bene and J. A. Pople, *J. Chem. Phys.* **52**, 4858 (1970).

i.e., less than four bonds for each oxygen. An oxygen making two bonds can only reorient, in water or alcohols, by the breakage of one bond, but a hydrogen from another molecule might move from one oxygen orbital site to another unoccupied one, and it has been suggested[178] that this asymmetrical process may require a similar energy to an orthodox bond breakage, although the entropy will be different. In aliphatic alcohols there are two relaxation processes widely separated in frequency, which at room temperature may be attributed to the normal and the asymmetric process. In water, however, there is a small spread of relaxation times ($\alpha = 0.015$), which is indistinguishable from two closely spaced relaxation times, perhaps attributable to the normal and asymmetric processes.

In both water and alcohols, the high-frequency limit ε_∞ is much larger than n^2 the square of the infrared refractive index, and in the submillimeter region there is a dispersion with superposed absorption bands. These might be regarded[179, 180] as subsidiary relaxation processes, at least in water and ethyl alcohol, for which the $\varepsilon''(\varepsilon')$ representations, with contributions from the principal relaxation subtracted, are shown in Fig. 16. The form of a resonance absorption in the $\varepsilon''(\varepsilon')$ plane should be a full circle, while that of a relaxation process is a half circle; the curves in Fig. 16 are composites of both types of processes.

4.4.4.4. Relaxation in Solids. Polar liquids normally freeze in such a way that their dipoles are locked in a lattice and required to absorb considerable energy in order to orient in an electric field; if orientation is not possible, the permittivity of the solid is much smaller than that of the liquid. Significant contributions can, however, arise from the mobility of ions, as well as from residual molecular rotation.

In an ionic crystal lattice the ions are able to vibrate in different modes, each mode being capable of absorbing electromagnetic energy at the appropriate frequency, which is usually in the far-infrared or submillimeter region of the spectrum. For example, the resonance absorption at 160 cm^{-1} of NaCl accounts for the bulk of its permittivity at lower frequencies. Translational motion of ions gives rise to absorption at lower frequencies, particularly in doped crystals (e.g., LiF doped with Mg^{2+} [181–183]) in which the impurity ion gives rise to an adjacent ionic vacancy which can migrate or

[178] J. B. Hasted, "Aqueous Dielectrics." Chapman & Hall, London 1973.

[179] M. S. Zafar, J. B. Hasted, and J. E. Chamberlain, *Nature Phys. Sci.* **243**, 6, 116 (1973).

[180] J. E. Chamberlain, M. N. Afsar, G. J. Davies, J. B. Hasted, and M. S. Zafar, *Nature (London)* **255**, 319 (1975).

[181] R. G. Beckenridge, *J. Chem. Phys.* **16**, 959 (1948).

[182] J. S. Dryden and N. D. A. A. S. Rao, *J. Chem. Phys.* **25**, 222 (1956); J. S. Dryden and J. S. Cook, *Aust. J. Phys.* **13**, 260 (1960).

[183] J. S. Dryden and R. J. Meakins, *Rev. Pure Appl. Chem.* **7**, 43 (1957).

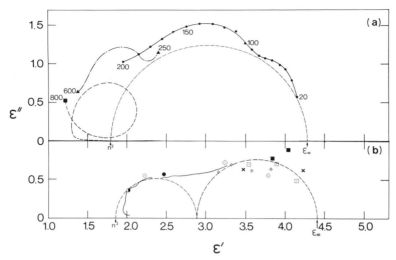

FIG. 16. $\varepsilon''(\varepsilon')$ representations of millimeter- and submillimeter-wave spectra of (a) water and (b) ethyl alcohol with calculated contributions from the principal relaxations subtracted [J. E. Chamberlain, M. N. Afsar, J. B. Hasted, M. S. Zafar, and G. J. Davies, *Nature* (*London*) **255**, 319 (1975). Broken line semicircles drawn from ε_∞ to n^2. Wavenumbers denoted in inverse centimeters. Full lines and various points represent experimental data.

diffuse. The energy barriers involved in this motion have characteristic magnitudes (in this particular instance 15 kcal/mole) which therefore correspond to dielectric dispersions of the Debye type. An additional dispersion can be found when the doping concentration is sufficiently large to give rise to the presence of different microphases (e.g., MgF_2) in which a characteristic ionic mobility is possible.

Another type of behavior, illustrated in Fig. 17, is to be found in hydrogen–halide crystals.[184] At the lowest temperatures the permittivity is small, but as the temperature is raised transitions occur to one or more solid phases possessing high permittivity which falls with increasing temperature. In these "rotator phases" dipole orientation can occur, with relaxation times similar to those in polar liquids. For HCl there is a single absorption with activation energy 2.6 kcal/mole, and preexponential constant 10^{-13} sec. In addition to the rotator phase there can exist other solid phases which are probably formed with association of the molecules in chains; treatment has been given using lattice defect models.[185]

[184] R. H. Cole and S. Havriliak, *Discuss. Faraday Soc.* **23**, 31 (1951); N. L. Brown and R. H. Cole, *J. Chem. Phys.* **21**, 1920 (1953); R. W. Swenson and R. H. Cole, *J. Chem. Phys.* **22**, 284 (1954).

[185] G. Adam, *J. Chem. Phys.* **43**, 662 (1965).

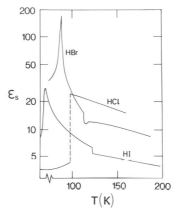

FIG. 17.   Temperature variation of static permittivities of solid hydrogen halides [R. H. Cole and S. Havriliak, *Discuss. Faraday Soc.* **23**, 31 (1951)].

Most rotator-phase solids[186, 100] consist of molecules whose shape is closer to spherical than to rod or disk. At the melting point the change in permittivity is not large, and it can also happen that the change in relaxation time is not large. In some instances the activation energies are smaller in the solid than in the liquid,[189] which might be due to the greater regularity in the solid of the region surrounding each molecule.

For solid camphane derivatives[190] the activation energy of relaxation increases with falling temperature, while the effective dipole moment decreases and the relaxation time distribution parameter increases; all of these suggest an increasing degree of molecular interaction as the temperature falls.

Relaxation spectra of long chain molecules in the solid state[183] can be interpreted with the aid of crystallographic studies. Usually there are several possible phases with a crystalline α-phase in which the active groups, both head and tail, assemble together in planes, with the inactive chains lying perpendicularly between them. Chains and groups can rotate about the long chain axis. As the temperature is reduced, a β-phase forms with tilted chains which are no longer able to rotate. There is a marked dependence of the activation energy on chain length.[191] The magnitudes of the permittivities

[186] C. P. Smyth, *in* "Dielectric Behavior and Structure," p. 114. McGraw-Hill, New York, 1955.

[187] A. H. White and W. S. Bishop, *J. Amer. Chem. Soc.* **62**, 16 (1940).

[188] C. Clemett and M. Davies, *Trans. Faraday Soc.* **58**, 1705 (1962).

[189] C. Clemett and M. Davies, *J. Chem. Phys.* **29**, 1347 (1960).

[190] C. Clemett and M. Davies, *Trans. Faraday Soc.* **58**, 1718 (1962).

[191] R. J. Meakins, *Progr. Dielect.* **3**, 151 (1961).

themselves depend, however, on the mechanical[192] and thermal[193] history of the specimens, as well as on impurities.[194]

High-frequency absorptions in the microwave band can arise in solids from intramolecular movements of polar bonds,[193, 195, 196] just as in liquids. For solid alcohols[197] the principal relaxations occur in the megahertz band, and the activation energies, which are independent of chain length, are as high as 15 kcal/mole for primary alcohols. In contrast to secondary alcohols for which the activation energies are much smaller (6 kcal/mole), the chain arrangement may be of the type

$$
\begin{array}{cccccc}
& \diagdown & & \diagdown & & \diagdown \\
& O{-}H \cdots & O{-}H \cdots & O{-}H \\
& \vdots & & \vdots & & \vdots \\
& H{-}O \cdots & H{-}O \cdots & H{-}O \\
& & \diagdown & & \diagdown & & \diagdown
\end{array}
$$

possibly with further cross linkages. The breakage of hydrogen bonds in orientation would be more extensive for such an arrangement than for the single arrangement

$$
\begin{array}{cccccc}
& \diagdown & & \diagdown & & \diagdown \\
& O{-}H \cdots & O{-}H \cdots & O{-}H
\end{array}
$$

Another group of solids for which the relaxation spectra are of interest is the group of clathrates. These are rigid lattices which form cages within which guest molecules, often polar, are free to move with only minimal disturbance of the lattice itself. In ice clathrates the crystal structure is quite different from that of pure ice; they contain cavities which may be pentagonal, dodecahedral, or hexakaidecahedral. In these cavities polar molecules such as ethylene oxide C_2H_4O can be trapped; it is found[198] that their presence raises the value of ε_∞ from 3.2 for pure ice to $\simeq 7.0$; the difference is ascribed to the reorientation of the ethylene oxide dipole.

4.4.4.5. Relaxation Processes in Solutions and Mixtures. The study of relaxation spectra in mixtures of polar liquids[199–203] provides a basis from

[192] J. W. Arnold and R. J. Meakins, *Trans. Faraday Soc.* **51**, 1667 (1955).

[193] J. S. Dryden and S. Dasgupta, *Trans. Faraday Soc.* **51**, 661 (1955).

[194] J. S. Dryden and H. K. Welsh, *Trans. Faraday Soc.* **60**, 2135 (1964).

[195] J. D. Hoffman and H. G. Pfeiffer, *J. Chem. Phys.* **22**, 132, 156 (1954).

[196] J. L. Farrands, *Trans. Faraday Soc.* **50**, 493 (1954).

[197] R. K. Meakins, *Trans. Faraday Soc.* **58**, 1953 (1962).

[198] G. J. Wilson and D. W. Davidson, *Can J. Chem.* **41**, 264, 1424 (1963); A. D. Potts and D. W. Davidson, *J. Phys. Chem. Ithaca* **69**, 996 (1965); R. E. Hawkins and D. W. Davidson, *J. Phys. Chem. Ithaca* **70**, 1889 (1961).

[199] A. Schallamach, *Trans. Faraday Soc.* **42A**, 180 (1946); P. Sixou, P. Dansas, and D. Gillot, *J. Chem. Phys.* **64**, 834 (1967).

[200] F. F. Bos, Thesis, Univ. of Leyden (1958); P. K. Kadara, *J. Phys. Chem. Ithaca* **62**, 887 (1958).

[201] D. J. Denney and R. H. Cole, *J. Chem. Phys.* **23**, 1767 (1955).

[202] E. Forest, and C. P. Smyth, *J. Phys. Chem. Ithaca* **69**, 1294 (1965).

[203] P. Daumézon, Thesis, Faculté des Sciences d'Orsay, Paris (1968).

which the molecular environment within the mixture can be envisaged. A mixture of two polar liquids, each with a single relaxation arising from single molecule rotation, might be expected if the mixing is imperfect to exhibit two dispersions corresponding to the two relaxation times of the pure liquids. If, however, the volume participating in the relaxation is sufficiently large to have a composition similar to the overall composition of the mixture, then a single dispersion is to be expected. Both situations are encountered. When the mixture is of two similar or "compatible" liquids, such as two alcohols or two ethers, then a single intermediate relaxation process is found; but a less compatible mixture, such as that of an alcohol with an ether, shows two separate relaxation times, indicating molecular segregation or aparthcid within the mixture. It appears that temperature and concentration are important factors in influencing the formation of "molecular phases" in a liquid mixture. In some instances the component of high concentration can impose its relaxation time on the entire mixture.

In aqueous solutions of nonelectrolytes such as glucose[204] it is possible to separate from the spectrum the relaxation times of the water, of the dissolved polar liquid and also that of the sheath of water molecules arranged around each dissolved molecule. These hydration or solvation sheaths may also have separate relaxation times when the solvent is a biomolecule[205] or an electrolytic ion.[206] In the latter case there are spreads of relaxation times and also some anomalies in the concentration dependence of principal relaxation times. There is a considerable diminution of the permittivity of water caused by the ionic electric field that clusters the dipoles in a spherical layer, thereby reducing their contribution to the polarization.[207]

Colloidal systems show an additional effect[208] that adds to the polarization rather than diminishes it. There is a sheath of ion pairs around the charged colloid particle, and their migration around the surface of the colloid particle is possible. An electric field will convert the particle with its sheath into a large dipole by virtue of the concentration of positive ions on one side and negative ions on the other. Distributions of relaxation times can arise from distributions of height of energy barriers hindering the mobility. A Cole–Cole distribution for a polystyrene latex solution is shown in Fig. 13b.

Biomolecules in aqueous solution exhibit relaxation spectra that include contributions from the molecular dipole, from the hydration sheath, and from the free liquid. Ionic conductivity also contributes, and the pH of the

---

[204] M. J. Tait, A. Suggett, F. Franks, S. Ablett and P. A. Quickenden, *J. Soln. Chem.* **1**, 131 (1972); F. Franks, D. S. Reid, and A. Suggett, *J. Soln. Chem.* **2**, 99 (1973).

[205] E. H. Grant, G. P. South, S. Takashima, and H. Ichimura, *Biochem. J.* **122**, 691 (1971); *J. Phys. Chem.* **72**, 4373 (1968).

[206] K. Giese, *Ber. Bunsenges. Phys. Chem.* **76**, 495 (1972).

[207] C. H. Collie, D. M. Ritson, and J. B. Hasted, *J. Chem. Phys.* **16**, 1 (1948).

[208] G. Schwarz, *J. Phys. Chem. Ithaca* **66**, 2637 (1962).

solution can distort the molecule in such a way that its resolved dipole moment is changed.[209]

4.4.4.6. Heterogeneous Materials. The relaxation spectra of heterogeneous materials differ from those of pure liquids and solids because of geometrical effects on the polarization. Expressions can be derived for the permittivity of a mixture of randomly arranged spherical or ellipsoidal particles of permittivity ε_i (the inner component) in a homogeneous medium of permittivity ε_o (the outer component).[210–214] When one or another component of the mixture exhibits a dielectric relaxation, then the apparent relaxation spectrum line shape of the mixture is modified, and can be calculated from these expressions.

The volumetric proportions of the mixture are v_i and v_o with

$$v_i + v_o = 1. \tag{4.4.249}$$

The electrical properties of the mixture are related to the average field E_i within the inner component particles. An approximation to this field can be made for ellipsoidal particles of axes a, b, and c. When a field E is applied along the axis a, the uniform field E_a within the ellipsoid is

$$E_a = \varepsilon_o E / \{\varepsilon_o + A_a(\varepsilon_i - \varepsilon_o)\}, \tag{4.4.250}$$

where the depolarization factor A_a along the a-axis is

$$A_a = \tfrac{1}{2}abc \int_0^\infty ds/(a^2 + s)^{3/2}(b^2 + s)^{1/2}(c^2 + s)^{1/2}. \tag{4.4.251}$$

Moreover

$$A_a + A_b + A_c = 1. \tag{4.4.252}$$

Numerical values of depolarization factors have been published.[215]

Along the a-axis the induced polarization P_a per unit volume of the ellipsoid is

$$P_a = (\varepsilon_i - \varepsilon_o)E_a; \tag{4.4.253}$$

the induced dipole moment is

$$\mu_a = \tfrac{4}{3}\pi abc P_a; \tag{4.4.254}$$

[209] E. H. Grant, G. P. South, and W. I. O. Walker, *Biochem. J.* **122**, 765 (1971).

[210] L.K.H. van Beek, *Progr. Dielec.* **7**, 69 (1967).

[211] G. P. de Loor, Thesis, Univ. of Leyden (1956).

[212] J. A. Stratton, "Electromagnetic Theory." McGraw-Hill, New York, 1941.

[213] D. Polder and J. H. van Santen, *Physica* **12**, 257 (1946).

[214] L. Landau and E. M. Lifshitz, "Electrodynamics of Continuous Media." Addison-Wesley, Reading, Massachusetts, 1960.

[215] K. W. Wagner, *Arch. Elektrotech.* **3**, 100 (1914); R. W. Sillars, *J. Inst. Elec. Eng.* **80**, 378 (1937).

and the polarizability is

$$\alpha_a = \tfrac{4}{3}\pi abc\varepsilon_o(\varepsilon_i - \varepsilon_o)/\{\varepsilon_o + A_a(\varepsilon_i - \varepsilon_o)\}. \tag{4.4.255}$$

The induced dipole moment polarizes the surroundings of the ellipsoid, so that the change ΔP in polarization owing to the introduction of N ellipsoids per unit volume is

$$\Delta P = N\alpha\bar{E}, \tag{4.4.256}$$

where $\bar{E}$ is the average field throughout the mixture. The dielectric constant ε_m of the mixture is related to ΔP by

$$(\varepsilon_m - \varepsilon_o)\bar{E} = P, \tag{4.4.257}$$

and N is related to the volume fraction by

$$v_i - \tfrac{4}{3}\pi abcN. \tag{4.4.258}$$

The calculation of E is carried out by deriving the polarization of the neighborhood of a particle by its induced dipole μ_a; this gives rise to a reaction field E_R and modifies the average field, giving

$$E = E_a/(1 - \alpha_a E_R). \tag{4.4.259}$$

The permittivity of the medium surrounding the particle is not necessarily equal to ε_o, but may be denoted for the time being by $\bar{\varepsilon}_o$. Thus

$$\bar{E}_a = \bar{\varepsilon}_o E/\{\bar{\varepsilon}_o + A_a(\varepsilon_i - \bar{\varepsilon}_o)\}, \tag{4.4.260}$$

which leads to a new expression for μ_a, from which ε_i is then eliminated:

$$E_R = A_a(1 - A_a)(\bar{\varepsilon}_o - \varepsilon_o)/\tfrac{4}{3}\pi abc\varepsilon_o\{\bar{\varepsilon}_o + A_a(\bar{\varepsilon}_o - \varepsilon_o)\}. \tag{4.4.261}$$

Finally one obtains

$$\varepsilon_m - \varepsilon_o = v_i(\varepsilon_i - \varepsilon_o)\bar{\varepsilon}_o/[\bar{\varepsilon}_o + A_a(\varepsilon_i - \bar{\varepsilon}_o)], \tag{4.4.262}$$

and, generalizing this expression for a random distribution of orientations of the ellipsoids, one obtains

$$\varepsilon_m - \varepsilon_o = \tfrac{1}{3}v_i(\varepsilon_i - \varepsilon_o)\bar{\varepsilon}_o \sum_{j=a,\,b,\,c} \{\bar{\varepsilon}_o + A_j(\varepsilon_i - \bar{\varepsilon}_o)\}^{-1}. \tag{4.4.263}$$

This equation is only valid for small v_i, but can be extended by analogy to larger v_i. Although $\bar{\varepsilon}_o$ is a function of v_i, it is often taken as equal to ε_o or ε_m for all values of v_i.

For spherical particles there is a single depolarization factor $A_j = \tfrac{1}{3}$, and using $\bar{\varepsilon}_o = \varepsilon_o$, which is applicable to high dilutions, one arrives at the Maxwell–Lewin formula:

$$(\varepsilon_m - \varepsilon_o)/3\varepsilon_o = v_i[(\varepsilon_i - \varepsilon_o)/(2\varepsilon_o + \varepsilon_i)]. \tag{4.4.264}$$

When the denominator $3\varepsilon_0$ is replaced by $\varepsilon_m + 2\varepsilon_0$, the equation is known as the *Rayleigh mixture formula*; a more exact form has been derived.[216]

For large values of v_i one may assume $\bar{\varepsilon}_0 = \varepsilon_m$ and obtain for spheres

$$(\varepsilon_m - \varepsilon_0)/3\varepsilon_m = v_i[(\varepsilon_i - \varepsilon_0)/(2\varepsilon_m + \varepsilon_i)] \qquad (4.4.265)$$

which is known as the *Böttcher mixture formula*; it is symmetrical in v_i and v_o.

An intermediate value of $\bar{\varepsilon}_0$ can be calculated as follows. The Rayleigh equation is written in differential form:

$$\frac{d\varepsilon_m}{\varepsilon_m} = \left(\frac{dv_i'}{1 - v_i'}\right)\left(\frac{\varepsilon_i - \varepsilon_m}{2\varepsilon_m + \varepsilon_i}\right) \qquad (4.4.266)$$

and integrated between the limits ε_0 and ε_m. This procedure leads to

$$1 - v_i = [(\varepsilon_i - \varepsilon_m)/(\varepsilon_i - \varepsilon_0)](\varepsilon_0/\varepsilon_m)^{1/3}, \qquad (4.4.267)$$

which is the Bruggeman–Niesel mixture formula. The procedure assumes $\varepsilon_m = \varepsilon_0$ at infinite dilution; for the addition of a small volume fraction δv_i, a change $\delta\varepsilon_m$ is calculated, and the procedure repeated. The solution is asymmetrical in v_i and v_o, but a more symmetrical procedure is as follows: On the addition of an incremental quantity of inner component, not only does ε_0 become $\varepsilon_m - \delta\varepsilon_m$ but also ε_i becomes $\varepsilon_m + \delta\varepsilon_m$; when applied to the Rayleigh equation this procedure yields

$$v_i' = \tfrac{1}{2} + [\delta\varepsilon_m/2(3\varepsilon_m - 2\delta\varepsilon_m)] \simeq \tfrac{1}{2} + \delta\varepsilon_m/6\varepsilon_m, \qquad (4.4.268)$$

$$v_i = (1 - v_i')v_i(\varepsilon_m + \delta\varepsilon_m) + v_i'v_i(\varepsilon_m - \delta\varepsilon_m). \qquad (4.4.269)$$

Expansion as the first two terms of Taylor's series leads to

$$v_i' = \tfrac{1}{2} - \tfrac{1}{4}\delta\varepsilon_m(d^2v_i/d\varepsilon_m^2)/(dv_i/d\varepsilon), \qquad (4.4.270)$$

$$3\varepsilon_m(d^2v_i/d\varepsilon_m^2) + (2dv_i/d\varepsilon_m) = 0. \qquad (4.4.271)$$

The solution which satisfies the boundary conditions $\varepsilon_m = \varepsilon_0$ for $v_i = 0$ and $\varepsilon_m = \varepsilon_i$ for $v_i = 1$ is the Looyenga equation[217, 218]

$$\varepsilon_m = (v_i\varepsilon_i^{1/3} + v_o\varepsilon_0^{1/3})^3. \qquad (4.4.272)$$

[216] R. E. Meredith and C. W. Tobias, *J. Appl. Phys.* **31**, 1270 (1960).
[217] H. Looyenga, *Physica* **31**, 401 (1965).
[218] L. D. Landau and E. M. Lifshitz, "Electrodynamics of Continuous Media." Addison-Wesley, Reading, Massachusetts, 1960.

For nonspherical particles[219, 220] the assumption $\bar{\varepsilon}_o = \varepsilon_m$ leads to

$$1 - v_i = \begin{cases} \left(\dfrac{\varepsilon_i - \varepsilon_m}{\varepsilon_i - \varepsilon_o}\right)\left(\dfrac{2\varepsilon_i + \varepsilon_o}{2\varepsilon_i + \varepsilon_m}\right) & \text{for disks,} \quad (4.4.273) \\[3mm] \left(\dfrac{\varepsilon_i - \varepsilon_m}{\varepsilon_i - \varepsilon_o}\right)\left(\dfrac{\varepsilon_i + 5\varepsilon_o}{\varepsilon_i + 5\varepsilon_m}\right) & \text{for rods.} \quad (4.4.274) \end{cases}$$

The Bruggeman-Niesel formula is

$$\varepsilon_m{}^2 = \frac{2(\varepsilon_i v_i + \varepsilon_o v_o) - \varepsilon_m}{(\varepsilon_i/v_i) + (\varepsilon_o/v_o)} \qquad \text{for disks} \qquad (4.4.275)$$

and

$$3\varepsilon_m{}^3 + (5\varepsilon_v' - 4\varepsilon)\varepsilon_m{}^2 - (v_i\varepsilon_i{}^2 + 4\varepsilon_i\varepsilon_o + v_o\varepsilon_o{}^2) - \varepsilon_i\varepsilon_o\varepsilon_v = 0 \qquad \text{for rods,} \qquad (4.4.276)$$

where

$$1/\varepsilon_v = (v_i/\varepsilon_i) + (v_o/\varepsilon_o) \qquad (4.4.277)$$

and

$$1/\varepsilon_v' = (v_i/\varepsilon_o) + (v_o/\varepsilon_i). \qquad (4.4.278)$$

Comparison of relaxation spectra with calculations from these equations has been made using artificial dielectrics: suspensions of spherical, rod-shaped, or disk-shaped particles in solids, liquids or waxes. For dilute mixtures, with $v_i < 0.2$, the agreement of the Maxwell–Lewin formula with experiment is satisfactory, but for $v_i \geqslant 0.2$, the agreement is less satisfactory. Figure 18 shows in the form $\varepsilon''(\varepsilon')$ the results of microwave experiments on Plexiglass containing water spheres. Natural heterogeneous dielectrics also provide tests of the calculations; Figure 19 illustrates the comparison of experimental spectra for snow with calculations made on the assumption that it consists of spheres of ice in air.[221, 222] It is seen that the calculations yield skewed arcs while the experiments yield semicircular arcs, and that the calculated effective relaxation time does not agree with the measured value.

Quite apart from the contributions to heterogeneous substance relaxation spectra arising from the relaxation of the components themselves, a low-frequency dispersion known as the *Maxwell–Wagner effect* is also known. Its origin may be understood by considering two parallel slabs of dielectric

[219] B. V. Hamon, *Aust. J. Phys.* **6**, 305 (1953).

[220] K. W. Wagner, *Arch. Elektrotech.* **3**, 100 (1914); R. W. Sillars, *J. Inst. Elec. Eng.* **80**, 378 (1937).

[221] Y. Ozawa and D. Kuroiwa, Microwave Propagation in Snowy Districts. Res. Inst. Appl. Elec., Hokkaido Univ., Japan, p. 31 (1958); D. Kuroiwa, *Un. Geod. Geophys. Int. Ass. Int. Hydrol. Sci.* **4**, 52 (1956).

[222] J. G. Paren, Thesis, Univ. of Cambridge (1970).

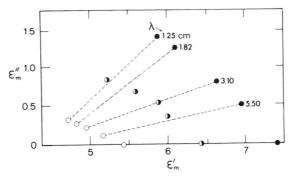

FIG. 18. $\varepsilon''(\varepsilon')$ representation of permittivity of Plexiglass containing spherical water droplets at $r_i = 0.25$. Data points (split circles) at indicated wavelengths (centimeters) lie midway between calculations on the assumptions $\bar{\varepsilon}_o = \varepsilon_o$ [Eq. (4.4.264)] (open circles) and $\bar{\varepsilon}_o = \varepsilon_m$ [Eq. (4.4.276)] (full circles). [J. B. Hasted, "Aqueous Dielectrics." Chapman and Hall, London, 1974.]

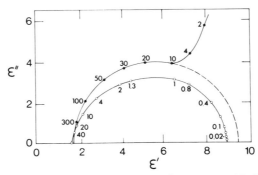

FIG. 19. $\varepsilon''(\varepsilon')$ representation of permittivity of snow [Y. Ozawa and D. Kuroiwa, Microwave Propagation in Snowy Districts. Res. Inst. Appl. Elec., Hokkaido Univ., Japan, p. 31 (1958); D. Kuroiwa, *Un Geodes. Geophys. Int. Ass. Int. Hydrol. Sci.* **4**, 52 (1956)]. Data points (filled circles) compared with calculations [J. G. Paren, Thesis, Univ. of Cambridge (1970)] for ice–air mixture using Eq. (4.4.272) (open circles). Frequencies indicated in kilohertz. Electrolytic conductivity enhances low-frequency data points at 2 or 4 kHz.

of thickness d_1 and d_2 situated in series between two condenser plates with permittivities ε_1 and ε_2 and conductivities σ_1 and zero; these will be assumed to be frequency independent. The dielectric polarization loss is zero, but the conductivity contribution is

$$\varepsilon_1'' = 4\pi\sigma_1/\varepsilon_0, \qquad\qquad (4.4.279)$$

where ε_0 is the permittivity of free space.

The complex capacitances of the two dielectric slabs are

$$C_1 = [\varepsilon_1 - (j4\pi\sigma_1/\varepsilon_0)]\varepsilon_0/4\pi d_1 \qquad\qquad (4.4.280)$$

and

$$C_2 = \varepsilon_2 \varepsilon_0 / 4\pi d_2. \tag{4.4.281}$$

The total capacitance is

$$C = \frac{C_1 C_2}{C_1 + C_2} = \frac{\varepsilon_0}{4\pi} \frac{\{[\varepsilon_1 - (j4\pi\sigma/\omega\varepsilon_0)]\varepsilon_2\} d_1 d_2}{\{[\varepsilon_1 - (j4\pi\sigma/\omega\varepsilon_0)]/d_1\} + (\varepsilon_2/d_2)}, \tag{4.4.282}$$

and the permittivity of the combined system is

$$\hat{\varepsilon} = \varepsilon_2[\varepsilon_1 - (j4\pi\sigma/\omega\varepsilon_0)]/\{[\varepsilon_1 - (j4\pi\sigma/\omega\varepsilon_0)]f_2 + \varepsilon_2 f_1\}, \tag{4.4.283}$$

where

$$f_1 = d_1/(d_1 + d_2) \quad \text{and} \quad f_2 = d_2/(d_1 + d_2) \tag{4.4.284}$$

as $\omega > 0$, $\varepsilon \to \varepsilon_2/f_2 = \varepsilon_s$ and as $\omega \to \infty$,

$$\varepsilon \to \varepsilon_1 \varepsilon_2/(\varepsilon_1 f_2 + \varepsilon_2 f_1) = \varepsilon_\infty. \tag{4.4.285}$$

At intermediate frequencies

$$(\hat{\varepsilon} - \varepsilon_\infty)/(\varepsilon_s - \varepsilon_\infty) - 1/(1 + j\omega\tau) \tag{4.4.286}$$

with

$$\tau = [(\varepsilon_1 f_2 + \varepsilon_2 f_1)/f_2 \sigma](\varepsilon_0/4\pi). \tag{4.4.287}$$

The system therefore shows a relaxation spectrum, and the relaxation time increases with decreasing conductivity.

Inhomogeneous dielectrics which are not of laminar structure can be interpreted similarly. Wagner[223] treated the case of conducting spherical particles (phase 2) dispersed at low concentration in a lossless dielectric (phase 1), the distances between the spheres being large compared with their radii.

The calculation yields a Debye relaxation with parameters

$$\varepsilon_\infty = \frac{\varepsilon_1\{2\sigma_1 + \sigma_2 + 2v_2(\sigma_2 - \sigma_1)\}}{2\sigma_1 + \sigma_2 - v_2(\sigma_2 - \sigma_1)}, \tag{4.4.288}$$

$$\varepsilon_s = \frac{3v_2\sigma_1\{(2\sigma_1 + \sigma_2)(\varepsilon_2 - \varepsilon_1) - (2\varepsilon_1 + \varepsilon_2)(\sigma_2 - \sigma_1)\}}{2\sigma_1 + \sigma_2 - v_2(\sigma_2 - \sigma_1)^2}$$

$$+ \frac{\varepsilon_1\{2\sigma_1 + \sigma_2 + 2v_2(\sigma_2 - \sigma_1)\}}{2\sigma_1 + \sigma_2 - v_2(\sigma_2 - \sigma_1)}, \tag{4.4.289}$$

$$\tau = \frac{\varepsilon_0}{4\pi}\left\{\frac{2\varepsilon_1 + \varepsilon_2 - v_2(\varepsilon_2 - \varepsilon_1)}{2\sigma_1 + \sigma_2 - v_2(\sigma_2 - \sigma_1)}\right\}. \tag{4.4.290}$$

[223] K. W. Wagner, *Arch. Elektrotech.* **3**, 100 (1914); R. W. Sillars, *J. Inst. Elec. Eng.* **80**, 378 (1937).

A similar treatment for ellipsoidal particles, due to Sillars[223] yields

$$\varepsilon_\infty = \frac{\varepsilon_1\{\varepsilon_1 + [A(1 - v_2) + v_2](\varepsilon_2 - \varepsilon_1)\}}{\varepsilon_1 + A(1 - v_2)(\varepsilon_2 - \varepsilon_1)}, \tag{4.4.291}$$

$$\varepsilon_s = \frac{\varepsilon_1\{\sigma_1 + [A(1 - v_2) + v_2](\sigma_2 - \sigma_1)\}}{\sigma_1 + A(1 - v_2)(\sigma_2 - \sigma_1)}$$

$$+ \frac{v_2\sigma_1\{[\sigma_1 + A(\sigma_2 - \sigma_1)](\varepsilon_2 - \varepsilon_1) + [\varepsilon_1 + A(\varepsilon_2 - \varepsilon_1)](\sigma_1 - \sigma_2)\}}{\{\sigma_1 + A(1 - v_2)(\sigma_2 - \sigma_1)\}^2}, \tag{4.4.292}$$

$$\tau = \frac{\varepsilon_0[\varepsilon_1 + A(1 - v_2)(\varepsilon_2 - \varepsilon_1)]}{4\pi[\varepsilon_1 + A(1 - v_2)(\varepsilon_2 - \varepsilon_1)]}. \tag{4.4.293}$$

Experimental tests of the Maxwell–Wagner effect have been carried out with polyethylene cylinders (axial ratio $a/b = 7$) completely filled with water and dispersed in paraffin wax. Figure 20 shows the fairly good agreement

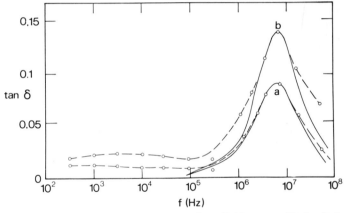

FIG. 20.   Frequency variation of loss angle $\tan \delta = \varepsilon''/\varepsilon'$ for water-filled cylinders in non-polar matrix [L. K. H. van Beek, Proc. Colloq. Ampère **11**, 229 (1962)]. Calculations (solid line) and experimental data (points and broken line) for volume fractions. Curve (a) is for $v_i = 0.0115$ (50 cylinder specimen) and curve (b) for $v_i = 0.0230$ (100 cylinder specimen).

between measured and calculated frequency dependence of the loss; the additional loss observed at lower frequencies may be caused by interactions between pairs of cylinders.

Many heterogeneous dielectric systems do not resemble suspensions of ellipsoids, but are more like insulators containing random conducting paths, surfaces, or filaments, which often connect with each other.[224] The extreme

[224] R. J. Meakins, *Progr. Dielec.* **3**, 151 (1961).

case of such a dielectric, namely, a material in which conducting paths stretch directly from condenser plate to condenser plate, would have a dielectric loss related to the conductivity by

$$\varepsilon'' = \sigma_2/\varepsilon_0\omega. \tag{4.4.294}$$

An inverse power proportionality of ε'' to frequency is characteristic of transport-conducting materials. Dielectrics which are believed to have conducting paths, such as animal fat, behave in this manner.[225] The variation may be expressed approximately as a power law $\varepsilon'' = Kf^{-k}$. For high concentrations of conducting material k approaches unity; it falls with decreasing concentration. There is some theoretical justification for a value $k = 0.5$. Frequently there are other processes besides transport conduction which complicate the observed behavior, but when the dc conductivity contribution is subtracted, a residual loss which is largely frequency independent will often remain. Alternatively, a Maxwell Wagner dispersion[226] will sometimes remain, but this interpretation is only free from suspicion when the loss maximum is clearly observed.

[225] L. K. H. van Beek, *Progr. Dielec.* **7**, 69 (1967)
[226] B. V. Hamon, *Aust. J. Phys.* **6**, 305 (1953).

5. RECENT DEVELOPMENTS

5.1. Beam-Foil Spectroscopy*

5.1.1. Introduction and History

"Beam-foil spectroscopy" (bfs) as an area of research now includes a wide range of experiments having the common feature that the radiation source is a fast beam of particles that have been excited by passage through a thin solid foil. It was recognized by Bashkin[1] at the beginning of the 1960's that ion beams such as have been routinely accelerated by nuclear physicists in Van de Graaff accelerators for years could become useful sources of radiation. It is largely due to the early publicity which he gave to the idea that it received a great deal of attention, especially from laboratories already possessing small accelerators whose use for nuclear and other research was not saturated.

In fact, the idea goes back well before the 1960's. As early as 1907 Wien[2] realized the importance of measuring the deexcitation time of excited atoms. He attempted to do so by measuring the spatial decay of light from "canal rays," atoms streaming through an aperture in the cathode of a gas-discharge tube. The canal rays were to stream from a high-pressure region, where they were collisionally excited, into a much lower-pressure region where their decay would not be collisionally induced. He was not successful until after the invention of the diffusion pump allowed him to reach a sufficiently low pressure in the observation region. Beginning in 1919[3] he performed a series of measurements of this type, determining decay constants in both the Lyman and Balmer series of hydrogen as well as in several other elements. It is interesting to note that, even for the very slow canal rays ($v \sim 10^7$ cm/sec) he was able to measure the average beam velocity using the Doppler shift of observed spectral lines. His technique was taken up by a number of other experimenters[4] in the 1920's, some of whom employed a gas jet to excite

[1] S. Bashkin, *Nucl. Instrum. Methods* **28**, 88 (1964).

[2] W. Wien, *Ann. Phys.* **30**, 3691 (1909); **23**, 428 (1907).

[3] W. Wien, *Ann. Phys.* **60**, 597 (1919); **66**, 229 (1921); **70**, 1 (1923); **73**, 483 (1924); **76**, 109 (1925); **83**, 1 (1927).

[4] A. J. Dempster, *Astrophys. J.* **57**, 193 (1923); J. S. McPetrie, *Phil. Mag.* **1**, 1082 (1926); R. d'E. Atkinson, *Proc. Roy. Soc. London Ser. A* **116**, 81 (1927).

* Chapter 5.1 is by C. Lewis Cocke.

the moving rays. One need only raise the beam energy and perhaps replace the gas exciter by a thin foil to obtain the beam-foil spectroscopy (bfs) so widely used today.

The first bfs publication of the renaissance was that of Kay in 1963,[5] who reported detection of some unidentified lines from foil-excited nitrogen and neon beams. Bashkin and Meinel[6] recorded spectra of beams of several elements from hydrogen to xenon at beam energies between 0.5 and 2 MeV and pointed out that the beam spectra were very similar to those obtained from novas, thus establishing early the potential usefulness of this laboratory source to those confronted with spectra from high-temperature astrophysical sources. Further data on similar beams were reported by Bashkin et al.[7] who showed clearly that the lifetimes of foil-excited states emitting visible radiation were recorded as spatially decaying intensities of the spectral lines as the beam proceeded beyond the exciting foil. Bashkin et al.[8] reported extension of this preliminary spectroscopy into the ultraviolet. They recorded spectra from a 5.1-MeV neon beam between 250 and 1300 Å and made rough measurements of the lifetimes of the $(2p^4)^2D$ state in Ne IV and the $(2p^2)^2D$ in Ne VI.

Bashkin et al.[9] reported the first bfs observation of quantum beats in the $n = 4$–7 levels of a foil-excited hydrogen beam. Berkner et al.[10] used foil-excited 0.97-MeV/amu beams from the Berkeley heavy ion linear accelerator (HILAC) to measure the lifetimes of the $(1s^22p)^2P^0_{1/2}$ and $(1s^22p)^2P^0_{3/2}$ states in the lithium isoelectric sequence between C and Ne. Kay[11] suggested that the charge state of the emitting species could be identified by comparing the dependence on beam energy of the yield of light from a spectral line with that of the equilibrium charge-state distribution. Very little was understood of the foil-excitation mechanism at the time, however, and several problems with this method of charge-state identification were discussed by Bashkin and Malmberg.[12] Clean separation of the charge states of the emitting ions was accomplished by Malmberg et al.[13] who used electrostatic analysis of the foil-excited beams.

It was fast becoming clear that the beam-foil radiation source indeed presented many control advantages over more conventional sources. Among these were isolation of beam type and charge state, knowledge of the velocity

[5] L. Kay, Phys. Lett. 5, 36 (1963).
[6] S. Bashkin and A. B. Meinel, Astrophys. J. 139, 413 (1964).
[7] S. Bashkin, A. B. Meinel, P. R. Malmberg, and S. G. Tilford, Phys. Lett. 10, 63 (1964).
[8] S. Bashkin, L. Heroux, and J. Shaw, Phys. Lett. 13, 229 (1964).
[9] S. Bashkin, W. S. Bickel, D. Fink, and R. K. Wangsness, Phys. Rev. Lett. 15, 284 (1965).
[10] K. Berkner, W. S. Cooper III, J. N. Kaplan, and R. V. Pyle, Phys. Lett. 16, 35 (1965).
[11] L. Kay, Proc. Phys. Soc. 85, 163 (1965).
[12] S. Bashkin and P. R. Malmberg, Proc. Phys. Soc. 87, 589 (1966).
[13] P. R. Malmberg, S. Bashkin, and S. G. Tilford, Phys. Rev. Lett. 15, 98 (1965).

of the emitter, and high equivalent source temperature. The number of contributions to the literature began to expand rapidly. By 1967 several laboratories had become actively engaged in doing spectroscopy on foil-excited beams from low-energy accelerators, and the first in a series of international conferences on beam-foil spectroscopy was held at the University of Arizona on November 20–22, 1967. It is interesting to notice that the contributors to this conference were not confined in their interests to "beam-foil spectroscopy," but rather had a broader common interest which might be grouped under the topic "atomic physics with low-energy accelerators." The development of beam-foil research programs coincided conveniently with a general redirection of many small Van de Graaff programs from the heavily worked areas of nuclear physics to interesting problems in atomic physics which had been largely overlooked during the earlier years of these accelerators. Thus the rapid growth of beam-foil spectroscopy has been aided by the immediate availability of the necessary accelerator technology. Further, development of the field has proceeded concurrently with many other new and interesting uses of accelerators to explore atomic collision problems.

Since the 1967 conference an explosion of literature on beam-foil spectroscopy has occurred. The greatest single area in which the bfs techniques have contributed is that of atomic lifetimes where measurements of a precision previously difficult to obtain have become straightforward. More unusual applications of bfs techniques to Lamb shift determinations and quantum beat experiments are now well known. More recent international conferences on beam-foil spectroscopy were held at Lysekil, Sweden (June 7–12, 1970) and Tucson, Arizona (October 2–6, 1972). The fourth such international conference was held at Gatlinburg, Tennessee (September 15–19, 1975).

In this chapter we do not pretend to cover in any exhaustive manner the great body of bfs work that has been done in the past decade. We attempt to present a condensed survey of some representative work from different areas of interest within the overall bfs heading. We hope that the interested reader will pursue the subject through the published conference proceedings[14–16] and through a number of excellent review articles.[17]

[14] S. Bashkin (ed.), "Beam-Foil Spectroscopy." Gordon & Breach, New York, 1968.

[15] I. Martinson, J. Bromander, and H. G. Berry (eds.), *Conf. Beam-Foil Spectrosc.*, *2nd* [*Nucl. Instrum. Methods* **90** (1970)].

[16] S. Bashkin (ed.), *Int. Conf. Beam-Foil Spectrosc. 3rd* [*Nucl. Instrum. Methods* **110** (1973)].

[17] S. Bashkin, *Appl. Opt.* **7**, 2341 (1968); W. S. Bickel, *Appl. Opt.* **7**, 2367 (1968); **6**, 1309 (1967); H. J. Andrä, *Physica Scripta* **9**, 257 (1974); I. Martinson, *Physica Scripta* **9**, 281 (1974); "Nuclear Spectroscopy and Reactions," Part C, p. 467. Academic Press, New York, 1974; H. G. Berry, *Physica Scripta* **12**, 5 (1975); I. Martinson and A. Gaupp, *Phys. Rep.* **15**, 13 (1974); S. Bashkin (ed.), "Beam-Foil Spectroscopy." Springer-Verlag, Berlin and New York, 1976.

5.1.2. General Characteristics of Radiation Source

Figure 1 shows a schematic of a beam-foil apparatus. An accelerated beam, selected by an analyzing magnet to be pure in mass and charge state, is directed onto a thin foil of a low nuclear charge (Z) material. The low Z minimizes Rutherford scattering and energy loss in the foil. Carbon is convenient. The reciprocal of the velocity of the beam is given nonrelativistically by $v^{-1} = 0.72(m/E)^{1/2}$ nsec/cm, where m is the projectile mass in atomic mass units and E the energy in megaelectron volts. If the foil is 10 $\mu g/cm^2$ carbon, its linear thickness is ~ 500 Å and the time spent by a 0.5-MeV/amu projectile in passage through this distance is $\sim 5 \times 10^{-15}$ sec. During passage the projectile experiences exchange of electrons with the foil material, some energy loss, and some angular deflection. It emerges into a vacuum surrounded by a distribution of electrons whose resolution into quantum states of the isolated ion or atom determines the ionization level and excitation state of the system. The subsequent decay of the free excited system in vacuum is then observed downstream of the exciting foil.

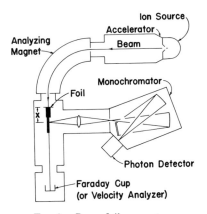

FIG. 1.   Beam-foil apparatus.

There are few sources of radiation about whose excitation distribution so little detail is known. This is due to the complexity and number of competing processes that are responsible for the final state of the merging system. It is certainly not possible to ascribe an excitation temperature to the excited ions. Nevertheless the fact that the velocity of the projectile may be large compared to characteristic velocities of electrons in even inner shells of the beam ions allows excitation of systems that would be present in thermal sources only at temperatures of the order of 10^5 to 10^6 K. For example, Fig. 2 shows the similarity between the charge-state distribution

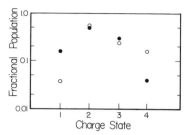

FIG. 2. Comparison of equilibrium charge-state fractions for carbon ions emerging from a carbon foil (●) at 1.156 MeV [P. L. Smith and W. Whaling, *Phys. Rev.* **188**, 36 (1969)] to those computed by Jordan for a solar coronal plasma (○) at 10^5 K in the absence of an equilibrium radiation field [C. Jordon, *Mon. Notices Roy. Astron. Soc.* **142**, 501 (1969)].

of carbon ions emerging from a thin carbon foil at the modest energy of 1.156 MeV[18] and a theoretical ionization equilibrium calculated for the solar corona at an effective electron temperature of 10^5 K.[19] It is clear that a foil-excited beam is indeed a "hot" source and is well suited to the study of highly ionized and excited systems.

Because the decay of the excited system takes place in high vacuum (typically $\sim 10^{-6}$ Torr) and the density of particles in the beam itself is low (typically $\leqslant 10^7/cm^3$), the radiation source is well below the densities for which self-absorption and secondary collision processes are important. One thus observes the decay of free and isolated systems. It is, furthermore, simple to couple evacuated detectors directly to the accelerator vacuum system, making straightforward observations into the far-ultraviolet and soft x-ray regions.

The low-particle density in the beam also renders the source rather weak, however, and high-efficiency low-background detectors are usually necessary. For example, photographic recordings of beam spectra with high-speed spectrographs may require several hours exposure time.[20] The excited beam is usually visible to the naked eye at beam currents above a few tenths of a microampere. While signal size will vary considerably from one beam type and spectral region to the next, the general weakness of the source has limited its usefulness to studies of systems in the beam, excluding its use as a more general-purpose radiation source.

Perhaps the most useful single characteristic of the source is that the emitting ions are excited at a rather precisely defined time zero and are moving at the known velocity of the accelerated beam, minus a slight

[18] P. L. Smith and W. Whaling, *Phys. Rev.* **188**, 36 (1969).
[19] C. Jordan, *Mon. Notices Roy. Astron. Soc.* **142**, 501 (1969).
[20] W. S. Bickel, *Appl. Opt.* **6**, 1309 (1967).

velocity loss in the foil. Thus the time behavior of the excited system is directly translated into a spatial behavior in the laboratory and mean lives of emitting states may be straightforwardly measured by observing the decay of the intensity of spectral lines with increasing foil-detector distance. Typical reciprocal velocities are in the nanosecond per centimeter range, allowing lifetimes between a few tenths of a nanosecond to several tens of nanoseconds to be measured.

Unfortunately, the motion of the ions represents a disadvantage when the source is used for spectroscopy. The finite angular acceptance of any detection system requires that photons emitted at different angles to the moving beam be accepted. Typical beam velocities are sufficiently high that the corresponding distribution in Doppler shifts usually limits the laboratory energy resolution $E/\Delta E$ to the order of 10^{-3}, corresponding to a resolving power of $10.^3$ Several efforts to minimize this effect are discussed later. Doppler spreading has limited the usefulness of the light source in the detailed classification of spectral lines. On the other hand, its use in low-resolution surveys of previously completely unstudied spectra has been high.

5.1.3. Beam-Foil Spectra

In some of the early beam-foil literature workers reported spectra in which many of the spectral lines were unidentified. This was the result of both a lack of previous experimental work on line classifications, especially on highly ionized spectra, and an unfamiliarity with general idiosyncrasies of the beam-foil source. Unusual features of the source which may result in unfamiliar spectra include the following:

(1) High ionization states are easily obtained even at modest beam energies.

(2) Multiple collision processes within the foil readily populate multiply excited systems.

(3) States at high excitation and with very high principal quantum number are copiously produced.

5.1.3.1. Techniques for Spectral Identification.

5.1.3.1.1. MASS PURITY. The beam-foil source has the advantage that the nuclear mass of the radiating system can be uniquely chosen by magnetic analysis of the accelerated beam. Ambiguity arises in some cases for which molecular ion beams have masses nearly equal to those of the desired elemental mass. For example, Bashkin et al.[21] found the use of $^{28}Si^+$ beams frustrated by contamination from $^{12}C^{16}O^+$ and $^{14}N_2^+$ beams. Nevertheless such contaminations can usually be controlled and mass purity established

---

[21] S. Bashkin (ed.), "Beam-Foil Spectroscopy," p. 5. Gordon & Breach, New York, 1968.

by attention to ion source operation and by selection of unambiguous isotopes or molecular combinations.

5.1.3.1.2. CHARGE-STATE IDENTIFICATION. Several methods for identification of the charge state of the emitting system have been employed. For sufficiently long-lived systems magnetic or electrostatic analysis may be performed on the foil-excited beam before it decays, thus separating the emitters into systems with well-defined charge states. Several variations on this theme have been employed, some of which are described by Bashkin.[22] For systems with short decay lengths electrostatic analysis after radiation of beam particles in coincidence with the emitted photons has been employed.[23]

A method far less rigorous but easier to employ[11, 24] consists of comparing the yield of light from an observed spectral line as a function of the beam energy (hereafter referred to as the "excitation function of the line") with the corresponding energy-dependent charge-state equilibrium curves.[25] This method relies on the tacit assumption that the energy dependence of excitation probability for populating a particular level is entirely contained in the variation of its parent charge-state yield with energy. Although this assumption is known to be quite poor in some cases, the method has nevertheless proved useful. For example, in Fig. 3 the excitation functions for several lines in foil-excited Ne are compared with the fractional yields of their parent charge states. The identification of the charges of the emitting ions on the basis of their excitation functions is clearly possible. It is interesting to note that doubly excited states appear to display characteristic excitation functions lying between the charge-state curves of their parents and the once further ionized system, a property that has helped identify these states.[26] The results of Andersen et al. are shown in Fig. 4.

5.1.3.1.3. DOPPLER LINE BROADENING. In the case of complex spectra the poor spectral resolution that results from the Doppler spread hinders line identifications by forcing line blending and inaccuracies in wavelength

[22] S. Bashkin, *Nucl. Instrum. Methods* **90**, 3 (1970); G. W. Carriveau and S. Bashkin, *Nucl. Instrum. Methods* **110**, 203 (1970); U. Fink, *J. Opt. Soc. Amer.* **58**, 937 (1968); J. A. Leavitt, J. W. Robson, and J. O. Stoner, Jr., *Nucl. Instrum. Methods* **110**, 423 (1973); see also P. R. Malmberg, S. Bashkin, and S. G. Tilford, *Phys. Rev. Lett.* **15**, 98 (1965).

[23] C. L. Cocke, B. Curnutte, J. R. Macdonald, and R. Randall, *Phys. Rev. A* **9**, 57 (1974).

[24] M. Dufay, *Nucl. Instrum. Methods* **90**, 15 (1970).

[25] I. S. Dmitriev, *Sov. Phys. JETP* **5**, 473 (1957); C. Zaidins, Thesis, California Inst. Technol. (1967); P. Hvelplund, E. Laegsgård, J. Ø. Olsen and E. H. Pedersen, *Nucl. Instrum. Methods* **90**, 315 (1970), P. L. Smith and W. Whaling, *Phys. Rev.* **188**, 36 (1969); C. D. Moak, L. B. Bridwell, H. O. Lutz, S. Datz, and L. C. Northcliffe, *in* "Beam-Foil Spectroscopy" (S. Bashkin, ed.), p. 157. Gordon & Breach, New York, 1968; H. Betz, *Rev. Mod. Phys.* **44**, 465 (1975).

[26] F. Gaillard, Thesis, Lyon (1969); N. Andersen, R. Boleu, K. Jensen, and E. Veje, *Phys. Lett.* **34A**, 227 (1971).

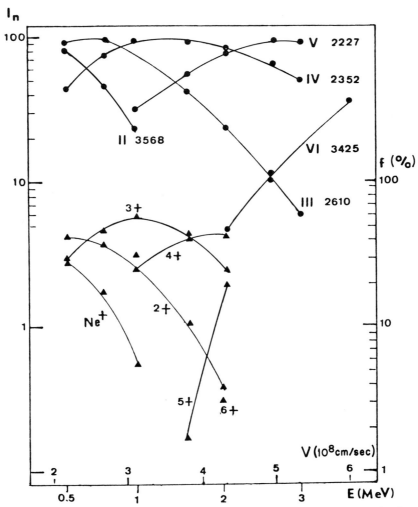

FIG. 3. Comparison of equilibrium charge-state fractions of neon ions emerging from a carbon foil with the yields of some neon line intensities in the beam-foil spectrum [M. Dufay, *Nucl. Instrum. Methods* **90**, 15 (1970)].

determinations. If radiation of wavelength λ from a beam of velocity v is observed at an angle θ to the beam through an optical system with an angular acceptance $\Delta\theta$, a first-order calculation of the spectral spread $\Delta\lambda$ due to Doppler effect is given by $\Delta\lambda/\lambda = v/c\,\Delta\theta\sin\theta$ (c is the speed of light). The worst spread is obtained at $90°$; the best at $0°$ or $180°$, but with a maximum Doppler shift. For example, light from a 0.5-MeV/amu beam, viewed

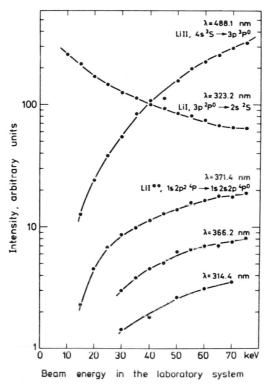

FIG. 4. Excitation functions for several lines in Li I and Li II showing the intermediate behavior of the doubly excited 1s2p² ⁴P state in Li I [N. Andersen, R. Boleu, K. Jensen, and E. Veje, *Phys. Lett.* **34A**, 227 (1971)].

at 90° and filling an *f*/5 optical system, will show a fractional spread $\Delta\lambda/\lambda = 6.5 \times 10^{-3}$.

Several methods have been used for reducing this spread. Leavitt *et al.*[27] have employed a "refocusing" technique which exploits the fact that Doppler-shifted rays passing through a grating spectrometer will reconverge to an image in a focal plane different from that for a normal monochromatic source. Axial viewing of the beam is often employed to give narrow lines but with large Doppler shifts. Roesler and Stoner[28] have used an off-axis Fabry–Perot interferometer to obtain narrow lines from a beam-foil source. Such techniques eventually are limited by an irreducible resolution contribution from the spread in vector velocity of the beam introduced by the

[27] J. A. Leavitt, J. W. Robson, and J. O. Stoner, Jr., *Nucl. Instrum. Methods* **110**, 423 (1973).
[28] F. L. Roesler and J. O. Stoner, Jr., *Nucl. Instrum. Methods* **110**, 465 (1973).

beam's passage through the foil.[29] The angular and energy spread attending the passage of heavy ions through foils is particularly severe at low energy.[30]

5.1.3.2. Spectral Features.   5.1.3.2.1. SINGLY EXCITED STATES. At energies below about 100 keV beam-foil spectra in the visible and near uv are often dominated by strong lines from the first and second spectra of the accelerated element for which line identifications and strengths from previous work with conventional light sources may be used in understanding the spectra. As the bombarding energy is raised, higher ionization states become dominant for which previous spectral classifications are rarer. Even very modest beam energies may result in the population of highly ionized spectra. This characteristic was partially responsible for the appearance of many unidentified lines in early beam-foil work. Figure 5 shows equilibrium

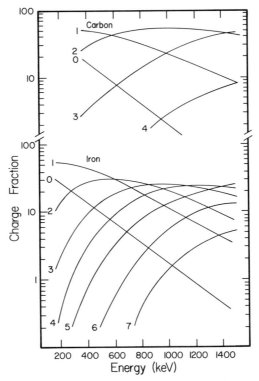

FIG. 5.   Equilibrium charge-state fractions for carbon and iron ions emerging from a carbon foil [P. L. Smith and W. Whaling, *Phys. Rev.* **188**, 36 (1969)].

[29] J. O. Stoner, Jr. and L. J. Radziemski, Jr., *Nucl. Instrum. Methods* **90**, 275 (1970).

[30] T. Andersen, K. A. Jessen, and G. Sørensen, *Nucl. Instrum. Methods* **90**, 41 (1970); G. Högberg, H. Nordén, and H. G. Berry, *Nucl. Instrum. Methods* **90**, 283 (1970).

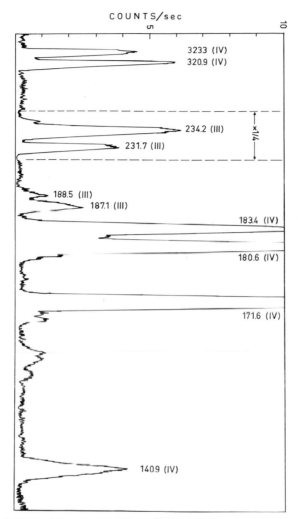

COUNTS/sec

FIG. 6.   Partial beam-foil spectrum of Mg at 330 keV [L. Lundin, B. Engman, J. Hilke, and I. Martinson, *Physica Scripta* **8**, 274 (1973)].

charge fractions from beams of C and Fe[25] emerging from carbon foils. It is clear that energies of several hundred kiloelectron volts are already sufficient for producing up to the fourth and fifth spectra of these elements. For example, a beam-foil spectrum in the uv of 330-keV Mg (Fig. 6) shows strong lines from Mg III–IV.[31]

[31] L. Lundin, B. Engman, J. Hilke, and I. Martinson, *Physica Scripta* **8**, 274 (1973).

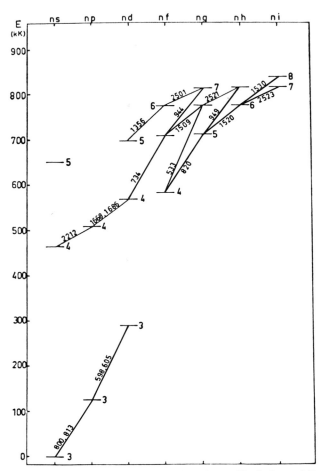

FIG. 7.   Partial level scheme of Cl VII showing lines observed in the beam-foil spectra. Term energies above $n = 3$ were assigned from these data [S. Bashkin and I. Martinson, *J. Opt. Soc. Amer.* **61**, 1686 (1971)].

In spite of the resolution limitations previously discussed a number of new line identifications and term energy measurements have been made from beam-foil spectra. For example, Fig. 7 shows transitions in Cl VII observed by Bashkin and Martinson[32] from 0.5–2 MeV chlorine beams from which first assignment of term energies above $n = 3$ were made.

5.1.3.2.2. MULTIPLY EXCITED STATES.   One of the major contributions of bfs to the literature of spectral classifications has been in the area of doubly excited states, i.e., states for which two electrons are excited above their

[32] S. Bashkin and I. Martinson, *J. Opt. Soc. Amer.* **61**, 1686 (1971); S. Bashkin, J. Bromander, J. Leavitt, and I. Martinson, *Physica Scripta* **8**, 285 (1973).

ground state configurations. That such states are produced much more strongly in foil excitation than in other light sources was first reported by Bickel et al.[33] in Stockholm and Buchet et al.[34] in Lyon (see Fig. 8). Simple arguments may be made to understand this result. Whereas in conventional light sources the time between excitational collisions is typically longer than the relaxation time of excited states, in passage through a foil the situation is reversed. The projectile may suffer many collisions in a time too short to allow an excited electron to relax. Thus production of multiply excited states via multiple collisions is expected. This picture is supported by the observation that doubly excited systems are only weakly excited in binary collision with gas targets.[34]

Many doubly excited systems are energetically capable of autoionization. Since this decay mode is intrinsically faster than the radiative one, photon emission will be observable only if the autoionization channel is forbidden. For example, in He I (see Fig. 9), of the first doubly excited unbound states, only the $(2p^2)^3P$ is forbidden to autoionize in LS coupling because emission to the p-wave $^1S_0$ continuum is required and this violates parity conservation. Thus allowed radiation to the $(1s2p)^3P^o$ is the dominant decay mode. This transition at 320 Å was observed in a helium spark discharge in 1928–1930[35] and was first observed in the beam-foil source by Berry et al.[36] along with several other lines identified as originating from doubly excited states. In general, autoionization is forbidden from $(2pnl)$ $^{3,1}L$ systems except as induced by spin–orbit and spin–spin parts of the Hamiltonian. Several transitions from these, and also from doubly excited systems for which autoionization rates are otherwise hindered sufficiently to allow competition from the radiative decay mode, have been observed by Knystautas et al.[37] and by Berry et al.[38]

Whereas transition between the doubly and singly excited systems are already in the far uv in He I, those between doubly excited systems will fall in the same region as the normal spectral lines of He I, above ~2000 Å. A number of such transitions have been observed and classified by Berry et al.[38] Such transitions between unbound states have opened an interesting opportunity for short lifetime determinations. If the upper level involved in the transition has an appreciable autoionization rate, the radiative decay

[33] W. S. Bickel, I. Bergström, R. Buchta, L. Lundin, and I. Martinson, Phys. Rev. 178, 118 (1969).

[34] J. P. Buchet, A. Denis, J. Desesquelles, and M. Dufay, Phys. Lett. 28A, 529 (1969).

[35] K. T. Compton and J. C. Boyce, J. Franklin Inst. 205, 497 (1928); P. G. Kruger, Phys. Rev. 36, 855 (1930).

[36] H. G. Berry, I. Martinson, L. J. Curtis, and L. Lundin, Phys. Rev. A 3, 1934 (1971).

[37] E. J. Knystautas and R. Drouin, Nucl. Instrum. Methods 110, 95 (1973).

[38] H. G. Berry, J. Desesquelles, and M. Dufay, Nucl. Instrum. Methods 110, 43 (1973); Phys. Rev. 6, 600 (1972); Phys. Lett. 36A, 237 (1971); N. Andersen, W. S. Bickel, R. Bolen, K. Jensen, and E. Veje, Physica Scripta 3, 255 (1971).

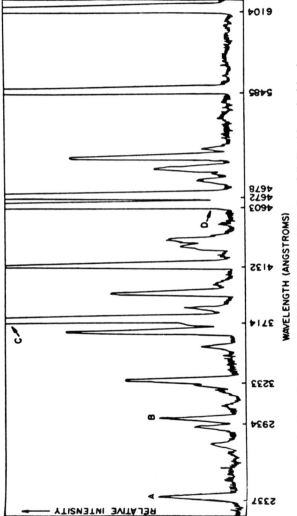

FIG. 8. Beam-foil spectrum of Li excited by a thin carbon foil. Doubly excited levels are denoted by A, B, C and D [W. S. Bickel, I. Bergström, R. Buchta, L. Lundin, and I. Martinson, *Phys. Rev.* **178**, 118 (1969)]. The incident particle energy is 56 keV.

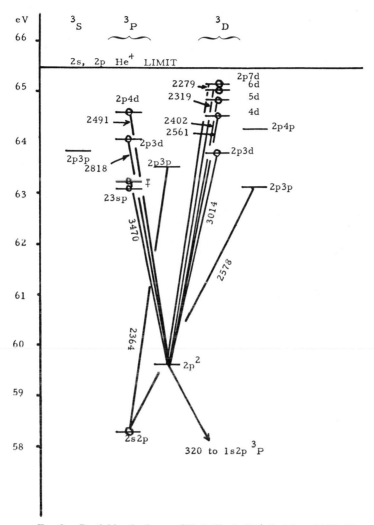

FIG. 9.   Partial level scheme of He I. The 1s He⁺ limit is at 24.58 eV.

will not be seen; however, radiative decay to a lower doubly excited system with a large autoionization width is possible. In the latter case the natural width of the spectral line will reflect the lifetime of the final state and may be large enough to measure, due to the speed of the autoionization process. Figure 10 shows such a broadening in the $(pp23)^3D \rightarrow (2s2p)^3P^o$ transition in He I at 2577.6 Å. From the observed width an autoionization lifetime of 7×10^{-4} sec is deduced for the 2s2p $^3P^o$ level.

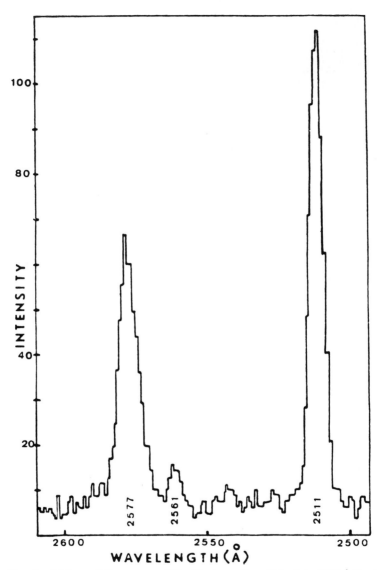

FIG. 10.   Partial beam-foil spectrum of He. The larger linewidth of the 2577 Å line in He I is due to the rapid autoionization of its lower term 2s2p ^{3}p°. [H. G. Berry, J. Desesquelles, and M. Dufay, *Phys. Rev.* **6**, 600 (1972)].

Some decays from doubly excited states of higher members of the He I isoelectronic sequence have been observed from beam-foil sources.[39] Of particular note are those seen in O, F, S, and Cl at high beam energy for which the transition energies are in the x-ray region (see Section 5.1.6).

The doubly excited states of the Li I isoelectronic sequences are among those most heavily studied using bfs. Quartet states of the type $(1s2snl)^4L$ and $(1s2pnl)^4L$ are spin forbidden to autoionize to the $(1s^2)^1S_0$ continuum. Classifications of lines in the quartet systems of Li I and Be II[40] are on rather good footing from beam-foil work and several transition in B III and C IV[41, 42] have been studied. For higher Z spin–orbit and spin–spin mixings allow observable electron emission from quartet systems to the $(1s^2)^1S_0$ systems, and extensive beam foil studies of these decays through the electron decay channel have been made (see Section 5.1.6). Figure 11 shows the quartet level scheme in Li I, indicating those transitions observed in beam-foil spectra. Many of the transitions have not been observed from other light sources. Several doubly excited states in the doublet system of Li I also possess observable radiative decays to both doubly and singly excited states and have been observed by Berry et al.[40] and Buchet et al.[39]

Multiply excited states in Be I and higher sequences have received further study using beam-foil techniques. For example, the Na I quartet system is homologous to the Li I system and one should thus expect a number of slowly autoionizing states whose radiative decay should be observable. Very little work in this area has yet been done. A single line in the bfs spectrum of Mg at 3480 Å has been attributed to a doubly excited state in Mg II on the basis of its excitation function,[31] and a similar transition has been observed in Na I at 3882.4 Å.[43] The transition is thought to be $(2p^53s3d)^4F^o_{9/2} \rightarrow$ $(2p^53s3p)^4D_{7/2}$. A number of unidentified lines from a foil-excited Na beam in the grazing incidence region are thought to arise from quartet systems.[43] Studies by Pegg et al.[44] of electron emission from such systems has been reported in Na I and Cl VIII. The identification of transitions in both radiative and electron decay channels suffers strongly from the serious lack of

---

[39] J. P. Buchet, M. C. Buchet-Poulizac, H. G. Berry, and G. W. Drake, *Phys. Rev. A* **7**, 922 (1973).

[40] H. G. Berry, E. H. Pinnington, and J. L. Subtil, *J. Opt. Soc. Amer.* **62**, 767 (1972); H. G. Berry, J. Bromander, I. Martinson and R. Buchta, *Physica Scripta* **3**, 63 (1971); N. Andersen, R. Bolen, K. Jensen, and E. Veje, *Phys. Lett.* **34A**, 277 (1971).

[41] I. Martinson, W. S. Bickel, and A. Öline, *J. Opt. Soc. Amer.* **60**, 1213 (1970); I. Martinson, *Nucl. Instrum. Methods* **90**, 81 (1970).

[42] H. G. Berry, M. C. Poulizac, and J. P. Buchet, *J. Opt. Soc. Amer.* **63**, 240 (1973).

[43] H. G. Berry, R. Hallin, R. Sjödin, and M. Gaillard, *Phys. Lett.* **50A**, 191 (1974).

[44] D. J. Pegg, I. A. Sellin, P. M. Griffin, and W. W. Smith, *Phys. Rev. Lett.* **28**, 1615 (1972); D. J. Pegg et al., *Phys. Lett.* **50A**, 447 (1975).

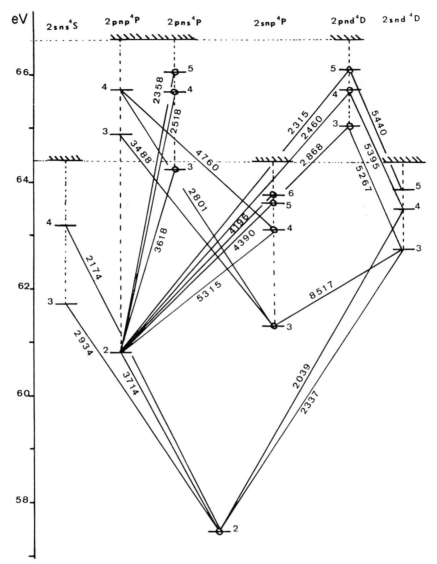

FIG. 11.   Partial level scheme for Li I showing the quartet states. The Li II ($^1S_0$) limit is at 5.39 eV.

theoretical level calculations especially for higher members of the Na I sequence. The reader is referred to the recent review article by Berry[17] for further discussion of recent theoretical and experimental work on multiply excited systems.

5.1.3.2.3. RYDBERG STATES. Many of the lines which went unidentified in early beam-foil spectra are now understood to originate from $\Delta n = 1$ and $\Delta n = 2$ transitions between states characterized by an electron with a very high principal quantum number circulating about a highly charged core. For example, even at the modest bombarding energy of 1.5 MeV the 3000–5000 Å spectra of Fe and Ni taken by Lennard et al.[45] are dominated transitions between Rydberg states with core charge $(q) \sim n$. The dominance of these transitions is due to the large production probabilities of such states in the foil-excitation process and to the selection of high n states made by the choice of the spectral region viewed. Transitions involving lower principal quantum numbers are at much shorter wavelengths for such highly ionized systems. Since the orbital radii for such states may be tens of angstroms it is difficult to imagine that they can survive creation from far inside the foil. A surface capture mechanism is suggested and appears roughly consistent with the data.[45, 46] That these excited states find themselves quickly free of perturbation from collisions after leaving the foil may largely account for their strong enhancement in beam-foil sources as opposed to plasma light sources. In the latter case the fragility of the states associated with large radii may destroy them before they can radiate.

Edlén[47] has discussed the energies of Rydberg states. The hydrogenic term energies T_{II} are given by

$$T_{II} = \frac{\text{Ry } \zeta^2}{n^2}\left[1 + \frac{\alpha^2 \zeta^2}{n^2}\left(\frac{n}{l + \frac{1}{2}} - \frac{3}{4}\right)\right],$$

where ζ is the core charge. For nonpenetrating orbits, the first-order corrections to these formulas are due to dipole and quadrupole polarization of the core by the orbiting electron. For the dipole case the shift Δp is given by $\Delta p = \alpha_d \zeta^4 \text{ Ry}\langle r_{nl}^{-4}\rangle$ where α_d is the dipole polarizability characteristic of the core state and values of $\langle r_{nl}^{-4}\rangle$ are tabulated by Edlen.[47] Observed transition energies can thus be used to deduce dipole and quadrupole core polarizabilities, as has been done by Berry et al.[48] and Lennard et al.[45, 46]

The study of such systems probably provides a better testing ground of the foil-excitation mechanism than for atomic structure theory. In some cases lines from the same n and different l can be resolved spectroscopically and thus information on the l-dependence of the foil-excitation probability deduced. Figure 12 shows the relative populations of the $(5l)^2L$ states in B III

[45] W. N. Lennard, R. M. Sills, and W. Whaling, *Phys. Rev. A* **6**, 884 (1972).

[46] W. N. Lennard and C. L. Cocke, *Nucl. Instrum. Methods* **110**, 137 (1973).

[47] B. Edlén, "Handbuch der Physik" (S. Flügge, ed.), Vol. 27, p. 80. Springer-Verlag, Berlin and New York, 1964.

[48] S. N. Bhardwaj, H. G. Berry, and T. Mossberg, *Physica Scripta* **9**, 331 (1974).

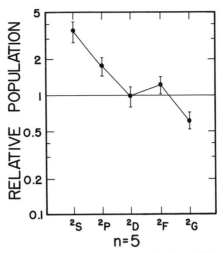

FIG. 12.   Relative populations of 5ℓ ^2L levels from the beam-foil spectrum of 1-MeV ^{11}B. The intensities have been divided by $2\ell + 1$ so statistical population corresponds to a horizontal line [I. Bromander, *Nucl. Instrum. Methods* **110**, 11 (1973)].

from a 1-MeV boron beam.[49] A clear enhancement of low l is seen, as is consistent with the formation of the states by direct electron capture. Population of higher l may result from Stark mixing of nearly degenerate l-states as the system departs the foil region.[50]

5.1.3.2.4.   HIGH-ENERGY BEAMS. Beam-foil spectra taken with higher velocity beams such as those available from Tandem Van de Graaffs and other linear heavy-ion machines show, in the visible and near-ultraviolet regions, essentially only Rydberg states.[51–53] The spectral lines observed are characteristic not of the nuclear charge but of the core charge of the emitting system. Thus beams of different nuclear charge, but similar charge-state distribution, will produce similar spectra, as is shown in the comparison of N and C spectra in Fig. 13.[53] Some lifetime work on these lines has been reported but is hard to evaluate due to severe cascading problems.[49] In general, Coulomb approximation calculations of the lifetimes will be more reliable than experimental beam-foil results.

More interesting transitions from high-velocity beams fall in the far-uv and grazing incidence regions. Berkner *et al.*[10] measured lifetimes of the

[49] I. Bromander, *Nucl. Instrum. Methods* **110**, 11 (1973).

[50] J. D. Garcia, *Nucl. Instrum. Methods* **110**, 245 (1973).

[51] M. Dufay, A. Denis, and J. Desesquelles, *Nucl. Instrum. Methods* **90**, 85 (1970).

[52] R. Hallin, J. Lindskog, A. Marelius, J. Pihl, and R. Sjödin, *Physica Scripta* **8**, 209 (1973).

[53] J. P. Buchet, M. C. Buchet-Poulizac, G. DoCao, and J. Desesquelles, *Nucl. Instrum. Methods* **110**, 19 (1973).

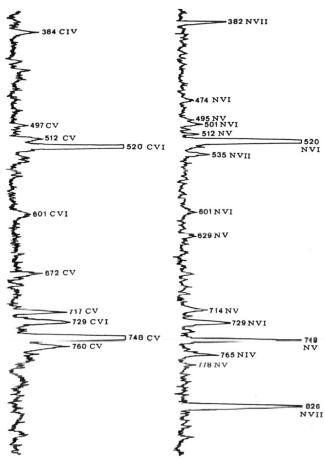

FIG. 13. Comparison of foil spectra from 1.15-MeV/amu C (bottom) and N (top) beams showing similar transitions for equally charged ions [J. P. Buchet, M. C. Buchet-Poulizac, G. DoCao, and J. Desesquelles, *Nucl. Instrum. Methods* **110**, 19 (1973)].

$(1s^22p)^2P$ state in C IV–Ne VIII using 0.97-MeV/amu beams from the Berkeley HILAC. The transition observed was to the $1s^22s$ and ranged from 1550 Å (C) to 770 Å (Ne). A number of transition in the 30–1200 Å region from 1.15 MeV/amu C, N, O, and Ne were studied by Buchet *et al.*[53] Transitions were observed in the He and Li isoelectronic sequences and measured lifetimes in the remarkably short region of tens of picoseconds were reported. A number of lines in the Li I and Be I isoelectronic sequences in Ne VII and Ne VIII were studied by Barrette *et al.*[54] below 150 Å, and upper limits of

[54] L. Barrette, E. J. Knystantas, and R. Drouin, *Nucl. Instrum. Methods* **110**, 29 (1973).

17 to 43 psec were found for several mean lives. Pegg et al.[55] have recently recorded with a grazing-incidence spectrometer spectra from fast foil-excited oxygen and fluorine beams. Many transitions in He and Li isoelectronic sequences were identified. Further examples of beam-foil spectra from systems with inner-shell vacancies are discussed in Section 5.1.6.

5.1.4. Transition Probabilities

5.1.4.1. Basic Technique. Perhaps the greatest single advantage of the beam-foil radiation source is that it allows an extremely simple method for the measurement of the lifetime of the emitting state. In the absence of cascading and background problems, the spatial decay of the intensity I of a spectral line will be given by $I(x) = I_0 e^{-x/v\tau}$, where v is the velocity of the post-foil beam, x the distance between foil and detector measured along the beam, and τ the lifetime of the emitting state. For example, the data of Martinson et al.[56] are shown in Fig. 14. Since the decay can often be traced over several mean lives, determination of τ with an accuracy of 10% is

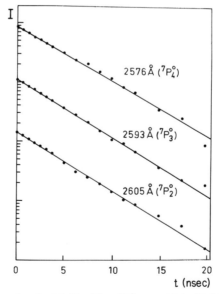

FIG. 14. Decay curves for the Mn II a $^7S_3$–z$^7P^0_{2,3,4}$ multiplet [I. Martinson, L. J. Curtis, J. Bryozowski, and R. Buchta, Physica Scripta 8, 62 (1973)].

[55] D. J. Pegg et al., Phys. Rev. A10, 745 (1974).
[56] I. Martinson, L. J. Curtis, J. Bryozowski, and R. Buchta, Physica Scripta 8, 62 (1973).

straight-forward, and with care, higher precision may be often obtained. The method does not require any knowledge of the absolute numbers of emitting states, the effective temperature of the source, absolute intensity calibration of the detection apparatus, etc. There is no problem with questions of optical thickness of the source or collisional line broadening. Since the foil-excitation mechanism is highly nonselective in the states that it populates, it allows mean-life measurement on states which are more difficult to obtain in other sources, such as doubly excited, highly ionized, and metastable systems.

On the other hand, if it is the transition probability per second A_{ij}[†] characterizing a transition from an initial state (i) to a final state (j) which is of interest, one must know, in addition to the mean life of the initial state, the fraction of decays of (i) which go to (j) [i.e., branching ratio from (i) to (j)]. For the case of simple spectra the branching ratios are often unity or may be calculated on first principles. In the case of more complex spectra they must be measured, and thus require additional experiments to compliment the beam-foil results. Branching ratio measurements are nevertheless considerably easier to perform accurately than are absolute oscillator-strength measurements because they require only measurements of relative intensities of lines originating from the same upper level, no knowledge of relative numbers of systems in different excited states being necessary.

We summarize next some of the complicating features that are often encountered in beam-foil lifetime measurements and survey some contributions made by the technique to the literature of transition probabilities.

5.1.4.1.1. POST-FOIL VELOCITY DETERMINATION. For high-velocity low-Z beams the post-foil velocity can be readily calculated by subtracting from the incident beam energy, known from the accelerator calibration to be typically a few parts in 10^3, the energy loss in the foil. Stopping powers for energetic ions passing through matter have been measured by numerous authors and a useful tabulation is given by Northcliffe and Schilling.[57]

While such a correction can often be made accurately, the procedure becomes somewhat treacherous for low-energy, high-Z beams, for which two problems arise. First, the energy loss comes both from energy transfer to electrons (electronic stopping) and, in hard collisions, to recoil nuclei (nuclear stopping).[58] A widely used formulation dealing with low-energy stopping powers is that of Lindhard et al.[59] A number of experimental determinations

---

[57] L. C. Northcliffe and R. F. Schilling, *Nucl. Data Tables A* **7**, 233 (1970); L. C. Northcliffe, *Ann. Rev. Nucl. Sci.* **13**, 67 (1963).

[58] N. Bohr, *K. Dan. Vidensk. Selsk. Mat. Fys. Medd.* **18**, 8 (1948).

[59] J. Lindhard and M. Scharff, *Phys. Rev.* **124**, 128 (1961).

[†] The oscillator strength f_{ij} for the transition (i) → (j) is related to A_{ij} by $f_{ij} = 1.5 \times 10^{-16}$ sec × $\lambda^2 A_{ij}$, where λ is the wavelength of the transition in angstrom units.

of low-energy electronic stopping powers are given by Hvelplund.[60] Calculations of the nuclear stopping contribution are given by Lindhard et al.[61] There is evidence that these calculations overestimate the nuclear stopping contribution for beam-foil experiments.[62] Unfortunately at very low energies, $\lesssim 100$ keV, such contributions may be as large as 10–20% of the beam energy and thus uncertainties in the size of the correction may lead to large uncertainties in post-foil velocity. Knowledge of the absolute foil thickness, necessary to making the correction, may be quite elusive. For example, Stoner et al.[63] find important thickness contributions from noncarbon materials on thin carbon foils. Further, foils generally thicken rapidly under bombardment unless the vacuum system is kept extraordinarily clean.

A second problem is that nuclear scattering in the foil may spread the beam appreciably at low energy, leading to loss of radiation from the beam at large foil-detector distances caused by scattering of the beam outside the acceptance region of the detector. The problem is discussed for example by Högberg et al.,[30] Stoner et al.,[64] and Andersen et al.[65] The problem is intrinsically quite complex, since both the energy loss and the excitation probability of a given level may depend on the angle at which the beam emerges from the foil.

Clearly it is advantageous to avoid energy-loss corrections altogether by a direct measurement of the post-foil beam velocity. To this end electrostatic analyzers (e.g., see Andersen[62]) and time-of-flight spectrometers have been used.[66] It is also possible to compare the apparent wavelengths of spectral lines observed at two different angles to the beam and to use the observed Doppler shift as a direct measure of the beam velocity. Such a method has the advantage of directly measuring the velocity of the excited component of the beam, but lacks precision at low energies because of the small size of the shift.

5.1.4.1.2. CASCADING. It is very rare that the decay profile from a beam-foil experiment is a pure single exponential. In many cases there is an

[60] P. Hvelplund, K. Dan. Vidensk. Selsk. Mat. Fys. Medd. **38**, 4 (1971); B. Fastrup, P. Hvelplund, and C. A. Sautter, K. Dan. Vidensk. Selsk. Mat. Fys. Medd. **35**, 10 (1966), and references contained therein.

[61] J. Lindhard, V. Nielsen, and M. Scharff, K. Dan. Vidensk. Selsk. Mat. Fys. Medd. **36**, 10 (1968); J. Lindhard, M. Scharff, and H. E. Schiøtt, K. Dan. Vidensk. Selsk. Mat. Fys. Medd. **33**, 14 (1963).

[62] T. Andersen, Nucl. Instrum. Methods **110**, 35 (1973); P. Hvelplund, E. Laegsgård, J. Ø. Olsen and E. H. Pedersen, Nucl. Instrum. Methods **90**, 315 (1970).

[63] J. O. Stoner, Jr., J. Appl. Phys. **40**, 707 (1969).

[64] J. O. Stoner, Jr. and L. J. Radzienmski, Jr., Nucl. Instrum. Methods **90**, 275 (1973).

[65] T. Andersen, K. A. Jessen, and G. Sørensen, Nucl. Instrum. Methods **90**, 41 (1970).

[66] R. M. Schectman, L. J. Curtis, C. Strecker, and K. Kormanyos, Nucl. Instrum. Methods **90**, 197 (1970).

important contribution from unresolved blends of many weaker lines from longer lived systems, or from scattered light in the detection apparatus. In such a case it may be possible to make a rough correction by subtracting background decay profiles adjacent to the line of interest. An illustration of this approach is given in Fig. 15.[67]

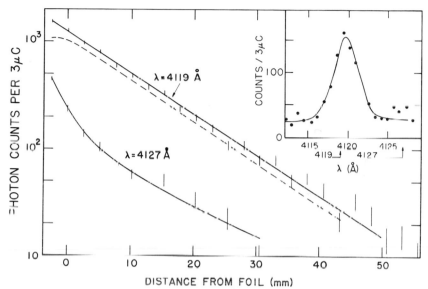

FIG. 15. Decay curve for the 4119 Å transition in Fe I at 500 keV. The curve obtained by background subtraction (dashed line) suggests nonexponential behavior at small distances. The lifetime was obtained from the slope of this curve at large distances [W. Whaling, R. B. King, and M. Martinez-Garcia, *Astrophys. J.* **158**, 389 (1969). By permission of the University of Chicago Press. Copyright 1969 by The University of Chicago.]

A more general problem is the feeding of the state of interest by cascading through higher excited states. The fitting of a decay curve by a sum of exponential decays attempts to take this effect into account. It is rarely useful or reliable to try to fit a curve with more than two exponential components. If the major decay is strong and much faster than levels cascading into the emitting state, then the overall effect of the slower cascades can be adequately described by a single slowly decaying component. An example is given in Fig. 16.[68] Such a procedure may, however, be dangerous unless some independent information on the size and origin of cascading contributions can be made, for example, from the strengths in the beam-foil spectrum of the

[67] W. Whaling, R. B. King, and M. Martinez-Garcia, *Astrophys. J.* **158**, 389 (1969).
[68] T. Andersen, O. H. Madsen, and G. Sørensen, *Physica Scripta* **6**, 125 (1972).

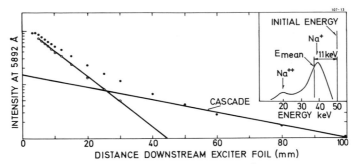

FIG. 16.   The decay downstream of the exciter foil for 5890 Å in Na I with a 50-keV Na ion beam, 5-μg/cm^2 carbon foil. The inset shows the energy distribution of post-foil particles. A cascade contribution to the decay curve is evident [T. Andersen, O. H. Madsen, and G. Sørensen, *Physica Script* **6**, 125 (1972)].

lines accompanying the feeding. A thorough analysis of this problem has been presented by Curtis *et al.*[69] who show that major errors in extracting mean lives may occur especially when one attempts to fit with a two-exponential decay a curve which in fact is composed of more than two strong components. They show that, by constraining the fit to be consistent with measured information on the strengths of the feeding cascades, a great gain in accuracy of the extracted mean lives may result. A similar approach has been used by Carre *et al.*[70] to analyze the decay schemes of 3s–3p transitions in Na II. They show that errors nearly as large as a factor of 2 may be made by the simpler two-exponential fit treatment.

Curtis *et al.*[71] proposed the use of a quantity which they call the *replenishment ratio R* as a figure of merit indicating the importance of cascading to a decay curve. The ratio R, a function of time after excitation, is defined as the ratio between the rate of feeding of the emitting level through cascades and the rate of its depletion by spontaneous decays; thus small R means little cascading. They further propose a treatment of observed decay curves which allows exact corrections for cascading and extraction of R if decay profiles for the feeding transitions can be measured. The method avoids the need to know the relative intensities of the cascading and principal transitions. Such a treatment is applied to several transitions in Si II–Si IV,[72] and in several cases shows considerable disagreement with the results from simple curve fitting of the same data. It is thus, in general, important to pay close attention

[69] L. J. Curtis, R. M. Schectman, J. L. Kohl, D. A. Chojnacki, and D. R. Schoffstall, *Nucl. Instrum. Methods* **90**, 207 (1970).

[70] M. Carré, M. L. Gaillard, and J. L. Subtil, *Nucl. Instrum. Methods* **90**, 217 (1970).

[71] L. J. Curtis, J. G. Berry, and J. Bromander, *Physica Scripta* **2**, 216 (1970).

[72] H. G. Berry, J. Bromander, L. J. Curtis, and R. Buchta, *Physica Scripta* **3**, 125 (1971).

to the energy-level network of the system being studied and to the intensities of observed spectral lines which accompany cascading in order to assess the importance of such effects in a particular situation.

Since the excitation process is highly nonselective in populating excited states, there is no consistent way to avoid cascading in normal beam-foil experiments. Several methods have, however, been suggested to produce cascade-unperturbed decay curves. Andrä et al.[73] have used a laser beam for resonant excitation of the $6\,^2S_{1/2} \rightarrow 6\,^2P_{3/2}$ transition in a 280-keV beam of Ba^+. The motion of the Ba^+ ions was used to allow Doppler tuning of the resonant excitation ($\lambda = 4554$ Å). Since only the $6\,^2P_{3/2}$ level was excited, its subsequent decay was entirely cascade free yielding a lifetime of 6.21 ± 0.06 nsec; a precision well beyond that possible in most beam-foil experiments. Using a beam-foil light source, Masterson and Stoner[74] have detected coincidences between photons feeding transitions in O II and those from subsequent decays. Such a process avoids cascades entirely, but the coincidence rates obtained are discouragingly small. Church and Liu[75] have measured the mean lives of a number of states in O II by subtracting decay profiles with and without a magnetic field B present. The resulting quantum beat curve (see Section 5.1.5) depends only on the mean life τ and the g-factor for the level. If alignment transfer by cascades is negligible, a far weaker requirement than that cascades be absent altogether, τ can be extracted unambiguously in terms of g and B. The precision of the method for three levels of O II is roughly 2–3% and compares favorably with results of normal beam-foil experiments. A somewhat similar method for excluding cascades by exploiting the polarization of light emitted by the primary decay from aligned states is discussed by Berry et al.[76]

5.1.4.2. Results. A survey of the NBS bibliographies[77] on atomic transition probabilities shows the enormous impact of bfs on this area of our literature. By the 1969 compilation nearly half of the experimental references for spectra of ionized systems are bfs measurements; there are very few entries at all above the first ionization state. The 1973 supplement shows the great majority of experimental entries for ionized systems are bfs and many of these are above the first ionization state.

Much early bfs work concentrated on lifetime measurements in all ionization states of low Z systems. Since the bfs method allows easy study of ionized systems, it has contributed much data that allow one to follow a

[73] H. J. Andrä, A. Gaupp, and W. Wittman, *Phys. Rev. Lett.* **31**, 501 (1973).

[74] K. D. Masterson and J. O. Stoner, Jr., *Nucl. Instrum. Methods* **110**, 441 (1973).

[75] D. A. Church and C. H. Liu, *Nucl. Instrum. Methods* **110**, 147 (1973).

[76] H. G. Berry, L. J. Curtis, and J. L. Subtil, *J. Opt. Soc. Amer.* **62**, 771 (1972).

[77] "Bibliography on Atomic Transition Probabilities." Nat. Bur. Std. Spec. Publ. 320, U.S. Govt. Printing Office, Washington, D.C., 1970; Suppl. 2, 1973.

particular transition along its isoelectronic sequence over an appreciable range in Z. A summary of systematic trends of oscillator strengths along isoelectronic sequences is given by Smith and Wiese,[78] and several papers[79] discuss developments taking place since this compilation appeared. Such presentations are useful for establishing the reliability of f-values on the basis of their consistency with the trends, and for the deduction by interpolation of f-values not directly measured.

In Fig. 17 we show a plot of oscillator strengths for the $(2s2p^2)^2D \rightarrow (2s^22p)^2P^o$ transition in the boron sequence versus $1/Z$.[78] The oscillator strength must go to zero for large Z since there is no change in principal quantum number. Four of the six experimental points are from beam-foil

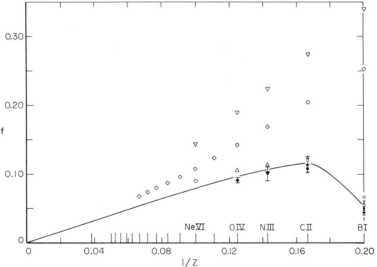

FIG. 17. Oscillator strengths versus $1/Z$ for the $2s^22p$ $^2P^o$–$2s2p^2$ ^2D transition in the B I sequence. Theoretical values are from: A. W. Weiss (○)—self-consistent field calculation with configuration interaction;(▽)—without configuration interaction [A. W. Weiss, *Phys. Rev.* **188**, 119 (1969)]; (△)—P. Westhaus and O. Sinanoglu, *Astrophys. J.* **157**, 997 (1969); (◊)—M. Cohen and A. Dalgarno, *Proc. Roy. Soc. London A* **286**, 510 (1964); (▲)—G. M. Lawrence and B. D. Savage, *Phys. Rev.* **141**, 67 (1966); (■)—A. Hese and H. P. Weise, *Z. Phys.* **215**, 95 (1968); (●)—J. Bromander, R. Buchta, and L. Lundin, *Phys. Lett.* **29A**, 523 (1969); (▼)—L. Heroux, *Phys. Rev.* **153**, 156 (1967); (♦)—W. S. Bickle, *Phys. Rev.* **162**, 7 (1967); (X)—I. Bergström, J. Bromander, R. Buchta, L. Lundin, and I. Martinson, *Phys. Lett* **28A**, 721 (1969). Experimental values of Bromander *et al.*, Heroux, Bickel, and Bergström *et al.* are bfs. [After M. Smith and W. Wiese, *Astrophys. J. Suppl. Ser.* **23**, 103 (1971). By permission of the University of Chicago Press. Copyright 1971 by The University of Chicago.]

[78] M. W. Smith and W. L. Wiese, *Astrophys. J. Suppl. Ser.* **23**, 103 (1971).
[79] M. W. Smith, G. A. Martin, and W. L. Wiese, *Nucl. Instrum. Methods* **110**, 219 (1973), and references cited therein.

work. Only the theoretical values of Weiss[80] and Westhaus and Sinanoğlu[81] come down to the experimental values for lower Z. These calculations differ from the other ones in that a careful handling of configuration interactions were taken into account, indicating the importance of this feature in this case.

Similar systematic trends may be followed along spectral series, as shown for example in Fig. 18. Here the oscillator strengths for the $(3snf)^3F \rightarrow$ $(3s3d)^3D$ transition in Al II has been plotted versus the effective principal

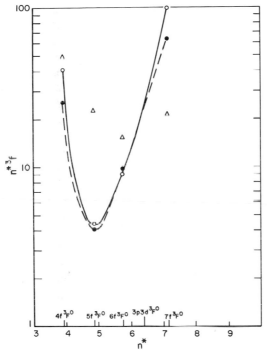

FIG. 18. Graph of $(n^* \, ^3f)$ versus n^* for the $3s3d \, ^3D$–$3snf \, ^3F^0$ series of Al II. The calculations are by A. W. Weiss: (○)—SCF with configuration interaction; (△)—SCF without configuration interaction. The experimental data (●) are from T. Andersen, J. R. Roberts, and G. Sørensen, *Physica Scripta* **4**, 52 (1971) for beam-foil with branching ratios from Weiss (○).

quantum number n^*. Since for high n^* the f-values should become hydrogenic and $n \, ^3f \rightarrow$ constant, the quantity $n^* \, ^3f$ is plotted. The calculations by Weiss including configuration interactions are seen to be in excellent agreement with the beam-foil results of Andersen *et al.*,[82] while the single configuration calculations are badly in error.

[80] A. W. Weiss, *Phys. Rev.* **188**, 119 (1969).
[81] P. Westhaus and O. Sinanoglu, *Astrophys. J.* **157**, 997 (1969).
[82] T. Andersen, J. R. Roberts, and G. Sørensen, *Physica Scripta* **4**, 52 (1971).

Lifetime work done on single-ended Van de Graaffs has been largely restricted to those elements that can be easily introduced in gaseous compounds into an rf ion source. The realization that a great deal of beam-foil work even on ionized spectra could be done at very low bombarding energies has resulted in heavy use of universal ion sources on accelerators without pressure vessels, typically $\lesssim 600$ kV accelerating potential. The versatility of universal ion sources and the convenience of not having them inside a pressure vessel has expanded the accessible range of elements to nearly the entire periodic table, and has resulted in a great deal of low-energy work for both light and heavy ions. The University of Aarhus and the Stockholm Research Institute for Physics have been especially productive in this respect. T. Andersen and G. Sorensen at Aarhus have extensively examined some of the experimental problems associated with foil energy loss and beam scattering at low energies and have concluded that much reliable lifetime work can be done at energies as low as 50 keV (for Na) and ions as heavy as Bi (200–800 keV).[62] Figure 19 shows a summary of an Aarhus study of trends

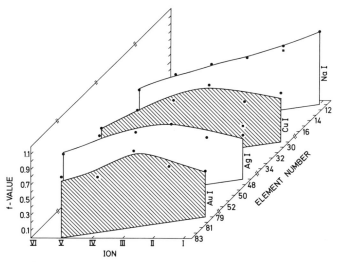

FIG. 19.   The f-values for lowest-lying ns ^{2}S–np ^{2}P transitions in the Na I, Cu I, Ag I, and Au I sequences [T. Andersen, A. K. Nielsen and G. Sørensen, *Nucl. Instrum. Methods* **110**, 143 (1973)].

both along isoelectronic sequences and for homologous transitions (i.e., similar transitions built around different closed shells) for the np ^{2}P $\rightarrow$ ns ^{2}S transition in Na I, Cu I, Ag I, and Au I sequences.[83] Of particular note is the work on the transition metals done both at Stockholm and Aarhus (see

[83] T. Andersen, A. K. Nielsen, and G. Sørensen, *Nucl. Instrum. Methods* **110**, 143 (1973).

next section) and extensions to the rare earths.[62] The reader is referred to the proceedings of the second[15] and third[16] international beam-foil conferences for further summaries of these results.

5.1.4.3. Astrophysical Applications. It has often been suggested that bfs is capable of making great contributions to astrophysical spectroscopic needs. While one might expect that the greatest value of bfs would be in the area of multiply ionized systems, in fact the most important concrete contributions of bfs to astrophysics has been in the determination of absolute f-value scales for the first and second spectra of the transition metals. The solar abundances of these elements are of particular astrophysical interest because of their nucleosynthesis in equilibrium processes in stellar interiors and their resulting large abundances in the universe. The latter characteristic makes them (especially Fe) important contributors to the solar opacity.

The deduction of the abundances of these elements from the strengths of their absorption lines in the solar photosphere, where neutral and singly ionized systems dominate, relies on having good laboratory values for the relevant oscillator strengths. A particularly heavily used source for these f-values has been the extensive work of Corliss and Bozman (CB).[84] Over the last few years a renewed interest in checking these and other previously accepted f-values for the transition elements has arisen. Serious errors in the CB values as well as in those of other workers have been found.

Beam-foil spectra of low-energy beams of these elements are dominated by the strongest lines of the neutral and singly ionized systems. Of these perhaps only a dozen or fewer will be satisfactory, in the sense that cascades and blending can be controlled, for mean-life measurements. If the lifetimes are supplemented by independent measurements of the branching ratios from the upper state to all lower states, many oscillator strengths may be deduced from a single mean lifetime. Further, f-values for the weak transitions, which are particularly useful in abundance analyses because the corresponding absorption lines are not saturated in the photospheric spectra, may be deduced.

The most important single element is iron, for which both mean lifetime and branching ratio work has been done at the California Institute of Technology. Both Fe I[67, 85] and Fe II[86] have been studied. The results in Fe I suggested that the previously accepted f-value scale for this spectrum was too high by roughly an order of magnitude and thus that the solar iron

[84] C. H. Corliss and W. R. Bozman, Nat. Bur. Std. Monograph 53, U.S. Govt. Printing Office, Washington, D.C., 1962.

[85] W. Whaling, M. Martinez-Garcia, D. L. Mickey, and G. M. Lawrence, *Nucl. Instrum. Methods* **90**, 363 (1970); M. Martinez-Garcia, W. Whaling, and D. L. Mickey, *Astrophys. J.* **165**, 213 (1971); T. Andersen and G. Sørensen, *Astrophys. Lett.* **8**, 39 (1971).

[86] P. L. Smith, W. Whaling, and D. L. Mickey, *Nucl. Instrum. Methods* **90**, 47 (1970); P. L. Smith and W. Whaling, *Astrophys. J.* **183**, 313 (1973).

abundance was really ten times that previously accepted. These results agree with those of several other workers,[87] as discussed by Bridges and Wiese.[88] Extension of this work to mean life measurements in Ti I and II,[89] V I and II,[89] Cr I–II,[90, 91] Ni I,[92, 93] Mn I–III,[56] Co I–III,[91] and Sc I–III[94] have now been reported. For Cr,[95] Ti,[89] and Ni[93] branching ratios have been measured. The CB f-values appear to contain serious normalization errors which vary both from one element to the next. The use of these and other recent f-value data have been used to perform much more reliable solar elemental abundance calculations than had previously been possible. For example, the Cr abundance was revised up by a factor of 5[95]; that for Ni by a factor of 2[93]; for Mn by a factor of 3.[96]

The contribution of bfs to the abundance determination problem has not been limited to the transition metals. Work on the rare earth region has been begun. Results of Curtis et al.[96] in Th II suggest a revision downward of the solar abundance by a factor of 2.4, for example. A review of specific applications of bfs to these problems is given by Smith.[97]

5.1.5. Quantum Beats

Because of spectral resolution limitations discussed in Section 5.1.3.1 it is not possible to study fine and hyperfine level structure spectroscopically using the beam-foil radiation source. Even if the initial "state" of an observed line is in fact a manifold of substates whose degeneracy is slightly removed under the Hamiltonian appropriate to the system beyond the foil, no line structure will be seen because of the low resolution of the experiment. Since, however, the time characterizing the excitation "pulse" is at least less than the foil traversal time, $\sim 10^{-14}$ sec, the various states of the manifold may be excited with well-defined relative phases. Under such conditions interference between coherent radiation from these states to a common final state may cause the line observed in low resolution to display variations in observed intensity, or "quantum beats," as a function of time after excitation.

[87] M. C. E. Huber and W. H. Parkinson, *Astrophys. J.* **156**, 1153 (1969), and references cited therein.

[88] J. M. Bridges and W. L. Wiese, *Astrophys. J.* **161**, L71 (1970).

[89] J. R. Roberts, T. Andersen, and G. Sørensen, *Astrophys. J.* **181**, 313 (1973).

[90] C. L. Cocke, B. Curnutte, and J. Brand, *Astron. Astrophys.* **15**, 299 (1971).

[91] E. H. Pinnington, H. O. Lutz, and G. W. Carriveau, *Nucl. Instrum. Methods* **110**, 55 (1973).

[92] J. H. Brand, C. L. Cocke, and B. Curnutte, *Nucl. Instrum. Methods* **110**, 127 (1973).

[93] W. N. Lennard, W. Whaling, R. N. Sills, and W. A. Zajc, *Nucl. Instrum. Methods* **110**, 385 (1973); W. N. Lennard, W. Whaling, J. M. Scalo, and L. Testerman, *Astrophys. J.* **197**, 517 (1975).

[94] R. Buchta, L. J. Curtis, I. Martinson, and J. Broyozowski, *Physica Scripta* **4**, 55 (1971).

[95] C. L. Cocke, A. Stark, and J. C. Evans, *Astrophys. J.* **184**, (1973).

[96] L. J. Curtis, I. Martinson, and R. Buchta, *Nucl. Instrum. Methods* **110**, 391 (1973).

[97] P. L. Smith, *Nucl. Instrum. Methods* **110**, 395 (1973).

Due to the rapid motion of the beam particles, a practically realizable spatial resolution of 10^{-4} m will give a time resolution of the order of 10^{-11} sec, which is sufficiently fast to allow studies of beat frequencies between levels split by fine and hyperfine interactions, as well as those split by externally applied electric and magnetic fields.

Quantum beats have been used to study the foil-excitation process, to measure g-factors and hyperfine coupling constants. We refer the reader to the recent review article of Andrä[17] for more detail than will appear here.

5.1.5.1. Stark Beats. In hydrogenlike systems states of opposite parity lie sufficiently close that beats may be observed between states which are Stark-mixed in an external electric field. The earliest observation of quantum beats in a foil excited light source was reported by Bashkin et al.,[9] who found oscillations on the decay curves from states with $n = 4–7$ in a foil-excited hydrogen beam when the decay proceeded in the presence of an electric field. They attributed the pattern to Stark beating between the ns and np states in hydrogen.

The simplest analysis of the time evolution of a foil-excited system consists in expanding the state prepared by the foil in terms of energy eigenstates of the system beyond the foil, noting that in general the amplitudes in such an expansion bear well-defined phase relations. If this coherence persists after averaging over many systems prepared by foil excitation, quantum beats may appear.

For the sake of illustration we follow a two-state analysis of the time evolution of a foil-excited $n = 2$ hydrogen system with $J = \frac{1}{2}$. Let us suppose that the foil populates only the $|2s\rangle$ state of hydrogen. The geometry of the experiment is shown in Fig. 20. We need consider only the $|2s\rangle$ and $|2p\rangle$

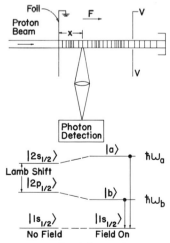

FIG. 20.   Stark beats schematic.

states. In the presence of an electric field F parallel to the beam the $|2s\rangle$ state is no longer an eigenstate of the full atomic Hamiltonian $\hat{H}$ but may be expanded in terms of the eigenstates $|a\rangle$ and $|b\rangle$ of $\hat{H}$. At $t = 0$, the moment of exit from the foil,

$$|2s\rangle = |a\rangle\langle a|2s\rangle + |b\rangle\langle b|2s\rangle. \qquad (5.1.1)$$

The coefficients $\langle 2s|a\rangle$, etc. are determined by the usual Stark mixing calculations.[98] The states $|a\rangle$ and $|b\rangle$ can also be expressed in terms of eigenstates of the field-free Hamiltonian by

$$
\begin{aligned}
|a\rangle &= |2s\rangle\langle 2s|a\rangle + |2p\rangle\langle 2p|a\rangle, \\
|b\rangle &= |2s\rangle\langle 2s|b\rangle + |2p\rangle\langle 2p|b\rangle.
\end{aligned}
\qquad (5.1.2)
$$

The state of the system $|\Psi(t)\rangle$ can now be found by allowing the states $|a\rangle$ and $|b\rangle$ in Eq. (5.1.1) to pursue their harmonic time dependences, damped by spontaneous decay constants γ_a and γ_b, respectively:

$$|\Psi(t)\rangle = |a\rangle\langle a|2s\rangle e^{-i\omega_a t - \gamma_a t/2} + |b\rangle\langle b|2s\rangle e^{-i\omega_b t - \gamma_b t/2} \qquad (5.1.3)$$

or, by using (5.1.2)

$$
\begin{aligned}
|\Psi(t)\rangle = &(|\langle a|2s\rangle|^2 e^{-i\omega_a t - \gamma_a t/2} + |\langle b|2s\rangle|^2 e^{-i\omega_b t - \gamma_b t/2})|2s\rangle \\
&+ (\langle 2p|a\rangle\langle a|2s\rangle e^{-i\omega_a t - \gamma_a t/2} + \langle 2p|b\rangle\langle b|2s\rangle e^{-i\omega_b t - \gamma_b t/2})|2p\rangle.
\end{aligned}
\qquad (5.1.4)
$$

The intensity of electric dipole radiation to the $|1s\rangle$ of hydrogen is then proportional to

$$
\begin{aligned}
|\langle 1s|\hat{E}1|\Psi(t)\rangle|^2 = |\langle 1s|\hat{E}1|2p\rangle|^2 \{&A^2 e^{-\gamma_a t} + B^2 e^{-\gamma_b t} \\
&+ 2AB e^{(\gamma_a + \gamma_b)t/2} \cos(\omega_a - \omega_b)t\},
\end{aligned}
\qquad (5.1.5)
$$

where $A = \langle 2p|a\rangle\langle a|2s\rangle$, $B = \langle 2p|b\rangle\langle b|2s\rangle$ (for A and B real) and $\hat{E}1$ is the appropriate electric dipole operator which will not connect $|1s\rangle$ and $|2s\rangle$ states. The beats, described by the third term of Eq. (5.1.5) may be thought of as arising from the oscillatory amplitude of the $|2p\rangle$ state in Eq. (5.1.4). Observed intensity peaks correspond to maxima in the intensity of this strongly radiating part of $|\Psi(t)\rangle$.

For the case of Ly_α radiation in hydrogen the two-level approximation is useful. When viewed with intermediate spatial resolution the perturbed $2s_{1/2}$–$2p_{1/2}$ splitting is responsible for the dominant beat frequency. The $2p_{3/2}$ state is separated too far in frequency for easy observation of either Stark-

[98] H. A. Bethe and E. E. Saltpeter, "Quantum Mechanics of One- and Two-Electron Atoms." Academic Press, New York, 1957; G. Lüders, *Ann. Phys.* **8**, 20 (1951).

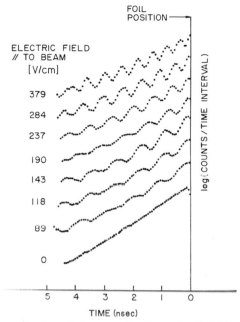

FIG. 21. Plots of $\log_{10}$ (Ly_α intensity) versus time in an electric field parallel to the beam [H. J. Andrä, *Phys. Rev. A* **2**, 2200 (1970)].

induced or field-free beats with the lower sublevels. Several authors[99–101] have studied this transition in the presence of either a static electric field or that induced by the motion of the atom through a magnetic field. The results of Andrä are shown in Fig. 21. The observed beat frequencies are in excellent agreement with those calculated from diagonalizing the Stark-perturbed hydrogen atom Hamiltonian[98] in the basis of $n = 2$ states. The phase of the beat pattern further gives information on the relative cross sections for formation of the 2s(σ_s) and 2p(σ_p) states at the foil. The maximum in the observed intensity at $t = 0$ indicates that the 2p state is populated more strongly than the 2s; Andrä finds $\sigma_p/\sigma_s = 4 \pm 1$ at 200 keV equivalent proton energy.[99]

For higher n the situation becomes rapidly very difficult to analyze because of the large number of levels which may be mixed. Weak field calculations are not applicable and one must do a full diagonalization of the perturbed Hamiltonian for each n.[98] Two-state mixings are not adequate as the eigenstates in the presence of the electric field generally contain large amplitudes

[99] H. J. Andrä, *Phys. Rev. A* **2**, 2200 (1970).
[100] W. S. Bickel, *J. Opt. Soc. Amer.* **58**, 213 (1968).
[101] I. A. Sellin, C. D. Moak, P. M. Griffin, and J. A. Biggerstaff, *Phys. Rev.* **188**, 217 (1969).

of states with several different l-values when expanded in the unperturbed-atom basis. Early observations of beats for high n suggested that the beat pattern might be so complex as to defy analysis.[100, 102] A correct analysis of the problem involves not only finding the perturbed eigenstates and energies but choosing initial conditions for the time dependent expansion of $|\Psi(t)\rangle$ which characterize the foil-excitation mechanism. Sellin et al.,[103] in a study of the Balmer series in hydrogen, argued that population of d-states might dominate the observed decay to the 2p. By following this suggestion with a rather complete analysis of decaying systems for $n = 3$–10 they were able to account for the major frequency components in Stark-perturbed H I and He II. Alguard and Drake[104] have studied Stark beats in both Ly_α and Ly_β, and have found the interesting result that population of s-states dominates that of p-states for $n = 2$–3 and proton energies between 150 and 508 keV, just the opposite of the results of Andrä[99] and Sellin et al.[101] Similar studies in He II and Li III have been carried out by Pinnington et al.[105]

Beating patterns discussed above will arise if nearly degenerate states of the same n but different l are unequally populated by the foil excitation. No true foil-produced coherence between states of different l is necessary. Using the beam direction as an axis of quantization, a diagonal density matrix representing the excited system is adequate to produce beats provided diagonal elements for different l are not all equal. It is certainly possible that nondiagonal elements might be present as well, however, if states of different l are excited with well-defined relative phases. Coherent population of states with different parity would correspond to the emergence of the atom from the foil with a nonzero electric dipole moment aligned along the beam axis. Burns and Hancock[106] analyzed the decay curve of H_β which displayed zero-field quantum beats and deduced a coherence between s and d states in the initial system. Eck[107] suggested a particularly simple experiment for detecting 2s–2p coherence from the Stark-perturbed decay of Ly_α. If the $n = 2$ part of the foil-excited hydrogen atom state $|\Psi(t = 0)\rangle$ is expanded in terms of the basis $|l, m\rangle$ with amplitudes $f_{lm} = |f_{lm}|e^{i\alpha_{lm}}$, Eck found that the beat signal between the $2p_{1/2}$ and $2s_{1/2}$ states in the presence of an electric field parallel to the beam is given by

$$(V/\omega)^2[\tfrac{1}{3}(\sigma_{p0} + 2\sigma_{p1}) - \sigma_s] \cos \omega t + (V\omega_0/\omega^2)\sqrt{\tfrac{1}{3}}\langle|f_s| \, |f_{p0}| \cos \alpha\rangle \cos \omega t$$
$$+ (V/\omega)\sqrt{\tfrac{1}{3}}\langle|f_s| \, |f_{p0}| \sin \alpha\rangle \sin \omega t.$$

[102] W. S. Bickel and S. Bashkin, *Phys. Rev.* **162**, 12 (1967).

[103] I. A. Sellin, C. D. Moak, P. M. Griffin, and J. A. Biggerstaff, *Phys. Rev.* **184**, 56 (1969).

[104] M. J. Alguard and C. W. Drake, *Phys. Rev. A* **8**, 27 (1973).

[105] E. H. Pinnington, H. G. Berry, J. Desesquelles, and J. L. Subtil, *Nucl. Instrum. Methods* **110**, 315 (1973).

[106] D. J. Burns and W. H. Hancock, *Phys. Rev. Lett.* **27**, 370 (1971).

[107] T. G. Eck, *Phys. Rev. Lett.* **31**, 270 (1973).

Here ω and ω_0 are the perturbed and unperturbed $2p_{1/2}$–$2s_{1/2}$ separations, respectively, $\alpha = \alpha_s - \alpha_{po}$, $\sigma_{lm} \equiv |f_{lm}|^2$, and V is the Stark matrix element coupling the $2s_{1/2}$ and $2p_{1/2}$ states. The first term depends only on there being a nonstatistical population of the substates and is independent of the sign of V. The second and third terms are nonzero only if the collision averaged values of $|f_s|\,|f_{po}|\{{\cos\alpha \atop \sin\alpha}\}$ are nonzero, representing average excitation coherence between 2s and 2p levels. Since these terms are odd in V, subtraction of the beat signals obtained with the electric field parallel and antiparallel to the beam direction will isolate them. Experimental detection of this difference signal has been made by Sellin et al.[108] and by Gaupp et al.[109] and the results of the former group are shown in Fig. 22. Physically the result corresponds to an initial distribution of the electron cloud which either leads or lags the proton at the moment of its emergence from the foil, or whose velocity distribution will cause it to do so at a later time. Beating between the perturbed $s_{1/2}$ and $p_{1/2}$ states will cause the cloud to ring back and forth as the system progresses downstream, reversing periodically the direction of the electric dipole. The direction of the applied electric field along the dipole will determine whether it will tend toward $|2s>$ or $|2p>$ character first in the ringing process, and roughly will determine the phase of the "coherence" contribution to the beat signal. Thus the beating associated with the initial coherence may be distinguished from that due to unequal but incoherent s–p population. Sellin et al.[108] conclude that the initial dipole moment of the atom is small but that the electron cloud does not match the proton in velocity. Gaupp et al.[109] deduce the density matrix describing the excited system and find that the size of the off-diagonal s–p term is nearly as large as it can be consistent with σ_s and σ_p. They also find this term to be largely imaginary, corresponding to a large velocity asymmetry in the initial electron cloud and deduce that the cloud leads the proton by 0.2 to 0.6 Bohr radii at the foil.

5.1.5.2. Zeeman Beats. For systems other than hydrogen nearly degenerate states with opposite parity are rare. Quantum beats from interference between radiation from different magnetic substates of a system Zeeman split by an external magnetic field may, however, be observed. If the foil excitation of a system characterized by angular momentum J leaves unequal populations of different $|M_J|$ states (i.e., the system is aligned), radiation from the subsequent decay of the state will not in general be isotropic. (Symmetry of the excitation Hamiltonian about the beam axis requires equal populations of $+M_J$ and $-M_J$ states.) If the system decays while precessing in an external magnetic field, the lighthouselike sweeping

[108] I. A. Sellin, J. R. Mowat, R. S. Peterson, P. M. Griffin, R. Laubert, and H. H. Haselton, *Phys. Rev. Lett.* **31**, 1335 (1973).
[109] A. Gaupp, H. J. Andrä, and J. Macek, *Phys. Rev. Lett.* **32**, 268 (1974).

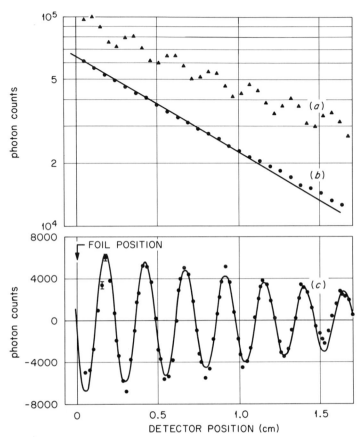

FIG. 22. Variation of Ly_α signal strength versus distance downstream at 525 and O V/cm and 186 keV beam energy. The straight line through the zero field data (b) corresponds in slope to τ_{2p}. The sum of signals for field parallel and antiparallel is given as (a), the difference of these signals as (c) [J. A. Sellin, J. R. Mowat, R. S. Peterson, P. M. Griffin, R. Laubert, and H. H. Haselton, *Phys. Rev. Lett.* **31**, 1335 (1973)].

of the radiation pattern past a detector fixed in space will result in the observation of beats. For example, we consider a p → s decay (we suppress spin coordinates) assuming only the $M_J (=m_l) = 0$ substate is populated, where the quantization axis is that of the beam (see Fig. 23). The system decays in the presence of an external magnetic field directed along the x-axis. The energy eigenstates of the system are those with well-defined angular momentum projection along the x-axis, denoted in the notation $|M_J{}^x\rangle$ as $|0\rangle$, $|+1\rangle$ and $|-1\rangle$ with eigenenergies $\hbar\omega_0$, $\hbar\omega_1$, and $\hbar\omega_{-1}$,

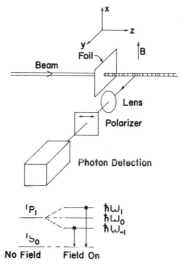

FIG. 23.   Zeeman beats schematic.

respectively. Expanding the initial state in terms of these states gives

$$|\Psi(t)\rangle = (1/\sqrt{2})|+1\rangle e^{-i\omega_1 t} + (1/\sqrt{2})|-1\rangle e^{-i\omega_{-1}t},   (5.1.6)$$

where the expansion of eigenstates of $\hat{J}_z$ in terms of those of $\hat{J}_x$ has been used. We omit damping factors due to spontaneous decay. If radiation from the system is viewed along the y-axis with polarization vector along z, the intensity I_z will be proportional to the square of the matrix element of $\hat{\varepsilon} \cdot \mathbf{r}$ (see Chapter 2.4) connecting $|\Psi\rangle$ with $|s\rangle$, or $I_z \propto |\langle s|\hat{z}|\Psi(t)\rangle|^2$. Since $|+1\rangle \propto -(y + iz)/r$ and $|-1\rangle \propto (y - iz)/r$, one obtains

$$I_z \propto (I_{z0}/4)|e^{-i\omega_1 t} + e^{-i\omega_{-1}t}|^2 = I_{z0} \cos^2 \tfrac{1}{2}(\omega_1 - \omega_{-1})t.   (5.1.7)$$

There will be no light emitted with polarization vector along x in this case, so no polarizer is necessary before the detector in order to see beats. Physically, this result corresponds to the precession around the x-axis of an electron distribution which initially has lobes extending along the z-axis. The beat frequency $\omega_1 - \omega_{-1}$ is related to the g-factor for the system by $\omega_1 - \omega_{-1} = 2\mu_B g B/\hbar$, where B is the magnetic field strength and μ_B the Bohr magneton. The expansion in Eq. (5.1.6) shows well-defined phases between $|+1\rangle$ and $|-1\rangle$ amplitudes. This coherence will persist if both states with $M_J = 0$ and $M_J = \pm 1$ systems are populated unequally and incoherently but is due entirely to the expansion of the eigenstates if $\hat{J}_z$ in terms of those of $\hat{J}_x$. The density matrix for the system, diagonal when the axis of quantization is that of the beam, may develop off-diagonal terms when the axis of quantization is chosen parallel to an external field. This is often referred to as "field-induced coherence" and is the simple consequence of the change of basis.

Experimentally, Zeeman beats may be used both to measure g-factors, as deduced from observed beat frequencies, and to determine the degree of alignment provided by the foil excitation.[110–112] The latter may be deduced from the amplitude of the beat signal. As long as the beat frequency depends linearly on B one may scan either the foil-detector distance x at fixed B or B at fixed x, the latter method eliminating the exponential damping term from the signal. Figure 24 shows an example of Zeeman beat signals obtained from the decays of the 4f 4D° in Ne II and the $2p_9$ in Ne I.[110] This data represented conclusive evidence for alignment of the initial states in the foil-excitation process and provided a previously unreported measurement of g_J for the 4f ^{4}D° state. The quantum-beat method is particularly interesting for obtaining g-values for ionized systems that are difficult to study in other light sources.

5.1.5.3. Zero-Field Quantum Beats. In an early study of Stark-induced beats in He II, Bashkin and Beauchemin[113] noted that some intensity variation persisted even at zero external field and suggested that it was caused by electric fields produced by beam-associated space charges. Macek[114] then showed that an external field is not necessary and that zero-field quantum beats are to be expected. A good physical discussion of how such beats arise is given by Andrä.[17] Since the foil-excitation time is quite short, parts of the atomic Hamiltonian responsible for fine and hyperfine structure do not have time to act during excitation. Coulomb forces should thus dominate the excitation and may selectively populate different l and $|m_l|$ but not $|m_s|$. If a system characterized by a well-defined l, m_l, and random spin orientation is expanded in states of well-defined J and M_J, which diagonalize the full Hamiltonian after excitation (ignoring for the moment hyperfine structure), the change in basis will establish well-defined phases between different J states and thus beats between unresolved fine-structure levels may be observed. Of course the excitation process will rarely populate a single l, m_l state. The coherence which attends

[110] C. H. Liu, S. Bashkin, and W. S. Bickel, *Phys. Rev.* **26**, 222 (1971).

[111] C. H. Liu and D. Church, *Phys. Lett.* **35A**, 407 (1971); D. A. Church and C. H. Liu, *Phys. Rev. A* **5**, 1031 (1972); *Physica* **67**, 90 (1973); *Nucl. Instrum. Methods* **110**, 267 (1973); M. Gaillard, M. Carré, and H. G. Berry, *Nucl. Instrum. Methods* **110**, 273 (1973); M. Druetta and A. Denis, *Nucl. Instrum. Methods* **110**, 291 (1973).

[112] T. Hadeishi, R. D. McLaughlin, and M. Michel, *J. Opt. Soc. Amer.* **61**, 653 (1971).

[113] S. Bashkin and G. Beauchemin, *Can. J. Phys.* **44**, 1603 (1966).

[114] J. Macek, *Phys. Rev. Lett.* **23**, 1 (1969).

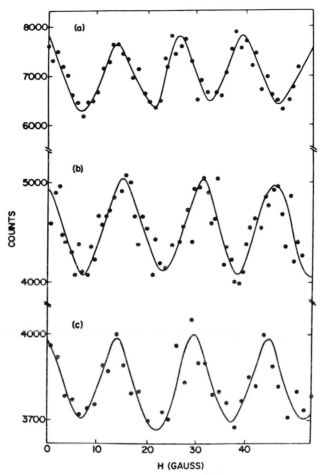

FIG. 24. Zeeman beat signals from the $2p_\nu$ level (6402 Å) (a and b) and the 4f $^4D^0$ level in Ne I (4220 Å) (c). Foil-detector distance d was fixed for each part and H scanned: (a) $d =$ 4.19 cm; (b) $d = 3.59$ cm; (c) $d = 3.59$ cm [C. H. Liu, S. Bashkin, and W. S. Bickle, *Phys. Rev.* **26**, 222 (1971)].

the basis change will, however, remain as long as states with different m_l are not populated equally.

An example of zero-field quantum beats in 3 $^3P \to$ 2 ^3S decay in ^{4}He I is shown in Fig. 25.[115] The major frequency observed corresponds to the 3 $^3P_1$–$^3P_2$ separation (658 MHz); higher spatial resolution reveals the 3 $^3P_0$–3 $^3P_2$ (8772 MHz) beats as well. From this data both the fine structure

[115] W. Wittman, K. Tillmann, and H. J. Andrä, *Nucl. Instrum. Methods* **110**, 305 (1973).

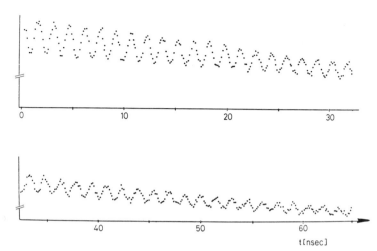

t [nsec]

FIG. 25.   Zero-field beats seen in the 3 ^{3}P → 2 ^{3}S decay of He I. The major frequency is 3 ^{3}P$_1$–^{3}P$_2$ beating (658 MHz) [W. Wittman, K. Tillmann, and H. J. Andrä, *Nucl. Instrum. Methods* **110**, 305 (1973)].

separations and the ratio between population cross sections for the $|m_l| = 0$ and 1 substates may be found.

Zero-field quantum beats have been used to determine both fine and hyperfine splittings and, from the latter, hyperfine coupling constants.[115–117] Quite complex spectra have been successfully analyzed by taking the Fourier transform of the observed beat pattern. An example of a beat pattern from the 4 ^{3}P → 3 ^{3}S transition in ^{7}Li II[117] is shown in Fig. 26 along with its Fourier transform. From the observed beat frequencies the hyperfine coupling constant was found to an accuracy of 1%.

5.1.5.4. Polarization in the Beam-Foil Source. All quantum beats in bfs rely on nonstatistical populations in the foil-excitation process, and indeed part of the interest in quantum-beat experiments has been in deducing initial population parameters from observed beat patterns. As long as the excitation geometry is symmetric about the beam axis, only alignment, i.e., equal population of $\pm M_J$ substates, in the source is possible.[118] It has

[116] H. J. Andrä, *Phys. Rev. Lett.* **25**, 325 (1970); D. J. Burns, C. W. Drake, M. J. Alguard, and C. E. Fairchild, *Phys. Rev. Lett.* **26**, 1211 (1971); P. Dobberstein, H. J. Andrä, W. Witmann, and H. H. Bukow, *Z. Phys.* **257**, 279 (1972); H. G. Berry and J. L. Subtil, *Phys. Rev. Lett.* **27**, 1103; K. Tillmann, H. J. Andrä, and W. Wittmann, *Phys. Rev. Lett.* **30**, 155 (1973); O. Poulson and J. L. Subtil, *Phys. Rev. A* **8**, 1181 (1973).

[117] H. G. Berry, J. L. Subtil, E. H. Pinnington, H. J. Andrä, W. Wittmann, and A. Gaupp, *Phys. Rev. A* **7**, 1609 (1973).

[118] U. Fano and J. Macek, *Rev. Mod. Phys.* **45**, 553 (1973).

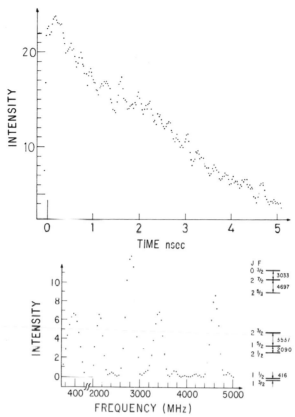

FIG. 26.   Intensity decay curve and its Fourier transform for the 3s ³S–4p ³P transition in
⁷Li II at 3684 Å [H. G. Berry, J. L. Subtil, E. H. Pinnington, H. J. Andrä, W. Wittman, and
A. Gaupp, *Phys. Rev. A* **7**, 1609 (1973)].

recently been shown that, by tilting the foil at an angle to the beam, polarization, i.e., unequal populations of $\pm M_J$ substates, of emitting states may be
obtained as well. Detection of this effect from observation of the polarization
state of light from a foil-excited $(3p)^1P_1^0$ state in ^{4}He I was first reported
by Berry *et al.*[119] An analysis of their data led them to the interesting conclusion that the electron cloud surrounding the helium nucleus leaves the
tilted foil possessing a nonzero average angular momentum about an axis
perpendicular to the beam but lying in the plane of the foil. Confirmation
of polarization of foil-excited states by tilted-foil exciters has been reported

[119] H. G. Berry, L. J. Curtis, D. G. Ellis, and R. M. Schectman, *Phys. Rev. Lett.* **32**, 751
(1974).

by Church *et al.*[120] and by Liu *et al.*,[121] who used quantum-beat signals from polarized states in He, O, and Ar to deduce the effect. These experiments have already stimulated considerable interest in understanding in more detail the foil-excitation processes,[122] although no entirely satisfactory explanation for the polarization has yet been given.[123]

5.1.6. High-Z Few-Electron Systems

While the bulk of beam-foil work has been done with single- ended Van de Graaffs and low-energy accelerators, the faster heavy ion beams available from Tandem Van de Graaffs and other higher-energy machines have proved especially useful for the production of systems with inner-shell vacancies. There are at least two assets of the higher-energy beams. First, the projectiles have a velocity comparable to that of inner-shell electrons, a situation which allows the collisional ejection of such electrons to be a probable event. Second, the outer shells of a swift projectile may be so heavily stripped by passage through the foil as to leave systems as simple as hydrogen- and heliumlike ones.

5.1.6.1. Lamb Shift Measurements. One of the classic atomic physics experiments of the century was the measurement by Lamb of the $2s_{1/2}-2p_{1/2}$ energy difference or Lamb shift S in hydrogen.[124] This energy difference is expected from quantum electrodynamics and may be thought of physically as arising from a smearing out of the electron charge due to emission and reabsorption of virtual photons. This smearing decreases the proton–electron attraction in the $2s_{1/2}$ state, allowing this state to be slightly less bound than the $2p_{1/2}$ state. The measurements of S in hydrogen, deuterium, and helium have used radio-frequency cavities to resonantly excite the (perturbed) $2s_{1/2}-2p_{1/2}$ transition,[125–129] a technique which becomes impractical for higher nuclear charge. Calculations of the Lamb shift have traditionally resulted in a double expansion in powers of $\alpha^i(Z\alpha)^j$, starting with $i = 1$ and $j = 4$.[130] Terms beyond the leading one account for less than 1% of

[120] D. A. Church, W. Kolbe, M. C. Michel, and T. Hadeishi, *Phys. Rev. Lett.* **33**, 565 (1974).

[121] C. H. Liu, S. Bashkin, and D. A. Church, *Phys. Rev. Lett.* **33**, 993 (1974).

[122] T. G. Eck, *Phys. Rev. Lett.* **33**, 1055 (1974).

[123] H. G. Berry, L. J. Curtis, and R. M. Schectman, *Phys. Lett.* **50A**, 59 (1974).

[124] W. E. Lamb, Jr. and R. C. Retherford, *Phys. Rev.* **72**, 241 (1947); **79**, 549 (1950); **81**, 222 (1951); **85**, 259 (1952); **80**, 1014 (1952).

[125] S. Triebwasser, E. S. Dayhoff, and W. E. Lamb, Jr., *Phys. Rev.* **89**, 98 (1953).

[126] R. T. Robiscoe and T. W. Shyn, *Phys. Rev. Lett.* **24**, 559 (1970).

[127] B. L. Cosens, *Phys. Rev.* **173**, 49 (1968).

[128] M. A. Narasimham and R. L. Strombotne, *Phys. Rev. A* **4**, 14 (1971).

[129] E. Lipworth and R. Novick, *Phys. Rev.* **108**, 1434 (1957).

[130] M. Leventhal, *Nucl. Instrum. Methods* **110**, 343 (1973); B. N. Taylor, W. H. Parker, and D. N. Langenberg, *Rev. Mod. Phys.* **41**, 375 (1969); T. Applequist and S. J. Brodsky, *Phys. Rev. Lett.* **24**, 562 (1970).

the shift at $Z = 1$ but almost 10% for $Z = 8$ (see Table 1). It has thus been of interest to test the calculation at higher Z in order to check the accuracy of higher terms in the expansion. Beam-foil techniques already have made possible measurements of S up to $Z = 8$.

TABLE I. Lamb Shift S

$$S = \sum_{i=1}^{\infty} \sum_{j=4}^{\infty} C_{ij}(\alpha)^i (Z\alpha)^j$$

Shift (GHz)	H	D	He	Li^{2+}	C^{5+}	O^{7+}
$C_{14}\alpha(Z\alpha)^4$	1.05055[a]	1.05135	13.82084	61.0708	733.59	1998.94
S (theory)	1.05791[a]	1.05917	14.0445	62.7391	782.87	2200.20
	1.05790[b]	1.059259	14.04478	62.7490	783.68	2205.20
S (exp.)	1.05790(6)[c]	1.05928(6)[e]	14.0454(12)[f]	63.031(327)[h]	780.1(8.0)[i]	2202.7(11.0)[j]
	1.05777(60)[d]	1.05900(6)[d]	14.0402(18)[g]			2215.6(7.5)[k]

[a] M. Leventhal, *Nucl. Instrum. Methods* **110**, 343 (1973); B. N. Taylor, W. H. Parker, and D. N. Langenberg, *Rev. Mod. Phys.* **41**, 375 (1969); T. Applequist and S. J. Brodsky, *Phys. Rev. Lett.* **24**, 562 (1970).

[b] G. W. Erickson, *Phys. Rev. Lett.* **27**, 780 (1971).

[c] R. T. Robiscoe and T. W. Shyn, *Phys. Rev. Lett.* **24**, 559 (1970).

[d] S. Triebwasser, E. S. Dayhoff, and W. E. Lamb, Jr., *Phys. Rev.* **89**, 98 (1953).

[e] B. L. Cosens, *Phys. Rev.* **173**, 49 (1968).

[f] M. A. Narasimham and R. L. Strombotne, *Phys. Rev. A* **4**, 14 (1971).

[g] E. Lipworth and R. Novick, *Phys. Rev.* **108**, 1434 (1957).

[h] C. Y. Fan, M. Garcia-Munoz, and I. A. Sellin, *Phys. Rev.* **161**, 6 (1967).

[i] D. E. Murnick, M. Leventhal, and H. W. Kugel, *Phys. Rev. Lett.* **27**, 1625 (1971).

[j] M. Leventhal, D. E. Murnick, and H. W. Kugel, *Phys. Rev. Lett.* **28**, 1609 (1972).

[k] G. P. Lawrence, C. Y. Fan, and S. Bashkin, *Phys. Rev. Lett.* **28**, 1613 (1972).

The basic technique was proposed by Fan *et al.* in 1967[131] and was used by them to measure the Lamb shift in ^{6}Li^{+2}. A Li beam is passed through a gas cell (replacing the foil exciter) where collisions leave a fraction of the beam in the $2s_{1/2}$ metastable state. Since this state is much longer lived than any other system capable of producing an eventual $n = 2 \rightarrow n = 1$ photon, it alone will survive to reach the detection region (see Fig. 27), where it passes between the plates of a parallel plate capacitor. Here a potential difference across the plates produces an electric field F. If the $2s_{1/2}$ state enters the field gently, the attendant adiabatic transition will leave the system in an eigenstate $|a\rangle$ of the Stark-perturbed Hamiltonian. This situation is to be compared with the sudden transitions which give rise to the quantum beats of Section 5.1.5.1. In first-order perturbation theory this state will be given by $|a\rangle = |2s_{1/2}\rangle + \alpha|2p_{1/2}\rangle$ where

$$\alpha = \langle 2p_{1/2}|e\mathbf{F}\cdot\mathbf{r}|2s_{1/2}\rangle/S = \sqrt{3}eFa_0/ZS. \qquad (5.1.8)$$

[131] C. Y. Fan, M. Garcia-Munoz, and I. A. Sellin, *Phys. Rev.* **161**, 6 (1967).

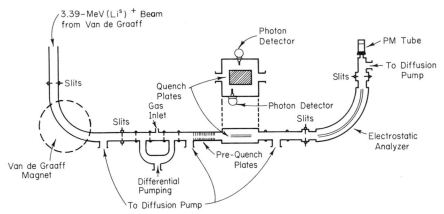

FIG. 27.   Apparatus used for Li Lamb shift experiment [C. Y. Fan, M. Garcia-Munoz, and
I. A. Sellin, *Phys. Rev.* **161**, 6 (1967)].

The state $|a\rangle$ will now decay to the $1s_{1/2}$ state via its $2p_{1/2}$ component with
a lifetime given by

$$\tau(a) = \tau(2p_{1/2})Z^2S^2/3e^2F^2a_0^2. \qquad (5.1.9)$$

The experiment consists in measuring $\tau(a)$ by the usual beam-foil method
using the quench radiation from the $|a\rangle \rightarrow 1s_{1/2}$ decay (135 Å in $^6Li^{2+}$).
Since all other quantities on the right-hand side of equation (5.1.9) may be
either measured or calculated, a value for S may be deduced. A more thorough
analysis including mixing with the $2p_{3/2}$ state and higher-order terms in
the perturbation expansion is found in Fan *et al.*[131] Sample decay curves
for two electric fields are shown in Fig. 28, from which a value for S of
$63,031 \pm 327$ MHz was deduced. The theoretical value is 62,749 MHz.[132]
   Similar measurements have now been made for $^{12}C^{5+}$ by Murnick *et al.*[133]
and $^{16}O^{7+}$ by Leventhal *et al.*[134] and by Lawrence *et al.*[135] (see Table I).
As one goes to higher Z several modifications to the above technique must
be made. The transition energies go into the x-ray region requiring detectors
such as proportional counters. The electric fields needed to appreciably
quench the $2s_{1/2}$ state become greater, the decay length scaling roughly as
Z^6/F^2 for fixed ion velocity. It becomes more convenient to use the motional
electric field seen by the ion moving through an applied magnetic field.
Beam-deflection problems which complicate the geometry of the spatial
distribution of decaying systems become important. It is nevertheless of

[131] G. W. Erickson, *Phys. Rev. Lett.* **27**, 780 (1971).
[132] D. E. Murnick, M. Leventhal, and H. W. Kugel, *Phys. Rev. Lett.* **27**, 1625 (1971).
[133] M. Leventhal, D. E. Murnick, and H. W. Kugel, *Phys. Rev. Lett.* **28**, 1609 (1972).
[134] G. P. Lawrence, C. Y. Fan, and S. Bashkin, *Phys. Rev. Lett.* **28**, 1613 (1972).

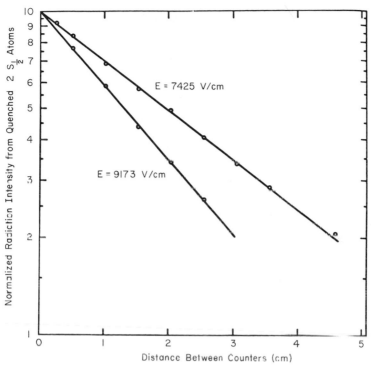

FIG. 28.   Decay curves for Li($2s_{1/2}$) in two different electric field strengths [C. Y. Fan, M. Garcia-Munoz, and I. A. Sellin, *Phys. Rev.* **161**, 6 (1967)].

interest and feasible to extend this technique to higher Z with the use of very large magnets and very-high-velocity beams.

An interesting alternative technique has been suggested for using foil-produced beams of $2s_{1/2}$ states in high-Z systems to measure the Lamb shift. There exist now powerful lasers in wavelength regions suitable for resonantly exciting the $2s_{1/2} \rightarrow 2p_{3/2}$ transition in certain high-Z ions. Since the $2p_{1/2}$–$2p_{3/2}$ splitting is accurately known, a measurement of the former transition energy can be used to deduce the Lamb shift. Variation of the effective laser frequency may be affected by some combination of laser tuning and Doppler tuning. Promising systems appear to be $Z = 9$ and $Z = 16$.

There is to date no significant discrepancy between theoretical and experimental values for the Lamb shift. The beam-foil values are not of particularly high precision. Nevertheless the simple fact that they were done at high Z has allowed a better test of higher powers in the expansion of Table I than was afforded by the high-precision values in H and He.

We note that the theoretical approach of Erickson[132] avoids the $Z\alpha$ expansion altogether.

5.1.6.2. Forbidden Decays. The dominant transition in the spectrum from a foil-excited beam will be, roughly speaking, those whose decay lengths are well matched to the spatial resolution of the detection apparatus. Decays that are too slow will not radiate appreciably within view of the detector, while those that are too fast will proceed before the systems depart the foil region. For the case of outer-shell transitions producing lines in the visible and ultraviolet, it is a fortunate coincidence that decay lengths for allowed electric dipole transitions are typically of the order of centimeters and match the typical spatial resolutions of detectors. For inner-shell transitions whose decay energies are in the x-ray region this match is rapidly lost, however. For example, a 2p → 1s transition rate is given by $6.25 \times 10^8 Z^4$ sec^{-1} and the corresponding decay length for a 1 nsec/cm beam becomes 0.012 mm for $Z = 6$. Electron emission channels reduce the lifetimes further, with the result that the allowed deexcitation of systems with inner-shell vacancies is generally too fast to allow such systems to be detected beyond the exciting foil.

Thus the study of higher-Z systems with inner-shell vacancies has focused on systems with small numbers of electrons which have a number of metastable states. The decay rates are often hindered by many orders of magnitude and are thus well suited to beam-foil study. In particular, many forbidden decays in hydrogen- and heliumlike systems, whose rates are inconveniently slow to observe in the neutral systems, have now been observed and their rates measured for high-Z systems. Figure 29 shows a plot of theoretical predictions for these decay rates versus Z along with some of the experimental values, most of which come from beam-foil work. A summary of experimental and theoretical lifetimes in hydrogenlike and heliumlike systems is given in Table II and III, respectively.

5.1.6.2.1. HYDROGENLIKE SYSTEMS. The $2s_{1/2}$ state in hydrogenlike systems decays via emission of two electric dipole photons to the $1s_{1/2}$ state with a theoretical rate given by $A = 6.22 Z^6$ sec^{-1}.[136] There is a competing relativistically induced M1 decay which scales as Z^{10} but which becomes important only for $Z \gtrsim 45$.[137] The two-photon decay mode was observed for the first time by Marrus and Schmieder in hydrogenlike argon.[138] They foil-excited a 412-MeV Ar beam from the Berkeley HILAC and looked for coincidences between the two photons (see Fig. 30). In argon the transition energy is 3.34 keV so that the two photons which share this energy are in

[136] J. Shapiro and G. Breit, *Phys. Rev.* **113**, 179 (1959).

[137] G. W. F. Drake, *Phys. Rev. A* **3**, 908 (1971); G. Feinberg and J. Sucher, *Phys. Rev. Lett.* **26**, 681 (1971).

[138] R. Marrus and R. W. Schmieder, *Phys. Rev. A* **5**, 1160 (1972).

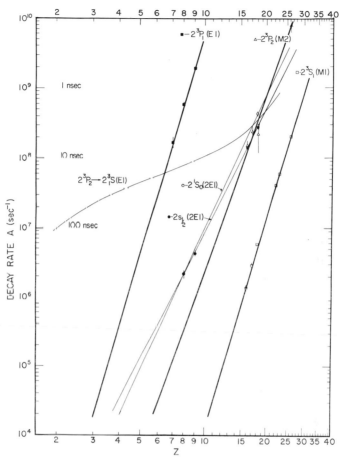

FIG. 29.   Decay rates for some forbidden decays in hydrogen- and heliumlike systems.

the soft x-ray region. The detectors used by Marrus and Schmieder were lithuim-drifted silicon detectors. The two-photon decay mode produces in either detector a continuum since the transition energy is shared between the two photons, but the sum of the coincident energy signals from the two detectors is the full transition energy, as seen in Fig. 31. By plotting the yield of continuum photons versus foil-detector distance the experimental lifetime of the $2s_{1/2}$ state in argon was found to be 3.54 ± 0.25 nsec. The theoretical value is 3.46 nsec.

The $2s_{1/2}$ state can easily be produced at lower Z by foil-exciting lower-energy beams. The photon energies quickly become awkwardly low, however, making the coincidence identification of the decay mode difficult. In an

TABLE II.   Lifetimes of the $2s_{1/2}$ State in Hydrogenlike Systems

Z	τ (exp) (nsec)	Reference	τ (theory)[f] (nsec)	Reference
1	$>2.4 \times 10^6$	[a]	121.5×10^6	[f]
2	$1.922(82) \times 10^6$	[b]	1.899×10^6	[f]
	$2.040 + 0.81 \times 10^6$	[c]		
	-0.34			
8	453(43)	[d]	463.6	[f]
9	237(17)	[d]	228.7	[f]
16	7.3(7)	[e]	7.11	[f, g]
18	3.54(25)	[e]	3.46	[f, g]

[a] W. L. Fite, R. T. Brackman, D. G. Hummer, and R. F. Stebbings, *Phys. Rev.* **116**, 363 (1959).
[b] M. H. Prior, *Phys. Rev. Lett.* **29**, 611 (1972).
[c] C. A. Kocher, J. E. Clendenin, and R. Novick, *Phys. Rev. Lett.* **29**, 615 (1972).
[d] C. L. Cocke, B. Curnutte, J. R. Macdonald, J. A. Bednar, and R. Marrus, *Phys. Rev. A* **9**, 2242 (1974).
[e] R. Marrus and R. W. Schmieder, *Phys. Rev. A* **5**, 1160 (1972).
[f] J. Shapiro and G. Breit, *Phys. Rev.* **113**, 179 (1959) (2E1).
[g] G. W. F. Drake, *Phys. Rev. A* **3**, 908 (1971); G. Feinberg and J. Sucher, *Phys. Rev. Lett.* **26**, 681 (1971) (M1).

TABLE III.   Lifetimes of Metastable Heliumlike States

State	Z	τ (exp) (nsec)	Reference	τ (theory) (nsec)	Reference
2^1S_0	2	$38(8) \times 10^6$	[a]	19.5×10^6	[o]
		$20(2) \times 10^6$	[b]		
	3	$503(26) \times 10^3$	[c]	513×10^3	[o]
	18	2.3(3)	[d]	2.53	[o]
2^3S_1	2	$3.8 \begin{Bmatrix} +10 \\ -2 \end{Bmatrix} \times 10^{12}$	[e]	7.9×10^{12}	[p]
				8.0×10^{12}	[q]
	16	706(83)	[f]	710	[p]
				698	[q]
	17	354(24)	[f]	381	[p]
				374	[q]
	18	172(12)	[g]	212	[p]
				208	[q]
	22	25.8(13)	[g]	27.4	[p]
				26.5	[q]
	23	16.9(7)	[h]	17.4	[p]
				16.8	[q]
	26	5.0(5)	[h]	5.0	[p]
2^3P_2	16	2.5(2)	[i]	2.66	[r]
	17	1.86(10)	[j]	2.01	[r]
	18	1.70(30)	[d]	1.51	[r]

TABLE III. Lifetimes of Metastable Heliumlike States (Continued)

State	Z	τ (exp) (nsec)	Reference	τ (theory) (nsec)	Reference
	23	(Four components[h])			
	26	0.11(2)	h	0.126	r
2^3P_1	7	4.2(5)	k	4.81	s
	8	1.7(3)	k	1.58	s
		1.52(11)	k		
		1.47(8)	l		
		1.47(2)	m		
	9	0.536(20)	n	0.515	s
		0.537(20)	m		

[a] A. S. Pearl, *Phys. Rev. Lett.* **24**, 703 (1970).

[b] R. S. Van Dyck, C. E. Johnson, and H. A. Shugart, *Phys. Rev. Lett.* **25**, 1403 (1970).

[c] H. H. Prior and H. A. Shugart, *Phys. Rev. Lett.* **27**, 902 (1971).

[d] R. Marrus and R. W. Schmieder, *Phys. Rev. A* **5**, 1160 (1972).

[e] H. W. Moos and J. R. Woodworth, *Phys. Rev. Lett.* **30**, 775 (1973).

[f] J. A. Bednar, C. L. Cocke, B. Curnutte, and R. Randall, *Phys. Rev. A* **11**, 460 (1975).

[g] H. Gould, R. Marrus, and R. W. Schmieder, *Phys. Rev. Lett.* **31**, 504 (1973).

[h] H. Gould, R. Marrus, and P. J. Mohr, *Phys. Rev. Lett.* **33**, 676 (1974).

[i] C. L. Cocke, B. Curnutte, and R. Randall, *Phys. Rev. A* **9**, 1823 (1974).

[j] C. L. Cocke, B. Curnutte, J. R. Macdonald, and R. Randall, *Phys. Rev. A* **9**, 57 (1974).

[k] I. A. Sellin, B. L. Donnally, and C. Y. Fan, *Phys. Rev. Lett.* **21**, 717 (1968), I. A. Sellin, M. Brown, W. W. Smith, and B. Donnally, *Phys. Rev. A* **2**, 1189 (1970).

[l] C. F. Moore, W. I. Braithwaite, and D. L. Matthews, *Phys. Lett. A* **44**, 199 (1973).

[m] P. Richard, R. L. Kauffman, F. F. Hopkins, C. W. Woods, K. A. Jamison, *Phys. Rev. Lett.* **30**, 888 (1973); *Phys. Rev. A* **8**, 2187 (1973).

[n] J. R. Mowat, J. A. Sellin, R. S. Peterson, and D. J. Pegg, *Phys. Rev. A* **8**, 145 (1973).

[o] G. W. F. Drake, G. A. Victor, and A. Dalgarno, *Phys. Rev.* **180**, 25 (1969); G. A. Victor and A. Dalgarno, *Phys. Rev. Lett.* **25**, 1105 (1967).

[p] G. W. F. Drake, *Phys. Rev. A* **3**, 908 (1971).

[q] W. R. Johnson and C. Lin, *Phys. Rev. A* **9**, 1486 (1974).

[r] G. W. F. Drake, *Astrophys. J.* **158**, 1199 (1969) (M2); H. Gould, R. Marrus, and P. J. Mohr, *Phys. Rev. Lett.* **33**, 676 (1974) (E1).

[s] W. L. Wiese, M. W. Smith, and B. M. Glennon, "Atomic Transition Probabilities," Vol. I. NSRDA–NBS.4, U.S. Govt. Printing Office Washington, D.C., 1966; G. W. F. Drake and A. Dalgarno, *Astrophys. J.* **157**, 459 (1969).

adaption of the Lamb shift experiment of Section 5.1.6.1, it is possible to Stark quench this state in a motional electric field and to use the resulting single-photon signal as a measure of the relative number of $2s_{1/2}$ states in the beam. In this way beam-foil lifetimes of this state in oxygen and fluorine have been obtained and the agreement with theory is again good.[139]

[139] C. L. Cocke, B. Curnutte, J. R. Macdonald, J. A. Bednar, and R. Marrus, *Phys. Rev. A* **9**, 2242 (1974).

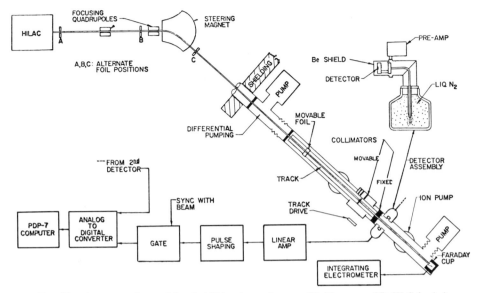

FIG. 30.   Apparatus for studying forbidden decays in argon [R. Marrus and R. W. Schmieder, *Phys. Rev. A* **5**, 1160 (1972)].

In addition to the beam-foil lifetime, experimental decay rates for the $2s_{1/2}$ state in HI[140] and He II[141, 142] have been determined and no discrepancy with theory has been found. Thus the 2 E1 decay mode for this system must be considered to be well understood.

5.1.6.2.2. HELIUMLIKE SYSTEMS. A partial level scheme for heliumlike chlorine is shown in Fig. 32 with energies taken from Doyle.[143] Only excited states with n = 2 are shown since higher n will quickly relax to these systems. There are five metastable heliumlike systems of which three may decay to the ground state via single-photon emission. The $2\,^1S_0$ state decays again via two-photon emission to the $1\,^1S_0$. Here there is no competing M1, since it is a $0 \rightarrow 0$ transition. The rate is given in the high-Z limit by $A = 16.4Z^6\ sec^{-1}$.[144] This transition was also observed by Marrus and Schmieder in heliumlike argon,[138] and an experimental lifetime for the state of 2.3 ± 0.3 nsec was found, in good agreement with the theoretical value of

[140] W. L. Fite, R. T. Brackman, D. G. Hummer, and R. F. Stebbings, *Phys. Rev.* **116**, 363 (1959).

[141] M. H. Prior, *Phys. Rev. Lett.* **29**, 611 (1972).

[142] C. A. Kocher, J. E. Clendenin, and R. Novick, *Phys. Rev. Lett.* **29**, 615 (1972).

[143] H. T. Doyle, *Advan. At. Mol. Phys.* **5**, 337 (1969); R. E. Scherr and C. W. Knight, *Rev. Mod. Phys.* **35**, 431 (1963).

[144] G. W. F. Drake, G. A. Victor, and A. Dalgarno, *Phys. Rev.* **180**, 25 (1969); G. A. Victor and A. Dalgarno, *Phys. Rev. Lett.* **25**, 1105 (1967).

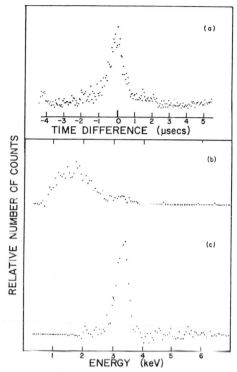

FIG. 31.   Coincidence signals from $2s_{1/2}$ decay in Ar^{17+}. (a) Distribution of time delays; (b) single Si[Li] detector energy spectrum; (c) spectrum of summed energy signals from two detectors [R. Marrus and R. W. Schmieder, *Phys. Rev. A* **5**, 1160 (1972)].

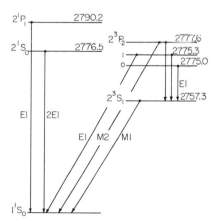

FIG. 32.   Partial level scheme for heliumlike chlorine.

2.35 nsec. There exists good agreement between the experimental rates in He I[145] and Li II[146] and corresponding theoretical values. The lifetime reported by Pearl[147] is apparently too long.

The $2\,^3P$ system has long been of interest in astrophysics. Gabriel and Jordan showed in 1969[148] that the relative strengths of the $2\,^3P \rightarrow 1\,^1S_0$ and $2\,^3S_1 \rightarrow 1\,^1S_0$ lines in the solar coronal spectrum could be used to deduce coronal electron temperatures. Although the theoretical decay rates should be expected to be reliable, several of them now have been verified experimentally using the beam-foil technique.

The $2\,^3P_1$ state decays via an allowed electric dipole decay to the 2^3S_1[149] and via spin–orbit mixing with the $2\,^1P_1$ state to the $1\,^1S_0$ ground state.[150, 151] The former decay rate scales as Z whereas the latter rate scales as Z^{10}. The K x-ray decay thus dominates for $Z \gtrsim 7$, and was first observed above helium by Sellin $et\ al.$ in N^{5+} and O^{6+}.[152] These early measurements were done by using proportional counters and a Bragg spectrometer. This decay has been studied more recently in fluorine by using a Si[Li] detector[153] in fluorine and in oxygen by using bent crystal spectrometers.[154, 155] Excellent agreement with theory is obtained.

The $2\,^3P_2$ state decays via an allowed E1 to the $2\,^3S_1$ and via a magnetic quadrupole transition to the $1\,^1S_0$. It is remarkable that the two rates could be competitive but they become nearly equal at $Z \simeq 18$. The former rate scales as Z, whereas the later scales as Z^8; for higher Z the M2 decay dominates.[156, 157] This decay has been observed in sulfur,[158] chlorine,[23] argon,[138] vanadium,[159] and iron,[159] and again the theoretical lifetimes[157, 159] are in adequate agreement with experiment.

The $2\,^3P_0$ state decays only to the $2\,^3S_1$ in helium. Hyperfine mixing may, however, allow it to emit a K x-ray as well. Effects of hyperfine mixing

[145] R. S. Van Dyck, C. E. Johnson, and H. A. Shugart, $Phys.\ Rev.\ Lett.$ **25**, 1403 (1970).

[146] M. H. Prior and H. A. Shugart, $Phys.\ Rev.\ Lett.$ **27**, 902 (1971).

[147] A. S. Pearl, $Phys.\ Rev.\ Lett.$ **24**, 703 (1970).

[148] A. H. Gabriel and C. Jordan, $Mon.\ Notices\ Roy.\ Astron.\ Soc.$ **145**, 241 (1969).

[149] W. L. Wiese, M. W. Smith, and B. M. Glennon, "Atomic Transition Probabilities," Vol. 1. NSRDA-NBS 4, U.S. Govt. Printing Office, Washington, D.C., 1966.

[150] G. W. F. Drake and A. Dalgarno, $Astrophys.\ J.$ **157**, 459 (1969).

[151] R. Elton, $Astrophys.\ J.$ **148**, 573 (1967).

[152] I. A. Sellin, B. L. Donnally, and C. Y. Fan, $Phys.\ Rev.\ Lett.$ **21**, 717 (1968); I. A. Sellin, M. Brown, W. W. Smith, and B. Donnally, $Phys.\ Rev.\ A$ **2**, 1189 (1970).

[153] J. R. Mowat, I. A. Sellin, R. S. Peterson, and D. J. Pegg, $Phys.\ Rev.\ A$ **8**, 145 (1973).

[154] P. Richard, R. L. Kauffman, F. F. Hopkins, C. W. Woods, and K. A. Jamison, $Phys.\ Rev.\ Lett.$ **30**, 888 (1973); $Phys.\ Rev.\ A$ **8**, 2187 (1973).

[155] C. F. Moore, W. J. Braithwaite, and D. L. Matthews, $Phys.\ Lett.\ A$ **44**, 199 (1973).

[156] R. H. Garstang, $Publ.\ Astron.\ Soc.\ Pac.$ **81**, 488 (1969).

[157] G. W. F. Drake, $Astrophys.\ J.$ **158**, 1199 (1969).

[158] C. L. Cocke, B. Curnutte, and R. Randall, $Phys.\ Rev.\ A$ **9**, 1823 (1974).

[159] H. Gould, R. Marrus, and P. J. Mohr, $Phys.\ Rev.\ Lett.$ **33**, 676 (1974).

on the decay of the 2 ^{3}P system in vanadium are discussed in Gould et al.[159]

The 2 ^{3}S$_1$ state is a particularly interesting one. This state was thought for years to decay via 2 E1 emission to the 1 ^{1}S$_0$ ground state until its single-photon decay was detected by Gabriel and Jordan[160] in the solar coronal spectrum for several heliumlike ions between carbon and silicon. This decay must be M1 and normally the rate would be zero because the M1 operator $\propto(\hat{\mathbf{l}} + 2\hat{\boldsymbol{\sigma}})$ has no radial dependence and thus cannot connect 2s and 1s states. It was suggested that configuration mixing of the type $(1s^2)^1S_0 \leftrightarrow (npn'p)^3P_0$ and $(1s2s)^3S_1 \leftrightarrow (npn'p)^3P_1$ might be responsible for the M1 decay. Griem[161] showed that these effects were too small by several orders of magnitude, but that finite wavelength corrections and relativistic treatment of the problem resulted in a nonzero M1 rate. A number of calculations of the rate have now been made.[162–164] Extensive calculations by Drake[162] and Johnson and Lin[163] are in essential agreement over a large range in Z. The predicted lifetime in helium is about $2\frac{1}{2}$ hr but, because of the Z^{10} scaling of the rate, comes down to 208 nsec in argon. The first measurement of the lifetime of this state was made by Marrus and Schmieder[138] using the beam foil method. They found a lifetime of 172 $\pm$ 30 nsec, and later reduced the error bar to 12 nsec,[165] indicating a clear disagreement with theory. A similar measurement in chlorine produced an experimental lifetime of 285 $\pm$ 25 nsec[166] shorter than the theoretical value of 374 nsec. Both of these experiments suffered from the problem that the decay lengths were quite long, being 9 and 5 m in the cases of argon and chlorine, respectively. The lifetimes were deduced from the slopes of the decay curves over less than one mean lifetime and, while strong arguments can be made that cascading cannot shorten the measured lifetime, the single-exponential character of the decay curves could not be verified experimentally. Measurements made at Berkeley on Ti $(Z = 22)$,[165] V $(Z = 23)$,[159] and Fe $(Z = 26)$,[159] for which the lifetimes are much shorter and the decay can be followed over more than one decay length, showed essential agreement between theory and experiment. Subsequent measurements in S $(Z = 16)$ and reexamination of chlorine, using an 8-m flight path were made by Bednar et al.[167] They found that at distances from the foil equal to or greater than one decay length the slope of the observed

[160] A. H. Gabriel and C. Jordan, Nature (London) 221, 947 (1969).

[161] H. R. Griem. Astrophys. J. 156, L103 (1969).

[162] G. W. F. Drake, Phys. Rev. A 3, 908 (1971).

[163] W. R. Johnson and C. Lin, Phys. Rev. A 9, 1486 (1974).

[164] G. Feinberg and J. Sucher, Phys. Rev. Lett. 26, 681 (1971).

[165] H. Gould, R. Marrus, and R. W. Schmieder, Phys. Rev. Lett. 31, 504 (1973).

[166] C. L. Cocke, B. Curnutte, and R. Randall, Phys. Rev. Lett. 31, 507 (1973).

[167] J. A. Bednar, C. L. Cocke, B. Curnutte, and R. Randall, Phys. Rev. A 10, (1975) (to be published).

decay curve approached that predicted by the theory. For shorter flight times an unexplained concave curvature to the decay was observed, emphasizing again the necessity for careful examination in any bfs experiment of the assumption that the exponential component corresponding to the mean life of interest has been isolated. It is likely that the discrepancy between theory and experiment found in the earlier experiments was due to this effect. It thus appears that the forbidden decay modes in the heliumlike system are all rather well understood. If interesting tests of the theory are to be made here they probably should be made at considerably higher Z than has been studied to date.

5.1.6.2.3. Systems with More Than Two Electrons. The addition of a third electron to a system with a K-vacancy immediately places the system in the continuum, since the binding energy of the last electron is less than the $n = 2 \rightarrow n = 1$ transition energy. Thus metastability is obtained only if autoionization as well as radiative decay is hindered. For example, in the lithiumlike system the (1s2s2p) and (1s2p^2) doublet states have either allowed x-ray or Auger-decay channels. The quartet states are spin forbidden to either radiate or autoionize, however, since either process requires $\Delta S = 0$. Decays from these metastable systems have been discussed in Section 5.1.3 for low Z. For higher Z spin–orbit mixing between the quartet and doublet systems becomes sufficiently strong that above $Z \sim 10$ most of the quartet states have taken on the decay characteristics of their doublet partners and become too short lived to proceed much beyond the exciting foil. The (1s2s2p)^{4}P$_{5/2}^\circ$ state is an exception to this, having no doublet counterpart, and has been extensively studied.

A great deal of the beam-foil spectroscopy that has been done on such systems has exploited the electron- rather than photon-decay channel. Sellin, Pegg et al.,[168] have taken electron spectra from the decay-in-flight of metastable lithiumlike systems for numerous elements between N and Ar. Some typical spectra are shown in Fig. 33. They have been able to identify a large number of their observed lines with the help of theoretical transition energy calculations by Holøien and Geltman[169] and by Junker and Bardsley.[170] In the cases of Cl and Ar, relativistic corrections to the transition energies are of the order of tens of electron volts and are detectable experimentally. For the most part the decay curves are complex, suggesting that the appearance of lines downstream may be due to cascading as well as to the intrinsic metastability of the emitting states.

[168] I. A. Sellin, *Nucl. Instrum. Methods* **110**, 477 (1972); D. J. Pegg et al., *Phys. Rev. A* **8**, 1350 (1973).

[169] E. Holøien and S. Geltman, *Phys. Rev.* **153**, 81 (1967).

[170] B. R. Junker and J. N. Bardsley, *Phys. Rev. A* **8**, 1345 (1973).

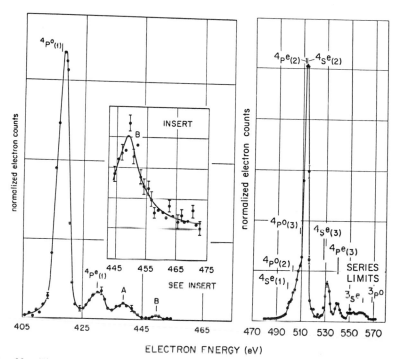

FIG. 33.   Electron spectra from decay-in-flight of foil-excited oxygen beam [D. J. Pegg *et al.*, *Phys. Rev. A* **8**, 1350 (1973)].

Lithiumlike lines appear in the x-ray spectra of foil-excited O, F, S, and Cl taken with bent-crystal spectrometers. In O and F a line presumably attributable to decay of the $(1s2s2p)^4P_{3/2,1/2}$ states to the $(1s^2 2s)^2S_{1/2}$ state appears and lifetime measurements on this line have been made by Richard *et al.*[154] and by Moore *et al.*[155] The situation is confused by uncertain indentification of the *j* of the decaying state and disagreement between workers on the lifetime.

The $(1s2s2p)^4P_{5/2}^o$ state is the longest-lived lithiumlike system with a 1s vacancy. Up to $Z \simeq 18$ this state decays predominantly via f-wave autoionization induced by spin–spin interaction terms in the Hamiltonian. There is a competing M2 radiative decay to the $(1s^2 2s)^2S_{1/2}$ state, whose rate is very nearly equal to that from the $2\,^3P_2$ heliumlike state. The radiative decay will become the larger branch above $Z \simeq 18$. The lifetime of the $^4P_{5/2}^o$ state has been studied over a large range of Z. Dmitriev *et al.*[171] measured the spontaneous charge change in foil-excited beams and attributed

[171] I. S. Dmitriev, V. S. Nikolaev, and Y. A. Teplova, *Phys. Lett.* **26A**, 122 (1969).

their result to decay from this state. Sellin, Pegg, and colleagues[168, 172] have measured its lifetime by observing the electron decay for many systems with $8 \leqslant Z \leqslant 18$. Their results are in good agreement with those of Cocke et al.[23, 158] who observed the M2 decay in sulfur and chlorine. These results were confirmed by Richard, Moore, and colleagues.[173] This last work is a good example of the high-resolution x-ray spectroscopy possible with a bent-crystal spectrometer on high-energy beams (Fig. 34). Levitt et al.[174]

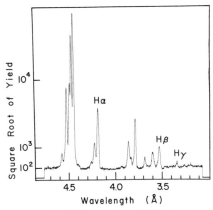

Fig. 34.   Beam-foil spectrum of chlorine taken with bent-crystal x-ray spectrometer with $E = 105$ MeV [P. Richard, private communication (1974)].

and Feldman et al.[175] measured the lifetimes in He⁻ and Li. A theoretical calculation of the lifetime of this state taking into account both radiative and autoionization channels has been made by Cheng, Lin and Johnson.[176] A comparison between their results and experiment shows rather good agreement, with small deviations appearing at large Z (Fig. 35).

The metastability achieved in low-lying heliumlike and lithiumlike systems is due largely to the very small number of final states which are energetically allowed. For example the K x-ray decay of the former and K-Auger decay of the latter system can go only to the $^1S_0$ heliumlike system. As one goes to systems with more electrons the number of decay channels grows rapidly in both density and complexity. Thus it becomes increasingly difficult to form an excited state with a K-shell vacancy whose decay is

[172] D. J. Pegg et al., Phys. Rev. A **9**, 1112 (1974); H. Haselton, R. S. Thoe, J. R. Mowat, P. M. Griffin, D. J. Pegg, and I. A. Sellin, Phys. Rev. A **11**, 468 (1975).

[173] P. Richard, private communication (1974).

[174] M. Levitt, R. Novick, and P. D. Feldman, Phys. Rev. A **3**, 130 (1971).

[175] P. D. Feldman, M. Levitt, and R. Novick, Phys. Rev. Lett. **21**, 331 (1968).

[176] K. Cheng, C. Lin, and W. R. Johnson, Phys. Lett. **48A**, 437 (1974).

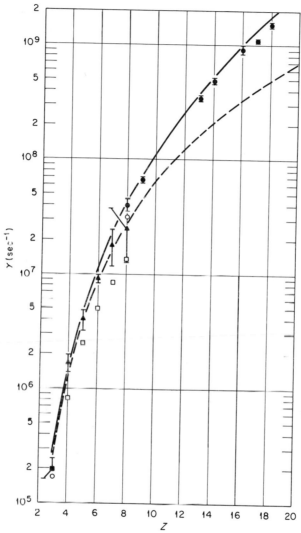

FIG. 35. Autionization rate of $(1s2s2p)^4P^o_{5/2}$ state versus Z. Data sources are: ●: H. Haselton, R. S. Thoe, J. R. Mowat, P. M. Griffin, D. J. Pegg, and I. A. Sellin, *Phys. Rev. A* **11**, 468 (1975); ▲: I. S. Dmitriev, V. S. Nikolaev, and Y. A. Teplova, *Phys. Lett.* **26A**, 122 (1969); ■: P. D. Feldman, M. Levitt, and R. Novick, *Phys. Rev. Lett.* **21**, 331 (1968); □: V. V. Balashov, V. S. Senashenko, and B. Tekou, *Phys. Lett.* **25**, 487 (1967); ○: S. T. Manson, *Phys. Lett.* **23**, 315 (1966); ––: M. Levitt, R. Novick, and P. D. Feldman, *Phys. Rev. A* **3**, 130 (1971); —: K. Cheng, C. Lin, and W. R. Johnson, *Phys. Lett.* **48A**, 437 (1974).

retarded by the several orders of magnitude necessary to leave it with a lifetime in the nanosecond region. Sellin *et al.*[168] have suggested that several lines in their delayed autoionization spectra may come from beryliumlike and higher systems, although no firm identifications are possible. Apart from cascade-fed levels, there appears to be no other evidence in delayed beam-foil spectra of metastable K x-ray emitters with more than three electrons.

Many of the metastable systems formed around a single hole in the 1s shell will possess analogs formed around single vacancies in higher shells. If the remaining shells are sufficiently emptied, one may again obtain metastable systems with inner shell vacancies whose decays may be either by electron or x-ray emission. For example, quartet sodiumlike systems with a single hole in the 2p shell and two particles in the $n = 3$ shell are analogous to the lithiumlike quartet systems discussed above. Electron decay from such systems was reported by Pegg *et al.*[44] in foil-excited chlorine, although lack of theoretical work on the level schemes for such highly ionized systems prevented identification of individual lines. Metastable M x-ray emitters from highly ionized iodine have been observed by Cocke *et al.*[177] Metastable states built on a single 3d vacancy, somewhat analogous to those built on the 1s vacancy in helium, of the type $3d^94s$, have been observed to decay via E2 radiation to the closed M shell in the presence of few spectator electrons. Again there is very little known about such highly ionized systems. Beam-foil work should prove useful in exploring such systems, emphasizing the metastable states. The allowed transitions are now receiving considerable study by using laser-produced and other very hot plasmas as light sources. One thus looks forward to a great improvement in the state of spectroscopy of highly ionized systems within the next few years.

ACKNOWLEDGMENTS

We thank T. Andersen, W. Whaling, I. Sellin, P. Richard, H. G. Berry, I. Martinson, and W. L. Wiese for providing help in assembling the manuscript and B. Curnutte for helpful comments.

[177] C. L. Cocke, S. L. Varghese, J. A. Bednar, C. P. Bhalla, B. Curnutte, R. Kauffman, P. Richard, C. Woods, and J. H. Scofield, *Phys. Rev. A* **12**, 2413 (1975).

5.2. Tunable Laser Spectroscopy*

5.2.1. Introduction

Frequency-tunable lasers, the classification for a new generation of optical instruments, were born in spectroscopy as children of the immediate past decade. Contrary to the present adolescence of these lasers, they are employed routinely and innovatively by spectroscopists in many investigations that provide new knowledge of the structure and of the optical properties of atoms, molecules, and condensed matter. In their maturity lasers of this type are certainly destined to be a very important subclass of the spectroscopic instruments of the future. It is appropriate then that this volume include a discussion of frequency-tunable lasers and their applications.

The first frequency-tunable laser, a Zeeman-tuned helium–neon system, was developed in 1963. Since then there has been continuously increasing activity in research and development related to tunable lasers. As a result there are now grating-tuned CO, CO_2, NO, HF, DF, and N_2O molecular gas lasers, tunable semiconductor lasers, the extremely practical dye laser, the spin–flip Raman laser, the four-wave parametric converter, and the parametric oscillator. Clever spectroscopic techniques also evolved during the decade, e.g., laser saturation spectroscopy, optoacoustic detection, laser heterodyne detection, intracavity enhanced absorption, double resonance, laser-stimulated fluorescence, and ultrahigh spectral resolution in the infrared. The wealth of scientific literature now available on these subjects would easily provide material for an entire volume. This chapter is only a fraction of the whole, however, and that singular constraint required a condensed discussion of the chosen subjects. Therefore, it is assumed that the reader possesses an understanding of the principles of optics and basic laser physics as presented collectively by several excellent books.[1–9]

[1] M. Born and E. Wolf, "Principles of Optics." Pergamon, Oxford, 1965.

[2] F. A. Jenkins and H. E. White, "Fundamentals of Optics." McGraw-Hill, New York, 1950.

[3] R. W. Ditchburn, "Light." Blackie, Glasgow and London, 1963.

[4] G. R. Fowles, "Introduction to Modern Optics." Holt, New York, 1968.

[5] B. A. Lengyel, "Lasers." Wiley, New York, 1971.

[6] A. Maitland and M. H. Dunn, "Laser Physics." North-Holland Publ., Amsterdam, 1969.

[7] W. V. Smith and P. P. Sorokin, "The Laser." McGraw-Hill, New York, 1966.

[8] A. Yariv, "Introduction to Optical Electronics." Holt, New York, 1971.

[9] H. G. Unger, "Introduction to Quantum Electronics." Pergamon, Oxford, 1970.

* Chapter 5.2 is by Marvin R. Querry.

TABLE I. Types of Frequency-Tunable Lasers and Their Optically Active Material, Ranges of Frequency Tunability, Spectral Output Width Γ, and Excitation Source ("Pump")

Type of laser	Active material	Range of tunability (cm^{-1})	Γ (cm^{-1})	"Pump"
He–Ne	Ne	vis.,[a] Ir, ± 90 GHz	< Doppler	Discharge
He–Xe	Xe	vis.,[a] Ir, ± 90 GHz	< Doppler	Discharge
N_2–CO–He	CO	1215.7–2012.7[b]	< Doppler	Discharge
N_2–CO_2–He	CO_2	907.7–2280[b]	< Doppler	Discharge
$GaAs_xP_{1-x}$	p–n junct.	11,905–15,625	$\leqslant 10^{-4}$ [c]	Current
$In_xGa_{1-x}As$	p–n junct.	3215–11,905	$\leqslant 10^{-4}$ [c]	Current
$InAs_xP_{1-x}$	p–n junct.	3215–11,110	$\leqslant 10^{-4}$ [c]	Current
$Pb_{1-x}Sn_xTe$	p–n junct.	<310–1515	$\leqslant 10^{-4}$ [c]	Current
$Pb_{1-x}Sn_xSe$	p–n junct.	$\leqslant$310–1175	$\leqslant 10^{-4}$ [c]	Current
$PbS_{1-x}Se_x$	p–n junct.	1175–2325	$\leqslant 10^{-4}$ [c]	Current
$Pb_{1-x}Ge_xTe$	p–n junct.	1665–2500	$\leqslant 10^{-4}$ [c]	Current
S–F Raman	n–InSb	1535–2000	$\sim 10^{-8}$	CO laser
S–F Raman	n–InSb	685–1110	<0.03	CO_2 laser
Organic dye	Dye molec.	13,890–27,780	$\lesssim 10^{-3}$	N_2 laser
SFM[d]	ADP	27,780–43,480	$\lesssim 10^{-3}$	Dye laser
SFM	Sr vapor	43,480–125,000	0.1	Dye laser
SFM	Te crystal	1725–1835	$\smallint$	CO_2 lasers
DFM[e]	Proustite	1770–3125	$\smallint$	Ruby and dye lasers
DFM	GaAs	5–145	$\smallint$	CO_2 lasers
DFM	$LiNbO_3$	20–190	$\smallint$	Dye lasers
DFM	$ZnGeP_2$	70–110	$\smallint$	CO_2 lasers
DFM	$AgGaSe_2$	665–1430	$\smallint$	ND:YAG ($LiNbO_3$)

[a] Tunability is ± 90 GHz about discrete frequencies of laser lines.
[b] Tunable to discrete vibrational–rotational lines.
[c] A Γ as narrow as 1.7×10^{-6} cm^{-1} has been observed for p–n junction lasers.
[d] Sum-Frequency mixing.
[e] Difference-Frequency mixing.
[f] Here, Γ is the same as for the pump laser with the narrowest spectral output width.

This chapter is divided into two sections. The first, tunable lasers, is a critique of the physical phenomena utilized in frequency tuning the individual infrared and visible lasers. Molecular lasing systems for the submillimeter region were reviewed in Volume 3 of this series.[10] The types of tunable lasers discussed here and their individual spectral regions of tunability are listed in Table I. In Section 5.2.3 spectroscopic applications are presented separately because they frequently are not unique to a single type of tunable laser, e.g., laser saturation spectroscopy is applicable to dye lasers and to spin–flip Raman lasers. The general intent of the discussion is to present a qualitative introduction to the methods of tunable laser spectroscopy along with many references to original publications.

[10] D. Williams, "Molecular Physics," Methods of Experimental Physics, Vol. 3, 2nd ed., Part A, Section 2.4. Academic Press, New York, 1974.

5.2.2. Tunable Lasers

5.2.2.1. Zeeman-Tuned He–Ne and He–Xe Lasers. During the decade 1963–1973 several spectroscopists independently constructed frequency-tunable helium–neon (He–Ne) and helium–xenon (He–Xe) lasers.[11-17] A schematic representation of a tunable laser of this type is presented in Fig. 1. The typical components of the tunable laser are as follows: The laser tube consists of a cylinderical, fused, quartz tube of about 6 mm i.d. and lengths

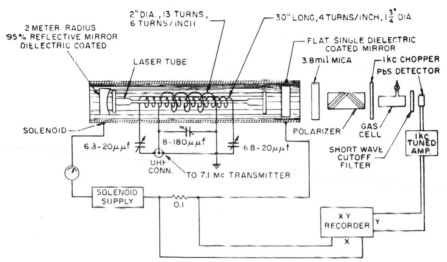

FIG. 1. Zeeman-tuned He–Ne laser spectrophotometer. Reproduced by permission of H. J. Gerritsen and M. E. Heller, *Appl. Opt. Suppl.* **2** 73 (1965).

that have ranged from 0.3 to 1.66 m. The tube contains helium and neon (or xenon) at pressure ratios varying from 5:1 to 7:1. Optium laser operation is achieved when the product of the tube diameter and the total gas pressure is 3.25 ± 0.35 Torr mm. The ends of the laser tube are sealed with fused quartz windows placed perpendicular to its cylindrical axis. Nontunable lasers most frequently have these windows placed so their normal is oriented at the Brewster angle relative to the axis of the laser tube. The sealed laser tube is placed inside a Fabry–Perot cavity formed by two flat or concave,

[11] R. L. Fork and C. K. N. Patel, *Appl. Phys. Lett.* **2**, 180 (1963).

[12] S. A. Ahmed, R. C. Kocher, and H. J. Gerritsen, *Proc. IEEE* **52**, 1356 (1964).

[13] W. E. Bell and A. L. Bloom, *Appl. Opt.* **3**, 413 (1964).

[14] H. J. Gerritsen and M. R. Heller, *Appl. Opt. Suppl.* **2**, 73 (1965).

[15] H. Brunet, *IEEE J. Quant. Electron.* **QE-2**, 382 (1966).

[16] K. Sakurai and K. Shimoda, *Jap. J. Appl. Phys.* **5**, 938 (1966).

[17] T. Kasuya, *Jap. J. Appl. Phys.* **11**, 1575 (1972).

first-surface, dielectric coated mirrors so that the axis of the tube is coincident with the axis of the cavity. One mirror is highly reflecting, 95% or greater, and the other, the output mirror, is usually selected to have a reflectance in the range 3–95%. The cylindrical tube is provided with electrodes appropriate for either dc or rf power sources. The entire laser, with the exception of the power source, is then placed inside a solenoid whose axis coincides with that of the laser tube. The solenoid is connected to a continuously variable, reversible polarity, dc power supply. A uniform, axial magnetic induction $\mathbf{B}$ is produced by passing a current through the solenoid; the components of the laser therefore must be made entirely from nonmagnetic materials.

The excess population of neon atoms in excited electronic states, which is necessary for the amplified stimulated emission of radiation from this system, is experimentally achieved through atomic *collisions of the second kind*. Helium atoms are excited directly or indirectly to the metastable singlet $2\ ^{1}S_{0}$ or the triplet $2\ ^{3}S_{1}$ electronic states by electron impact in either the dc or rf discharge created inside the laser tube. The term values $T = -E/hc$ for these states of He are listed in Table II; here E is the negative binding

TABLE II.[a]   Selected Electronic States, Corresponding
Term Values T, and Lande g-factors for
Helium, Neon, and Xenon

Atom	Racah	Paschen	$T\,(\text{cm}^{-1})$	g
He		$2s\ ^{3}S_{1}$	159,850.31	
He		$2s\ ^{1}S_{0}$	166,271.70	
Ne	$5s'[1/2]_{1}^{\circ}$	$3s_{2}$	166,658.48	1.295
Ne	$4s'[1/2]_{1}^{\circ}$	$2s_{2}$	159,536.6	
Ne	$4p'[3/2]_{2}$	$3p_{4}$	163,710.6	1.184
Ne	$3p'[3/2]_{2}$	$2p_{4}$	150,860.5	1.301
Ne	$3p'[1/2]_{0}$	$2p_{1}$	152,972.7	
Xe	$5d[7/2]_{3}^{\circ}$	$3d_{4}$	80,970.73	1.121
Xe	$6p[5/2]_{2}$	$2p_{4}$	78,120.30	1.106

[a] Data from C. E. Moore, "Atomic Energy Levels," Nat. Bur. Std. Circular 467. U.S. Govt. Printing Office, Washington, D.C., 1952.

energy of the electron, h Planck's constant, and c the speed of light in vacuum. The excited helium atoms then inelastically collide with, and very efficiently transfer their electronic excitation energy to, the neon atoms. After the collision the neon atom is in an excited electronic state and the helium atom in its ground electronic state.

Neon, is characterized by a filled electronic p subshell: the electronic configuration is $1s^{2}\ 2s^{2}\ 2p^{6}$. During the collision with the excited helium atom one of these electrons undergoes a transition to either a 3s, 3p, 3d, 4s, 4p, 4d, 4f, or 5s, . . . orbital of higher energy. The electronic configuration of the resultant neon atom in the excited electronic state is thus $1s^2$, $2s^2$, $2p^5$ for the electronic core plus the single electron in the higher-energy orbital. It was established from the now historic investigations of emission spectra that noble gases exhibit the phenomenon known as pair coupling, i.e., the total angular momentum $\mathbf{J}_c$ of the electronic core couples with the orbital angular momentum $\mathbf{L}$ of the single electron in the higher-energy orbital yielding an intermediate angular momentum $\mathbf{K} = \mathbf{J}_c + \mathbf{L}$. The vector $\mathbf{K}$ then couples with the spin angular momentum s of the single electron yielding the total angular momentum $\mathbf{J} = \mathbf{K} + \mathbf{S}$ of the atom. The multiplicity, or the degree of the degeneracy, of the excited electronic states of neon is $2J + 1$.

We resrict our attention for the moment to the inverted population density of neon atoms remaining in the $5s'[1/2]_1^0$ Racah state ($3s_2$ Paschen state) after the energy transfer process of the second kind. The term value for this electronic state is also listed in Table II. In the Racah notation the 5 and the s denote the principal quantum number and the orbital (s, p, d, f, . . .), respectively, of the single electron in the higher-energy orbital. The prime symbol on the $5s'$ denotes antiparallel orientation of the net angular momentum $\mathbf{L}_c$ and the net spin angular momentum $\mathbf{S}_c$ of the core electrons: $\mathbf{J}_c = \mathbf{L}_c + \mathbf{S}_c$. Conversely, the absence of the prime symbol denotes parallel alignment of $\mathbf{L}_c$ and $\mathbf{S}_c$. Furthermore, the $[1/2]_1^0$ part of the Racah notation signifies that the magnitude of $\mathbf{K}$ is $K = 1/2$, the number inside the bracket, that the magnitude of $\mathbf{J}$ is $J = 1$, the value of the subscript, and that the state is of odd parity. States of even parity are represented by deleting the small letter o (odd) as a superscript. An excellent comparative discussion of the Racah and Paschen notations as applied to the noble gases has been prepared by Lengyel.[18] The stronger He–Ne laser emissions at wavelengths 632.8 nm, 1.523 μm, and 3.391 μm are due to the electronic transitions $5s'[1/2]_1 - 3p'[3/2]_2$, $4s'[1/2]_1 - 3p'[1/2]_0$ and $5s'[1/2]_1 - 4p'[3/2]_2$ in the Racah notation or $3s_2 - 2p_4$, $2s_2 - 2p_1$, and $3s_2 - 3p_4$ in the Paschen notation, respectively. The term values for these states are given in Table II.[19]

In the presence of the uniform axial magnetic induction $\mathbf{B}$ produced by the solenoid, the $(2J + 1)$-fold degeneracy of each electronic state is removed by the Zeeman interaction of the magnetic moment $\boldsymbol{\mu}$ of the excited neon atom with the magnetic induction $\mathbf{B}$. The interaction energy $\Delta\xi = -\boldsymbol{\mu} \cdot \mathbf{B}$ is further provided by the expression $\Delta\xi = +m_j g \mu_B B$, where g is the Lande g-factor, $\mu_B = eh/2m$ the Bohr magnetron, B the magnitude of the magnetic

[18] B. A. Lengyel, "Lasers," pp. 294–302. Wiley, New York, 1971

[19] C. E. Moore, "Atomic Energy Levels," Nat. Bur. Std. Circular 467. U. S. Govt. Printing Office, Washington, D. C., 1952.

induction, and m_j represents all the integer or odd half-integer values in the range $-J \leqslant m_j \leqslant +J$. The change in term value due to the Zeeman interaction is

$$\Delta T = -m_j g B \mu_B / hc. \tag{5.2.1}$$

The Zeeman interaction splits the $5s'[1/2]_1{}^\circ$ and $p'[3/2]_2$ states of neon into three and five fine structure states, respectively. The allowed transitions for emission of electric dipole radiation from neon are those for which the total angular momentum changes by $\Delta J = 0, \pm 1$ units and the magnetic quantum number changes by $\Delta m_j = 0, \pm 1$. Electromagnetic radiation emitted during transitions for which $\Delta m_j = 0, +1, -1$ is linearly π, right-circular σ^+, and left-circular σ^- polarized, respectively. The π-component is not observed in the direction parallel or antiparallel to the magnetic field.

The frequency of electromagnetic radiation emitted during an electronic transition is

$$v = c[T_{0i} - T_{0f} - \mu_B B(m_{ji} g_i - m_{jf} g_f)/hc], \tag{5.2.2}$$

where i and f indicate initial and final states. The change Δv of frequency produced by varying the magnetic induction by $|\Delta \mathbf{B}|$ is

$$\Delta v = -(\mu_B / h)(m_{ji} g_i - m_{jf} g_f)|\Delta \mathbf{B}|. \tag{5.2.3}$$

Restricting (5.2.3) to the three σ^+ transitions, $\Delta m_j = +1$,

$$\Delta v = -(\mu_B / h)[m_{ji}(g_i - g_f) - g_f]|\Delta \mathbf{B}|. \tag{5.2.4}$$

Further, the relative intensities I of the transitions for which $\Delta J = +1$ are about $1:3:6$ for $m_{ji} = -1, 0, +1$, respectively, as provided by the relation[20]

$$I = k(J_i + m_{ji} + 1)(J_i + m_{ji} + 2), \tag{5.2.5}$$

where k is a constant proportionality factor. The frequencies of the three σ^+ components are separated by less than the Doppler width of either component when the magnetic induction is below 0.22 T (2200 G). Thus the average shift in frequency must be calculated with respect to the weighted average of the relative intensities of the three individual components:

$$\langle \Delta v \rangle = [(0.1)(\Delta v_{m_{ji}=-1}) + (0.3)(\Delta v_{m_{ji}=0}) + (0.6)(\Delta v_{m_{ji}=1})]$$
$$= (1.5 g_f - 0.5 g_i)\mu_B |\Delta \mathbf{B}|/h = 15.79 \quad \text{GHz} \cdot \text{T}^{-1} |\Delta \mathbf{B}|, \tag{5.2.6}$$

where we used the Lande g-factors for the $5s'[1/2]_1{}^\circ - 4p'[3/2]_2$ transition as listed in Table II. Increasing the magnetic induction produces plasma effects that decrease the gain per unit length of the laser and thus normally quench the laser action at about 0.2 T. In one He–Ne laser tube 55 cm long and with a solenoid producing 55.8×10^{-4} T A^{-1}, multimode laser emission

---

[20] H. E. White, "Introduction to Atomic Spectra." McGraw-Hill, New York, 1934.

was sustained for solenoid currents in the range ± 23 A.[16] This provided continuous frequency tuning throughout a ± 2 GHz range centered about $8.840 \times 10^{+13}$ Hz, the frequency of radiation emitted during the $5s'[1/2]_1{}^0 - 4p'[3/2]_2$ transition in the absence of the magnetic induction. The relative output power from the laser decreased continuously with increases in the magnitude of the solenoid current, thus falling to about 20% at ± 23 A relative to 100% at zero current. In another He–Ne laser tube 160 cm long multimode laser emission was sustained for magnetic induction to ± 0.2 T providing a tuning range of ± 3.16 GHz for the same laser transition.[14] The spectral width δv of radiation emitted by the laser operating in multimode is essentially the Doppler width 100 ± 15 MHz. The resolving power $v/\delta v$ is therefore 8.84×10^5 at the central frequency and correspondingly increases or decreases as the laser frequency is increased or decreased. An order of magnitude of greater resolving power can be achieved by locking the laser to single-mode operation. Care must also be taken to isolate the laser from all thermal or acoustical disturbances so that frequency stability is maintained.

Although midway through the preceding paragraph attention was restricted to the σ^+-polarization component, a similar analysis leading to the negative of (5.2.6) and to the same range of tunability is valid for the σ^--polarization component. The laser emission therefore consists of both the σ^+ and σ^- components which are of higher and lower frequency, respectively. The σ^+ and σ^- components are separated into two perpendicular linear polarization components by passing the laser radiation through a thin sheet of mica which acts as a quarter-wave plate. It is traditional to then use a polarizer to select the linear component of σ^+ origin. The polarizer may be the transmission type consisting of a stack of arsenic trisulfide glass plates or a stack of silver chloride or silver bromide plates placed at the Brewster angle relative to the laser beam, or it may be of the reflection type consisting of a highly polished silicon surface placed at the Brewster angle relative to the laser beam. For spectroscopic applications the cw laser beam is usually chopped, allowed to interact with the sample, and then observed with lead sulfide or gold-doped germanium detectors[21] coupled to phase-sensitive electronic components. It is very convenient to plot simultaneously with an $x–y$ recorder the amplified signal from the detector as the y input and the solenoid current, which is directly proportional to the frequency of the laser radiation, as the x input.

In the foregoing discussion of the tunable He–Ne laser attention was restricted to the $5s'[1/2]_1{}^0 - 4p'[3/2]_2$ transition in neon. Similar frequency

[21] For a comprehensive list of commerically available detectors and lasers and their manufacturers see "Laser Focus Buyers Guide" published annually by Advanced Technology Publ., Inc., Newton, Massachusetts.

tunability was, however, observed for both the 632.8-nm and 1.15-μm wavelengths and for nearly all the discrete He–Ne laser transitions in the 3–9-μm wavelength range.[15] Frequency tunability has also been observed for several He–Xe laser wavelengths in the infrared: 3.36, 3.50, and 5.57 μm. In principle there are hundreds of laser transitions in neon, xenon, argon, and krypton that can be frequency tuned. Frequencies and corresponding electronic transitions for noble gas lasers are tabulated in the literature.[22, 23]

A very interesting extension of the range of frequency tunability for the He–Xe laser was made recently by Kasuya.[17] The electronic transition of interest here was from the $5d[7/2]_3^\circ - 6p[5/2]_2$ states of xenon which are also listed in Table II. It is fortuitous that (5.2.6) is also valid for this particular electronic transition in xenon. In the case of weak magnetic induction one expects the Zeeman effect to dominate. The tuning range was extended by use of superconducting magnets generating maximum axial induction of 6 T. The magnetic induction was uniform to one part in 10^4 over the 0.3-m length of the laser tube. Plasma effects were eliminated by use of side-arm gas vessels attached to the laser tube. The resultant tuning range was about ± 90 GHz centered at 8.54×10^{13} Hz (3.50 μm). The output intensity of the laser, as previously expected, continuously declined as the induction was increased from 0 to 1.3 T. Surprisingly, the intensity then increased continuously, but at a more rapid rate than it had declined, as the induction was further increased from 1.3 to 2.3 T. Finally, the intensity slowly declined as the induction was again further increased from 2.3 to 6 T. The laser was used for investigations of the absorption spectrum of CH_3F.

5.2.2.2. Molecular CO and CO_2 Lasers. Continuous-wave stimulated emission in the infrared from linear CO_2 molecules was observed in 1964.[24] From that and subsequent investigations[25–34] of stimulated emission from

---

[22] R. J. Pressley, "CRC Handbook of Lasers." Chemical Rubber Co., Cleveland, Ohio, 1971.

[23] W. L. Faust, R. A. McFarlane, C. K. N. Patel, and C. G. B. Garret, *Phys. Rev.* **133A**, 1476 (1964).

[24] C. K. N. Patel, *Phys. Rev.* **136**, A1187 (1964).

[25] C. K. N. Patel, *Phys. Rev. Lett.* **13**, 617 (1964).

[26] C. K. N. Patel, *Appl. Phys. Lett.* **6**, 12 (1965).

[27] C. K. N. Patel, *Appl. Phys. Lett.* **7**, 15 (1965).

[28] T. J. Bridges and C. K. N. Patel, *Appl. Phys. Lett.* **7**, 244 (1965).

[29] C. K. N. Patel, *Appl. Phys. Lett.* **7**, 246 (1965).

[30] C. K. N. Patel, P. K. Tien, and J. H. McFee, *Appl. Phys. Lett.* **7**, 290 (1965).

[31] C. K. N. Patel, *Phys. Rev.* **141**, 71 (1966).

[32] R. M. Osgood, Jr. and W. C. Eppers, Jr., *Appl. Phys. Lett.* **13**, 409 (1966).

[33] T. S. Fahlen, *Appl. Opt.* **12**, 2381 (1973).

[34] An extensive bibliography for several types of molecular lasers was presented by K. N. Rao and C. W. Mathews, "Molecular Spectroscopy: Modern Research," Section 3.3, pp. 141–178. Academic Press, New York, 1972.

molecular gases has evolved the present large selection of commercial[21] pulsed and cw lasers of this type. Two investigations[24, 31] were of amplified stimulated emission at low power (milliwatt) from pure CO_2 and pure CO, respectively. Although those two investigations of pure gas lasers were crucial to the present understanding of the physical processes associated with molecular lasers, this section is primarily directed to a discussion of mixed gas molecular lasers of moderate output power (watts) which were obtained when proper amounts of gaseous nitrogen and helium were included with CO_2 or CO inside the laser cavity.

The configuration and energetics of molecular N_2-CO_2 and N_2-CO lasers bear great similarity to those of the atomic He–Ne lasers described in Section 5.2.2.1. A schematic representation of a continuous flow molecular laser developed by Patel appears in Fig. 2. The fused quartz or Pyrex laser tube of 25 mm inside diameter (i.d.) had two parallel side arms; one provided

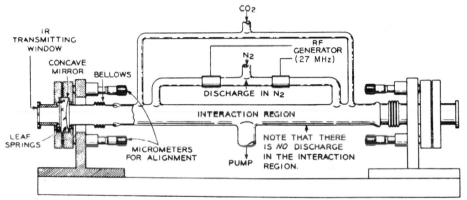

FIG. 2. Experimental apparatus for obtaining laser action in N_2-CO_2. Reproduced by permission of C. K. N. Patel, *Phys. Rev. Lett.* **13**, 617 (1964).

a CO_2 inlet port at each end of the interaction region, and the other provided a region where N_2 was excited by electron impact in a 27-MHz rf discharge and then flowed into each end of the interaction region of the laser tube. The flow rates were adjusted nearly equal for both N_2 and CO_2 so that 0.8 Torr was the total pressure in the interaction region. The laser cavity, which includes the interaction region, consisted of two confocal concave gold mirrors placed 1.3 m apart. The mirrors were coated opaque with gold and thus radiant output, in the milliwatt range for slightly less than 100 W of input rf power, was through a 0.5-mm aperture centered in one of the mirrors. In higher-power lasers the mirror substrates must be cooled to prevent damage. Micrometers provided for careful alignment of the mirrors.

A mechanical vacuum pump removed the mixed gases through a central port at the bottom of the laser tube. The ends of the laser tube were sealed with infrared transmitting windows.

Subsequently, excitation of the N_2–CO_2 mixture itself was achieved in a 2-m-long, 400-W dc discharge in the interaction region only,[27] thus eliminating the need for the side arms. The laser cavity of 2.4 m length was terminated by gold mirrors of focal lengths $+5$ and -5.5 m. Radiant output power observed from an 8-mm diameter hole centered in the concave mirror was 4.5 W when $P_{N_2} = P_{CO_2} \simeq 0.4$ Torr, and was 11.9 W for an air–CO_2 mixture when P_{air} and P_{CO_2} were 1.8 and 0.4 Torr, respectively. The remaining power was partly dissipated in a ballast resistor connected externally between the electrodes of the laser tube. Currents in the milliampere range and potentials in the kilovolt range sustained the discharge. Continuous wave (cw) laser emission from carbon monoxide was also obtained in a N_2–CO mixture with a similar laser; the temperature of the laser-tube walls, however, was reduced this time by pumping cooled methanol through a surrounding jacket.[29]

Further efficiency of the laser process was achieved with another continuous flow, similarly cooled, Pyrex laser tube of 1 m length and 10.5 mm i.d. The ends of this tube, however, were closed with vacuum flange mountings for 5-mm-thick potassium chloride windows oriented at the Brewster angle relative to the tube axis. Excitation of the N_2–CO_2 mixture was by cold cathode discharge operating at about 7.5 mA and 9 kV between the ends of the tube. The laser cavity consisted of gold-coated mirrors of focal lengths $+1.5$ and -5 m placed 1.65 m apart. Radiant output of 4.5 W was observed through a 3-mm-diameter hole centered in the convex mirror. Optimum output was derived when the partial pressures of N_2 and CO_2 were equal and when the total pressure was 7.9 Torr. The output power also increased as the tube walls were cooled from 313 to 213 K.

Additional efficiency of the stimulated emission process, up to 9%, was soon obtained by adding helium as a third gas to either the N_2–CO_2 or the N_2–CO mixture.[30, 32] Partial pressures of the three-gas mixtures that yielded optimum laser power are listed in Table III.

The inverted population of the higher vibrational–rotational energy levels of CO or CO_2 in the N_2–CO or N_2–CO_2 mixed gas laser proceeds by collisions of the second kind. The importance of energy transfer from N_2 to either CO or CO_2 during these collisions was evident from the 10^3 times increase in output power of the mixed gas lasers relative to the pure gas lasers. A qualitative explanation of the physics pertinent to the wavelength tunability of the N_2–CO_2–He and N_2–CO–He lasers will be more easily understood if we digress briefly to review the energetics of diatomic and linear triatomic molecules. The reader is referred to Sections 2.3.2.2, 2.4.4,

TABLE III.^{a} Partial Pressures (Torr) for Optimum
Laser Action, Relative Power P, and
Relative Efficiency E

| Partial pressure (Torr) | | | | | | |
N_2	CO	CO_2	He	Air	P	E
	0.3		8.8	1.4	1	1
1.7		5.0	4.8		0.5	0.7
1.0	0.2		5.2		0.3	0.4

^{a} Data from R. M. Osgood, Jr. and W. C. Eppers,
Jr., *Appl. Phys. Lett.* **13**, 409 (1966).

and 2.5.4 for a more complete discussion. In the remainder of this section
we depart from SI units in favor of the more traditional units of molecular
spectroscopy, e.g., frequencies in wavenumber units of reciprocal centimeters.

Solution of the Schroedinger equation for diatomic molecules,[35, 36] subject to conditions of the Born–Oppenheimer approximation, leads to product
wave functions

$$\psi = \psi_e \psi_v \psi_r \tag{5.2.7}$$

and total energy

$$E = E_e + E_v + E_r. \tag{5.2.8}$$

The subscripts e, v, r designate parts that arise from electronic, vibrational,
and rotational motions, respectively. The individual energies represent bound
states and are therefore negative. Term values T in wavenumber units of
reciprocal centimeters are provided by multiplying (5.2.8) by $(-hc)^{-1}$; the
result being

$$T = T_e + G + F, \tag{5.2.9}$$

where G and F traditionally symbolize vibrational and rotational term values,
respectively.

The anharmonic vibrating–rotator model of the diatomic molecule, represented by the Morse potential energy function, further provides

$$G = \omega_e[(v + \tfrac{1}{2}) - X_e(v + \tfrac{1}{2})^2 + Y_e(v + \tfrac{1}{2})^3 - Z_e(v + \tfrac{1}{2})^4 + \cdots] \tag{5.2.10}$$

and

$$F = B_v J(J + 1) - D_v J^2(J + 1)^2 + \cdots, \tag{5.2.11}$$

where v and J are, respectively, the vibrational and rotational quantum

[35] G. Herzberg, "The Spectra and Structures of Simple Free Radicals: An Introduction to
Molecular Spectroscopy." Cornell Univ. Press, Ithaca, New York, 1971.
[36] G. Herzberg, "Molecular Spectra and Molecular Structure," Vol. I, Spectra of Diatomic
Molecules. Van Nostrand-Reinhold, Princeton, New Jersey, 1950.

numbers, ω_e the classical frequency (in reciprocal centimeters) for harmonic oscillations of small amplitude about the equilibrium position, and X_e, Y_e, Z_e, ... are constants providing for a mathematical description of the anharmonic oscillations. The coefficients B_v and D_v are given by

$$B_v = B_e - \alpha_e(v + \tfrac{1}{2}) + \gamma_e(v + \tfrac{1}{2})^2 \qquad (5.2.12)$$

and

$$D_v = D_e + \beta_e(v + \tfrac{1}{2}), \qquad (5.2.13)$$

where D_e is the dissociation energy (in reciprocal centimeters) and $B_e = h/(4\pi CI_e)$ with $I_e = \mu r_e^2$ being the equilibrium value for the moment of inertia. The relatively small constants α_e, β_e, γ_e mathematically provide for interaction between vibrational and rotational motions. We note that coefficients bearing the e or v subscripts can have different values for each of the electronic and vibronic states, respectively, of a single diatomic molecule. Values for these constants are usually determined by careful analysis of molecular spectra obtained in the laboratory. In Table IV we list molecular

TABLE IV.[a]   Molecular Constants for Ground Electronic States of N_2 and CO

Molecule	N_2 (X $^1\Sigma_g^+$) (cm^{-1})	CO (X $^1\Sigma^+$) (cm^{-1})
T_e	0	0
W_e	2359.61	2169.82
$W_e X_e$	14.456	13.292
$W_e Y_e$	+0.00751	0.01082
$W_e Z_e$		0.0000572
B_e	2.010	1.93141
α_e	0.0187	0.017520
γ_e		2.96×10^{-6}
D_e		6.18×10^{-6}
β_e		-1.76×10^{-9}
r_e	1.094×10^{-10} m	1.1281×10^{-10} m
μ	7.00377 amu	6.85841 amu

[a] Compiled from Refs. 28 and 36.

constants for the X $^1\Sigma_g^+$ and X $^1\Sigma^+$ ground electronic states[37] of N_2 and CO, respectively. The term value T_e is traditionally set equal to zero for the ground electronic state with all other term values then stated relative to this arbitrary choice.

[37] For further explanation of the notation for electronic states see G. Herzberg, "Molecular Spectra and Molecular Structure," Vol. 1, Chapter 5. Van Nostrand-Reinhold, Princeton, New Jersey, 1950.

Spectroscopic notation such as $X \, {}^1\Sigma_g^+$ for the electronic states of molecules is very similar to that for atoms. The Greek letters Σ, Π, Δ, $\Phi, \ldots$, respectively, designate a total orbital angular momentum of magnitude $\Lambda = 0, 1, 2, 3, \ldots$ units of $\hbar$ about the internuclear axis. The numerical superscript on the left specifies the multiplicity $2S + 1$, where S is the quantum number associated with the total electronic spin angular momentum. The plus or minus sign as a superscript on the right denotes that ψ_e is symmetric or antisymmetric, respectively, to reflection about any plane through the internuclear axis. Homonuclear molecules such as N_2 possess a center of symmetry. The right subscript g or u, for "gerade" or "ungerade," designates the even or odd character, respectively, of the product electronic wave function for reflection about this center of symmetry. The g or u notation is not used for heteronuclear molecules such as CO which do not possess the center of symmetry. The arabic symbol on the left is used as follows: X denotes the ground electronic state, A, B, C, $\ldots$ denote increasingly higher-energy states of the same multiplicity, and a, b, c, $\ldots$ denote increasingly higher-energy states of multiplicity different from that of the ground state.

Similar quantum-mechanical analyses for linear triatomic molecules, CO_2 being an example, provides

$$G(v_1, v_2, v_3) = \sum_{i=1}^{3} \omega_i(v_i + \tfrac{1}{2}d_i) + \sum_{i=1}^{3} \sum_{k>i}^{3} X_{ik}(v_i + \tfrac{1}{2}d_i)(v_k + \tfrac{1}{2}d_k)$$
$$+ \cdots + g_{22}l_2^2 \tag{5.2.14}$$

and

$$F_v(J) = B_v[J(J + 1) - l_i^2] \tag{5.2.15}$$

for the vibrational and rotational term values, respectively. There is also an electronic term value T_e analogous to that for diatomic molecules. There are three vibrational degrees of freedom with normal frequencies $\omega_1, \omega_2, \omega_3$ and with v_1, v_2, v_3 as vibrational quantum numbers for the symmetric stretch, bending, and asymmetric stretching notions, respectively. The bending and asymmetric stretch vibrations of frequencies ω_2 and ω_3 are infrared active. The ω_1- and ω_3-modes are nondegenerate, while the ω_2-mode is doubly degenerate. The parameters d_i equal 1 or 2 correspond to the nondegeneracy or double degeneracy, respectively, and the X_{ik} are anharmonicity constants. The quantum numbers $l_1 = l_3 = 0$, but

$$l_2 = v_2, v_2 - 2, \ldots, 1 \quad \text{or} \quad 0 \tag{5.2.16}$$

corresponding to angular momentum $l_2\hbar$ about the internuclear axis as generated by the doubly degenerate bending mode; g_{22} is a corresponding molecular constant. The value of B_v is

$$B_v = B_e - \sum_{i=1}^{3} \alpha_i(v_i + \tfrac{1}{2}d_i) + \cdots. \tag{5.2.17}$$

Equations (5.2.14)–(5.2.17) do not include Fermi-resonance perturbations[38] which are observed in CO_2 spectra because $\omega_1 \approx 2\omega_2$. The vibrational state of the linear triatomic molecule is denoted as $(v_1 v_2{}^{l_2} v_3)$; thus the ground vibrational state is (0000).

In the spectroscopy of diatomic and polyatomic molecules, transitions from one energy state of the molecule to another are written by first listing the symbol for the state of higher energy, say H, and second the symbol for the state of lower energy, say L. Thus $H \rightarrow L$ and $H \leftarrow L$ designate emission and absorption of radiation, respectively. Electric dipole selection rules for electronic transitions are

$$\Delta\Lambda = 0, \pm 1, \qquad \Delta S = 0, \qquad (5.2.18)$$

and

$$\Sigma^+ \leftarrow|\rightarrow \Sigma^-, \qquad \Sigma^+ \leftrightarrow \Sigma^+. \qquad (5.2.19)$$

Further for homonuclear molecules

$$g \rightarrow u, \qquad g \leftarrow|\rightarrow J, \qquad u \leftarrow|\rightarrow u. \qquad (5.2.20)$$

The symbol $\leftrightarrow$ ($\leftarrow|\rightarrow$) is read "does (does not) combine with." There are no rigid selection rules for vibrational transitions within anharmonic molecules, but spectra show great preference for

$$\Delta v = v' - v'' = \pm 1, \qquad (5.2.21)$$

where v' and v'' designate quantum numbers for the higher- and lower-energy states, respectively. Rotational transitions are provided by

$$\Delta J = J' - J'' = 0, \pm 1, \qquad (5.2.22)$$

except that $J' = 0 \leftarrow|\rightarrow J'' = 0$. The transitions $\Delta J = 0, +1, -1$ provide for the Q-, R-, and P-branches, respectively. Lines in the P-branch are denoted as $P(J'')$.

We now return our attention to molecular lasers. Energy transfer processes within the N_2–CO_2–He laser, which we will discuss first, are currently better defined than those for the N_2–CO–He laser. An energy-level (term value) diagram showing the pertinent vibrational levels of N_2 and CO_2 is shown in Fig. 3.[39] During the electric discharge inside the laser tube, N_2 molecules are excited either directly or indirectly by electron impact from the $N_2(v = 0)$ vibrational level to the metastable $N_2{}^*(v = 1)$ level of the X $^1\Sigma_g{}^+$ electronic state. The population densities of $N_2{}^*(v - 1)$ in the absence of CO_2 easily range from 10 to 30% of the total N_2 population in the discharge. Through

[38] Correction for the Fermi-resonance phenomenon is found in G. Herberg, "Molecular Spectra and Molecular Structure," Vol. 1, pp. 191–193. Van Nostrand-Reinhold, Princeton, New Jersey, 1950.

[39] C. K. N. Patel, *Appl. Phys. Lett.* **6**, 12 (1965).

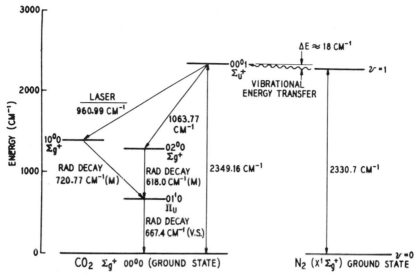

FIG. 3. Energy-level diagram of selected vibrational levels of CO_2 and N_2. Rotational levels are not shown in the diagram. Reproduced by permission of C. K. N. Patel, *Phys. Rev. Lett.* **13**, 617 (1964).

collisions of the second kind, N_2 transfers this vibrational energy to CO_2 which is then excited to the $00^0 1$ vibrational level. The energy discrepancy of ~ 18 cm^{-1} in this reaction is provided by thermal energy; $kT \simeq 210$ cm^{-1} at ordinary room temperature. A small number of CO_2 molecules are also excited to the $00^0 1$ level by electron impact in the discharge. The radiative transitions $\sum_u{}^+(00^0 1) - \sum_g{}^+(10^0 0)$ and $\sum_u{}^+(00^0 1) - \sum_g{}^+(02^0 0)$ are the source of laser action near wavelengths 10.4 and 9.4 μm, respectively. The $10^0 0$ and $02^0 0$ levels each have short radiative lifetimes and thus CO_2 undergoes further transitions to the $\prod_u(01^1 0)$ state, which has a relatively long radiative lifetime. The CO_2 molecules then radiatively decay as shown in the diagram but predominantly nonradiatively decay back to the ground state by colliding with the cold walls of the laser tube or with He atoms in the N_2–CO_2–He laser, whereupon they are once again available to begin the entire process. Patel's calculations[24, 31] show the optical gain per unit length is significantly greater for the P-branch than for the R-branch in both diatomic and linear triatomic molecules. Although many R-branch lines exhibit a negative gain, laser action was observed in the R-branch of both CO_2 and N_2O vibrational spectra.[40] Laser emission from CO_2 has been observed for the discrete vibrational–rotational lines listed in Table V. The spectral width

[40] G. Moeller and J. Dane Rigden, *Appl. Phys. Lett.* **8**, 69 (1966).

TABLE V.[a]   Discrete Vibrational–Rotational Laser Lines from CO_2
and Outermost[b] Frequencies

Vibrational band	Rotational lines		Frequencies (cm^{-1})
00^01-10^00	$R(J = 2K)$,	$2 \leqslant K \leqslant 27$	964.7–994.18
00^01-10^00	$P(J = 2K)$,	$2 \leqslant K \leqslant 28$	957.8–907.7
00^01-02^00	$R(J = 2K)$,	$2 \leqslant K \leqslant 26$	1067.5–1095.7
00^01-02^00	$P(J = 2K)$,	$2 \leqslant K \leqslant 30$	1060.5–1005.4
01^11-11^10	$\cdot P(J)$,	$19 \leqslant J \leqslant 45$	911.3–886.4
10^02-10^01	$P(J = 2K + 1)$,	$3 \leqslant K \leqslant 12$	2296.2–2280.5

[a] Summarized from K. N. Rao and C. W. Mathews, "Molecular Spectroscopy-Modern Research," pp. 153–154. Academic Press, New York, 1972.
[b] For example, in first entry the outer most frequencies are $R(4) = 964.7$ cm^{-1} and $R(54) = 994.18$ cm^{-1}.

of these lines at low pressures is just the Doppler width which is about 2.5×10^{-3} cm^{-1}(~ 75 MHz) at 300 K. There have been recent attempts to pressure-broaden the rotational lines from TEA and higher-pressure TE lasers in order to obtain broad-band spectral emission. TEA lasers are transverse excitation atmospheric pressure N_2–CO_2–He systems in contrast to the systems of low pressure and longitudinal excitation described here.

Continuous wave emission from CO lasers has been obtained simultaneously, or in cascade from pulsed lasers, for the P-branch lines of several vibrational bands of the X $^1\sum^+$ electronic state. Table VI presents a summary of these laser lines and their outermost frequencies. The spectral distribution of the radiant output, i.e., the number of lines, is a function of pressure, temperature, and the mode of excitation. The relatively complex energy transfer processes of the N_2–CO–He laser remain subjects of current scientific inquiry. Several plausible processes, which may be operating individually or in concert, have, however, been proposed in the literature.[29, 32, 41–44] Among these is the vibrational–vibrational transfer of the type

$$N_2^*(v) + CO(v') \rightarrow N_2(v - 1) + CO^*(v' + 1),$$

which occurs during collisions of the second kind; in this manner CO could be successively excited to higher and higher vibrational levels. Deposition of carbon, frozen CO_2, and solid N_2O_3 on the cold walls of the CO laser tube,

[41] F. Legay, C. R. Acad. Sci. Paris 266, 554 (1968).
[42] F. Legay and N. Legay-Sommaire, C. R. Acad. Sci. Paris 257, 2644 (1963).
[43] N. Legay-Sommaire and F. Legay, J. Phys. Radium 25, 917 (1964).
[44] N. Legay-Sommaire, L. Henry, and F. Legay, C. Acad. Sci. Paris 260, 3339 (1965).

TABLE VI.[a]   Discrete Vibrational–Rotational Laser Lines from
CO and Outermost[b] Frequencies

$v'-v''$	$P(J)$	Frequencies (cm^{-1})	$v'-v''$	$P(J)$	Frequencies (cm^{-1})
4–3	$13 \leqslant J \leqslant 16$	2012.7–2000.1	21–20	$5 \leqslant J \leqslant 15$	1611.8–1577.1
5–4	$9 \leqslant J \leqslant 18$	2003.2–1965.8	22–21	$5 \leqslant J \leqslant 14$	1587.1–1556.3
6–5	$7 \leqslant J \leqslant 19$	1985.1–1936.0	23–22	$5 \leqslant J \leqslant 14$	1562.4–1532.0
7–6	$7 \leqslant J \leqslant 20$	1959.2–1906.3	24–23	$5 \leqslant J \leqslant 13$	1537.9–1511.2
8–7	$7 \leqslant J \leqslant 22$	1933.4–1872.3	25–24	$5 \leqslant J \leqslant 12$	1513.3–1490.4
9–8	$7 \leqslant J \leqslant 22$	1907.7–1847.2	26–25	$5 \leqslant J \leqslant 13$	1489.0–1462.8
10–9	$7 \leqslant J \leqslant 22$	1882.1–1822.1	27–26	$4 \leqslant J \leqslant 14$	1467.7–1436.0
11–10	$6 \leqslant J \leqslant 20$	1860.2–1805.3	28–27	$5 \leqslant J \leqslant 11$	1440.4–1421.4
12–11	$6 \leqslant J \leqslant 19$	1834.6 1784.3	29–28	$5 \leqslant J \leqslant 11$	1416.1–1397.4
13 12	$6 \leqslant J \leqslant 19$	1809.1–1759.3	30–29	$5 \leqslant J \leqslant 10$	1391.9–1376.6
14–13	$6 \leqslant J \leqslant 16$	1783.7–1746.3	31–30	$5 \leqslant J \leqslant 11$	1367.9–1349.5
15–14	$6 \leqslant J \leqslant 16$	1758.4–1721.3	32–31	$5 \leqslant J \leqslant 10$	1343.8–1328.7
16–15	$5 \leqslant J \leqslant 14$	1736.7–1704.1	33–32	$6 \leqslant J \leqslant 10$	1316.9–1304.9
17–16	$7 \leqslant J \leqslant 16$	1704.6–1671.7	34–33	$6 \leqslant J \leqslant 10$	1293.0–1281.1
18–17	$7 \leqslant J \leqslant 15$	1679.6–1650.8	35–34	$6 \leqslant J \leqslant 9$	1269.0–1260.4
19–18	$7 \leqslant J \leqslant 15$	1654.7–1626.1	36–35	$5 \leqslant J \leqslant 9$	1248.0–1236.5
20–19	$5 \leqslant J \leqslant 15$	1636.6–1601.8	37–36	$6 \leqslant J \leqslant 8$	1221.3–1215.7

[a] Summarized from K. N. Rao and C. W. Mathews, "Molecular Spectroscopy-Modern Research," pp. 144–145. Academic Press, New York, 1972.

[b] For example, in the first entry the outermost frequencies are $P(13) = 2012.7$ cm^{-1} and $P(16) = 2000.1$ cm^{-1}.

and the formation of NO indicate that both CO and N_2 are dissociated during the discharge. Thus an exothermic reaction was proposed:

$$C + NO \rightarrow CO^* + N + \Delta\varepsilon_K.$$

A visible glow emanating from the laser tube also indicates that excited electronic states of N_2 and CO may also play a role through collisions or cascades to lower levels whereby the correct population inversion is achieved. it is believed that He and the cold walls aid in depopulating the lower laser levels.

Now that we understand some of the physical properties of the molecular lasers, the methods of frequency tunability can be summarized in a single paragraph. We have seen that CO and CO_2 lasers, as well as other types of molecular lasers, emit at many discrete infrared frequencies characteristic of their radiative vibrational–rotational transitions. The lasers are frequency tuned to a particular P-branch or R-branch line by varying the pressure, by varying the temperature of the tube walls, by adjusting the length of the laser cavity, by inserting a frequency selective element in the cavity so that the gain is reduced for all but one of the laser lines, or by a combination of two or more of these methods. The frequencies v of allowed laser oscillations are inversely proportional to the optical path length L of the laser tube; Δv is proportional to $\Delta L/L^2$. This method of varying the cavity length or the pressure, thus varying the index of refraction, is used to frequency-tune commercial CO_2 lasers to different P- and R-branch lines. Insertion of a frequency-selective element is commonly accomplished by replacing one of the laser-cavity end mirrors with a reflective diffraction grating which has sufficient dispersion, or resolving power, to tune the cavity to a single rotational line. The cavity is then tuned to a frequency given by $v = n \csc \theta/2d$, where n is the diffraction order, θ the angle between the axis of the laser cavity and the normal to the grating, and d the spacing between lines on the grating. There is actually a small spread in frequencies because the inside diameter of the laser tube or the mirror geometry provides a limiting aperature of finite dimensions. By rotating the grating, which is usually in a Littrow mount, one changes θ and thereby tunes the laser from one P- or R-branch line to another. The laser must also be stabilized against fluctuations in temperature, pressure, and power supply voltage if the output is to be most useful for spectroscopic applications.

5.2.2.3. Diode Injection Lasers. The first operational semiconductor lasers that were of gallium arsenide (GaAs) emitting at 0.84 μm in the near infrared were developed independently and simultaneously in 1962 by three research groups.[45–47] Very soon other type III–V semiconductors, e.g., InP, InSb, InAs[48–50] were made to sustain laser action at 0.9, 5.18, and 3.11 μm, respectively. Stoichiometrically mixed type III–V semiconductor lasers such as $GaAs_xP_{1-x}$, $In_xGa_{1-x}As$, and $InAs_xP_{1-x}$, one by one, subsequently provided laser emission,[51–54] dependent on the value x, at wavelengths within the regions 0.64–0.84, 0.84–3.11, and 0.9–3.11 μm. In 1964 type IV–VI semiconductors such as PbSe and PbTe, and shortly thereafter PbS,

[45] R. N. Hall, G. E. Fenner, J. D. Kingsley, T. J. Soltys, and R. O. Carlson, *Phys. Rev. Lett.* **9**, 366 (1962).

[46] M. I. Nathan, W. P. Dunike, G. Burns, F. H. Dill, and G. J. Lasher, *Appl. Phys. Lett.* **1**, 61 (1962).

[47] T. M. Quist *et al.*, *Appl. Phys. Lett.* **1**, (1962).

[48] K. Weiser and R. S. Lewitt, *Appl. Phys. Lett.* **2**, 178 (1963); *Bull. Amer. Phys. Soc.* **8**, 29 (1963).

[49] R. J. Phelan, A. R. Calawa, R. H. Rediker, R. J. Keyes, and B. Lax, *Appl. Phys. Lett.* **3**, 143 (1963).

[50] I. Melngailis, *Appl. Phys. Lett.* **2**, 176 (1963); I. Melngailis and R. H. Rediker, *J. Appl. Phys.* **37**, 899 (1966).

[51] N. Holnyak, Jr., and S. F. Bevacqua, *Appl. Phys. Lett.* **1**, 82 (1962).

[52] I. Melngailis, A. J. Strauss, and R. H. Rediker, *Proc. IEEE* **51**, 1154 (1963).

[53] F. B. Alexander *et al.*, *Appl. Phys. Lett.* **4**, 13 (1964).

[54] For a review of semiconductor lasers see H. Rieck, "Semiconductor Lasers." Macdonald, London, 1970.

yielded laser emission at 8.5, 6.6, and 4.3 μm, respectively.[55–57] Next, the stoichiometrically mixed type IV–VI semiconductors, $Pb_{1-x}Sn_xTe$, $Pb_{1-x}Sn_xSe$, and $PbS_{1-x}Se_x$, provided laser emission, again dependent on the value x, in the corresponding wavelength regions[58–64] 6.6->30, 8.5->30, and 4.3–8.5 μm. These semiconductors lasers whether n-type, p-type, or p–n junctions, provide laser action when external ("pump") energy is properly supplied in either optical, electron beam, or electrical current form. The physical properties of type III–V and IV–VI semiconductors are very similar. The type III–V semiconductor lasers were previously reviewed in detail by Rieck[54] and by Lengyel.[5] We therefore restrict the remainder of this discussion to the type IV–VI semiconductor lasers, and furthermore, to p–n junction diode-type lasers which presently appear the most easily adaptable to spectroscopic applications.

The fabrication of a p–n junction diode-injection laser is unlike that of the gas lasers described in the two preceding subsections. An artist's representation of a p–n junction diode laser appears in Fig. 4. Crystalline type IV–VI semiconductors are grown either from the melt by use of the Bridgman technique[65] or from the vapor by various recrystallization techniques.[66] Growth from the vapor seems to be preferred because it yields well-faceted crystals, some as large as 1 cm^3, from which lasers are most easily fabricated. An excess of either the nonmetal or metal atomic constituent renders the crystal p-type or n-type, respectively. A physical description of p- and n-type materials is deferred to a later paragraph in this subsection. Properly cleaved, polished, and etched bulk p-type or n-type crystals provide laser action when pumped optically or by electron beams. A p–n junction crystal, however, must be fabricated if the material is to provide laser action when subjected to electrical current excitation. The p–n junction is prepared by proper

[55] J. F. Butler, A. R. Calawa, R. J. Phelan, Jr., A. J. Strauss, and R. H. Rediker, *Solid State Commun.* **2**, 303 (1964).

[56] J. F. Butler, A. R. Calawa, R. J. Phelan, Jr., T. C. Harman, A. J. Strauss, and R. H. Rediker, *Appl. Phys. Lett.* **5**, 75 (1964).

[57] J. F. Butler and A. R. Calawa, "Physics of Quantum Electronics" (P. L. Kelley, B. Lax, and P. E. Tannenwald, eds.), pp. 458–466. McGraw-Hill, New York, 1966.

[58] J. F. Butler, A. R. Calawa, and T. C. Harman, *Appl. Phys. Lett.* **9**, 427 (1966).

[59] J. O. Dimmock, I. Melngailis, and A. J. Strauss, *Phys. Rev. Lett.* **16**, 1193 (1966).

[60] J. F. Butler and T. C. Harman, *Appl. Phys. Lett.* **12**, 347 (1968).

[61] I. Mengailis, *J. Phys.* **29**, C4–84 (1968).

[62] G. A. Antcliffe and J. S. Wrobel, *Appl. Phys. Lett.* **17**, 290 (1970).

[63] G. A. Antcliffe and S. G. Parker, *J. Appl. Phys.* **44**, 4145 (1973).

[64] J. F. Butler, Semiconductor Lasers and Emitters for the Infrared. Available from Arthur D. Little Inc., Cambridge, Massachusetts 02140 (1974).

[65] J. F. Butler, A. R. Calawa, R. J. Phelan, Jr., A. J. Strauss, and R. H. Rediker, *Solid State Commun.* **2**, 303 (1964).

[66] A. R. Calawa, T. C. Harman, M. Finn, and P. Youtz, *Trans. AIME* **242**, 374 (1968).

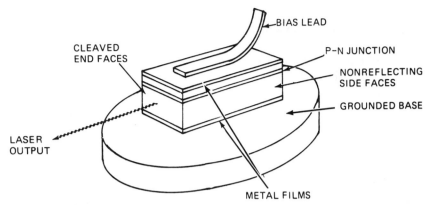

FIG. 4. An artists drawing of a Pb–salt p–n junction diode laser. Further details are provided in the text. Reproduced by permission of Arthur D. Little, Inc., Cambridge, Massachusetts.

application of combined diffusive–annealitive procedures whereby n-type or p-type layers about 30 μm deep are formed in p-type or n-type grown crystals, respectively. The type IV–VI semiconductor crystals have the NaCl face-centered cubic structure. After preparation of the p–n junction, rectangular parallelepiped structures as shown in Fig. 4 are obtained by cleaving the crystal both parallel and perpendicular to the natural (100) faces. The two longer crystal faces perpendicular to the plane of the p–n junction are etched, thus reducing their flatness and reflectance. The two shorter faces perpendicular to the plane of the p–n junction therefore form the Fabry–Perot cavity of the laser. These two shorter faces are etched in an electrolytic solution to remove about a 10-μm thick layer of material damaged in the cleaving process. These faces are then polished to increase their flatness and reflectance. The two remaining crystal faces, which are parallel to the plane of the p–n junction, are electroplated with gold and indium layers on the p- and n-surfaces, respectively, thus providing ohmic contacts. The gold layer is then coated with a buffer metal such as platinum. Indium then bonds this surface of the diode to an electrically grounded copper base which also acts as a heat sink. A copper lead is attached to the top of the diode, which is then mounted in a low-inductance assembly. The dimensions of a finished diode laser are typically 200-μm separation between the two planes with the ohmic contacts, 500-μm separation between the two planes forming the Fabry–Perot cavity, and 250-μm separation between the remaining pair of planes.

Quantum mechanically, pure crystalline intrinsic semiconductors are characterized by continuous, finite, eigenenergy bands that electrons are either permitted or forbidden to occupy.[67-69] A graphical representation of the energy band structure is achieved by plotting regions of both permitted and forbidden eigenenergy E against the wavevector k of the electron. The lowest band of permitted eigenenergies appearing on a graph of this type is known as the *valence band*. This band contains the electrons that participate in the chemical bonds between the atoms of the crystal. The maximum eigenenergy within the valence band is known as the *valence-band energy* E_v. Just above the valence band there occurs a finite continuous band of forbidden eigenenergies, and above this then occurs the second finite continuous band of permitted eigenenergies. The second permitted band is known as the *conduction band*. Electrons occupying the conduction band are not bound to a particular atom but are free to move about the crystal lattice. External electric fields will cause these free electrons to move within the crystal, thus producing a current. The minimum eigenenergy within the conduction band is known as the *conduction band energy* E_c. The width E_g of the intermediate forbidden band, in energy units, is thus $E_g = E_c - E_v$ which is referred to as the *band-gap energy*. The probability $f(E)$ that an electron occupies an eigenstate of energy E follows the Fermi–Dirac distribution law when thermodynamic equilibrium exits:

$$f(E) = [1 + \exp(E - E_F)/k_0 T]^{-1}, \qquad (5.2.23)$$

where k_0 is Boltzmann's constant, T the absolute temperature, and E_F the Fermi energy. For a pure intrinsic semiconductor $E_F \simeq (E_v + E_c)/2$. At low temperatures the conduction band will be void of electrons. At relatively higher temperatures (5.2.23) predicts that a few electrons will occupy the conduction band, thus leaving an equal number of electronic vacancies, i.e., holes which are regarded as positive charge carriers, in the valence band. The holes migrate to the highest energy levels within the valence band and the electrons migrate to the lowest energy levels within the conduction band. Electrons in the conduction band can either directly or indirectly, depending on the precise energy-band structure of the semiconductor, transfer to the valence band, recombine with a hole, and thus emit a photon or a photon plus phonon, respectively. During the direct transition the linear momentum of the electron remains unchanged, whereas the momentum of the electron changes significantly during the indirect transition with vector momentum

[67] J. S. Blakemore, "Solid State Physics." Saunders, Philadelphia, Pennsylvania, 1969. See Chapters 3 and 4 for a good introductory discussion of energy band theory.

[68] C. Kittel, "Introduction to Solid State Physics," 4th ed., Wiley, New York, 1971. See Chapters 9–11 for discussion of energy band theory.

[69] J. Callaway, "Quantum Theory of the Solid State," Part A. Academic Press, New York, 1974. See Chapter 4 for a more advanced theoretical discussion of energy band theory.

being conserved by the accompanying phonon. The inverse process, absorption of a photon or a photon plus phonon, also occurs with an electron going to the conduction band and the hole thus being created in the valence band. During periods of thermal equilibrium the spectral distributions of photons and of phonons are each quantified by use of the Bose–Einstein statistics which lead to the Planck radiation formula. During thermal equilibrium there exists within the crystal a balanced exchange of emitted and absorbed photons for the direct transitions, or photons plus phonons for the indirect transitions. Germanium and silicon are examples of semiconductors characterized by indirect optical transitions. Semiconductors that produce laser action, e.g., GaAs, PbTe, PbSe, PbS, etc., are characterized by direct optical transitions.

Semiconductor diodes that produce laser action are extrinsic materials, whereas in the preceding paragraph we primarily discussed the physical properties of intrinsic semiconductors. Extrinsic semiconductors are fabricated by doping crystals with atoms that must either *accept* an extra electron from the crystal or *donate* an extra electron to the crystal thus allowing natural participation in the valence bonds between adjacent atoms. When an excess of either acceptor or donor atoms is present, the crystal is said to be p-type or n-type, respectively. Here we recall that inclusion of an excess of the nonmetal or metal atomic constituents during the growth process correspondingly renders IV–VI semiconductors p-type or n-type. The most favorable acceptor atoms for semiconductor applications have outermost electrons whose eigenenergies are just above the valence-band energy E_v. Similarly, the most favorable donor atoms have outermost electrons whose eigenenergies are just below the conduction-band energy E_c. Thus the eigenenergies of the outermost electrons of acceptor and donor atoms provide discrete localized energy states that lie within the forbidden energy band. Concentrations of 10^{15} to 10^{16} cm^{-3} of either acceptor or donor atoms removes the localization and discreteness of these eigenenergy levels. The energy E_{Fc} of the Fermi level in n-type material is shifted into the conduction band and electrons thus partially occupy this band. Similarly, the energy E_{Fv} of the Fermi level in p-type materials is shifted into the valence band and thus holes partially occupy energy states within this band. Here we recall that planar p–n junctions are normally fabricated in laser diodes by taking as grown n-type crystals such as PbSe and forming a p-type layer by heating the crystals in vapor rich in Se, the nonmetal acceptor atom.

At thermal equilibrium the Fermi energy level E_{Fv} applicable to the p-type part of the crystal must coincide, across the p–n junction, with the Fermi energy level E_{Fc} of the n-type part of the crystal. Thus the energy-band structure of the n-type material is displaced downward relative to that of the p-type material, as shown in Fig. 5. By connecting a positive voltage V to the p-type face of the crystal, e.g., the bias lead shown in Fig. 4, current is produced between this lead and the grounded base. The current causes electrons from the n-type material and holes from the p-type material to be simultaneously injected into the region of the p–n junction. The recombination of electrons and holes provides for emission of photons. Absorption and emission of photons are competitive processes within the crystal. Stimulated emission of photons exceeds the combined rates of spontaneous emission and absorption provided that, first, $E_{Fc} - E_{Fv} > E_g$ and, second, that $eV > E_g$. The first condition provides for the population inversions for electrons and holes. The second condition, $eV > E_g$, provides for external work sufficient for recombination and also provides for energy to compensate for ohmic losses. The series resistance of a laser diode at room temperature must be less than 0.2 ohm. The ohmic losses physically manifest themselves as Joule heat which can be used to control the temperature of the diode-injection laser. It must be stressed that a state of thermal equilibrium does not prevail in the region of the p–n junction during the laser-action process.

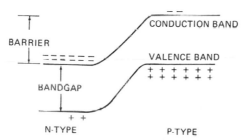

FIG. 5. Diagram of energy bands across a p–n junction for conditions near thermal equilibrium. The Fermi energy level is shifted upward into the conduction band in the n-type material and downward into the valence band in the p-type material. Reproduced by permission of Arthur D. Little, Inc., Cambridge, Massachusetts.

Each diode-injection laser is also characterized by a threshold current density J_t where laser action first dominates the combined processes of spontaneous emission and internal absorption. The threshold current density J_t and the radiant output power P_o from the laser are[8]:

$$J_t = 8\pi v^2 n^2 ed \, \Delta v[\alpha L - \ln(R)]/LC^2 n_i, \tag{5.2.24}$$

$$P_o = [(I - I_t)\eta_i hv/e][\ln(R)/(\ln(R) - \alpha L)], \tag{5.2.25}$$

where v is the spectral frequency of laser emission (in hertz), n the refractive index of p–n junction region (dimensionless), e the electronic charge (in coulombs), d the width of p–n junction region (in meters), Δv the width of the p–n junction spontaneous emission (in hertz), η_i the internal quantum efficiency for radiative recombination (dimensionless), c the speed of light in vacuum (in meters per second), α the absorption loss factor (in reciprocal meters), L the distance between reflective faces (in meters), $R \equiv [(n - 1)/(n + 1)]^2$ the reflectance of end faces (dimensionless), I the current through the diode (in amperes), I_t the threshold current for laser action (in amperes), and h the Planck's constant (in Joule · seconds). Laser emission occurs at frequencies $v = E_g/h$. The band-gap energy E_g of the IV–VI semiconductors is dependent on their chemical composition and on temperature. This dependence furnishes physical means for controlling the radiant output frequency of the lasers. The dependence of laser frequency, in wavenumber units of reciprocal centimeters, on chemical composition is demonstrated graphically in Fig. 6 for four different Pb–salt lasers. The temperature and pressure dependence is described in the next paragraph. Continuous wave operation at temperatures as great as 65 K has been achieved with a few IV–VI semi-conductors, but cw operation is generally restricted to temperatures near that for liquid helium. Semiconductor lasers frequently operate simulta-

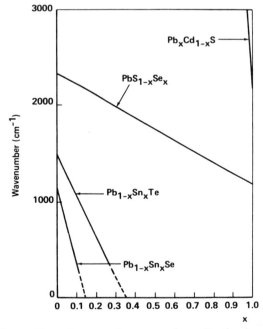

FIG. 6. Dependence of laser frequency, in wavenumber units of reciprocal centimeters on the chemical composition of four Pb–salt diode lasers. The solid lines indicate spectral regions and corresponding chemical compositions for which laser emission has been observed in the laboratory. Reproduced by permission of Arthur D. Little, Inc., Cambridge, Massachusetts.

neously on several modes of the optical cavity. The spectral width of a single mode is stated to be 3×10^{-5} cm^{-1} or less for commercially available lasers, but a width as narrow as 1.7×10^{-6} cm^{-1} was observed from a $Pb_{0.88}Sn_{0.12}Te$ cw laser operating at 943.3 cm^{-1} (10.6 μm). The spatial distribution of the laser emission is directly dependent on the geometrical dimensions of the active laser region of the diode, and thus, in the plane of the p–n junction the full beam divergence has been observed to vary from 0.02 to 0.14 rad.

Several physical parameters, controlled individually or in combination, are used to frequency tune the radiant output from semiconductor lasers. Among these parameters are temperature, forward bias current, magnetic induction, uniaxial pressure, and hydrostatic pressure. The pressure coefficient for the energy gap in PbTe is about $dE_g/dP = -(7.5 \pm 0.3) \times 10^{-6}$ eV/bar. Uniaxial pressure, which changes the length of the laser cavity and the crystalline parameters, is the most easily applied method for frequency tuning, but the lasers are often fractured and thus destroyed because of nonuniformity in the external pressure. Production of uniform, large magnitude, hydrostatic pressures requires very bulky specialized equipment which renders this method of frequency tuning generally impractical for spectroscopic applications. The band gap E_g, and thus the laser frequency, is dependent on temperature. For example,[63] the change in the band gap for PbTe in the temperature range 7–35 K obeys the quadratic relation

$$E_g(T) - E_g(7 \text{ K}) = [(0.72 \pm 0.10) \times 10^{-4} \text{ eV/K}^2]T^2, \quad (5.2.26)$$

while a linear relation

$$E_g(T) - E_g(35 \text{ K}) = [4.85 \times 10^{-4} \text{ eV/K}](T - 35 \text{ K}) \quad (5.2.27)$$

is obeyed in the 35–80 K range. Additionally, the refractive index of the material decreases as the energy gap increases with temperature. The use of forward bias current for frequency tuning is simply a manifestation of the temperature tuning concept. Semiconductors of type IV–VI possess a relatively low thermal conductivity, and therefore, in the vicinity of the p–n junction the Joule heat can raise the temperature several tens of degrees above that of the liquid helium heat sink. An example of current tuning is shown in Fig. 7. Note that as the current increases the laser shifts from one mode to another. Also note that one mode continuously spans a spectral interval 0.3–0.7 cm^{-1} wide, that the spacing between modes is about 2 cm^{-1}, but mode shifting produces a discontinuity in frequency of only about 0.05 cm^{-1}, and that the full spectral region of tunability is about 43 cm^{-1} wide. Present commercially available lasers will span 1–5 cm^{-1} in a single current-tuned optical mode.

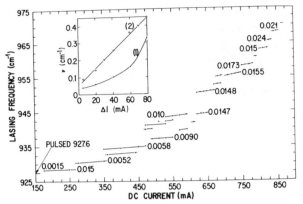

FIG. 7. Lasing frequencies (reciprocal centimeters) versus forward bias current (milliamperes) for a $Pb_{1-x}Sn_xTe$ laser of the etched mesa configuration. Emission is single mode near I_t, the threshold current, then multimode above $3I_t$, and at higher currents single-mode behavior reappears. The detailed tuning curve is shown on the inset for the mode at threshold (1) and the next mode (2) to indicate the nonlinearity. Only the mode at threshold (1) was nonlinear. The tuning rate (reciprocal centimeters) per milliamperes is noted for selected modes. Reproduced by permission of G. A. Antcliffe and S. G. Parker, *J. Appl. Phys.* **44**, 4145 (1973).

Placement of the semiconductor in a region of uniform magnetic induction **B** introduces additional quantum-mechanical phenomena which are also utilized to frequency tune the laser. The Pb IV–VI semiconductors possess four direct minimum, anisotropic, energy gaps at the L point in the Brillouin zone.[57, 70, 71] Surfaces of constant energy for both the valence and conduction band edges are prolate ellipsoids oriented with major axes in the $\langle 111 \rangle$ direction. This band structure is strongly influenced by spin–orbit and by relativistic effects. Both the valence and the conduction bands possess a twofold spin or time-reversal degeneracy. The presence of a magnetic induction **B** removes this twofold degeneracy and provides spin–split Landau levels. Quantum-mechanically, the energies of the lth order Landau sublevels for the conduction (c) and valence (v) bands are

$$E_{cl} = E_c + (l_c + \tfrac{1}{2})\hbar\omega_c + m_{sc}g_c\mu_B B, \qquad (5.2.28)$$

$$E_{vl} = E_v - (l_v + \tfrac{1}{2})\hbar\omega_c + m_{sv}g_v\mu_B B, \qquad (5.2.29)$$

respectively, where E_c and E_v are as previously defined, $\omega_c = eB/m_c^*$, m_s the electron spin quantum number $\pm\tfrac{1}{2}$, g the electron g-factor, μ_B the Bohr magnetron, and $l = 0, 1, 2, 3, \dots$. The effective mass parameters m_c^* and

[70] J. O. Dimmock and G. B. Wright, *Phys. Rev.* **135**, A821 (1965).

[71] J. Callaway, "Quantum Theory of the Solid State," Part A and also Part B. Academic Press, New York, 1974.

m_v^* for the conduction and valence bands, respectively, are nearly the same magnitude for Pb IV–VI semiconductors, while g_v and g_c are approximately equal in magnitude but are thought to be opposite in sign. Electric dipole selection rules require $\Delta l = 0$ during an electronic transition across the band gap. Transitions between states with $l = 0$ are the most probable according to experimental observations.[57] Furthermore, radiation associated with a spin-reversal electronic transition is linearly polarized perpendicular to **B** when propagating perpendicular to **B**. Radiation associated with spin-conserving transitions is linearly polarized in the direction of **B**. Spin-conserving $(+ +, - -)$ and spin-reversing $(+ -, - +)$ transitions between zero-order Landau levels provide the following tuning rates (in reciprocal centimeters per tesla):

$$\Delta \nu / \Delta B = (e/2\pi m^* c) + (m_{sc} g_c - m_{sv} g_v)(\mu_B / hc), \qquad (5.2.30)$$

where $m_s = \pm \frac{1}{2}$ is the quantum number for electron spin. Observation of magnetoemission yielded for PbSe, $m^* = 0.022 m_o$, and $|g_v| = |g_c| = 29$; for PbS, $m^* = 0.05 m_o$, $|g_v| = 7.0$, and $|g_c| = 11.5$; and for PbTe, $m^* = 0.028 m_o$, $|g_v| = |g_c| = 19$. The magnetic induction also effects the refractive index n of the material:

$$(\partial n / \partial B_{[100]})_v \simeq 1.25 \times 10^{-10} \quad T^{-1}. \qquad (5.2.31)$$

The mode equation $\lambda_q = 2nL/q$ is thus effected, but is a smaller influence than the shifting of the Landau levels. A combination of magnetic and current tuning is recommended when continuous frequency tuning is desired over a 40–50 cm^{-1} spectral interval. Spectroscopic applications of semiconductor diodes are presented in Hinkley,[72] Nill et al.,[73] Hinkley et al.,[74] Nill et al.,[75] Antcliffe et al.,[76] and Aronson et al.[77]

5.2.2.4. Spin–Flip Raman Lasers. Spontaneous Raman scattering of infrared radiation from magnetic spin–split Landau states of conduction-band electrons in n-type semiconductors was theoretically predicted in 1966 for the two transitions[78, 79] $\Delta l_c = 2$, $\Delta m_{sc} = 0$ and $\Delta l_c = 0$, $\Delta m_{sc} = 1$, the

[72] E. D. Hinkley, *Appl. Phys. Lett.* **16**, 351 (1970).

[73] K. W. Nill, F. A. Blum, A. R. Calawa, and T. C. Harman, *Appl. Phys. Lett.* **19**, 79 (1971).

[74] E. D. Hinkley, A. R. Calawa, and P. L. Kelley, *J. Phys.* **43**, 3222 (1972).

[75] K. W. Nill, F. A. Blum, A. R. Calawa, and T. C. Harman, *Appl. Phys. Lett.* **21**, 132 (1972).

[76] G. A. Antcliffe, S. G. Parker, and R. T. Bate, *Appl. Phys. Lett.* **21**, 505 (1972). Pb$_{1-x}$Ge$_x$Te laser were used to investigate the spectrum of No near 5.3 μm.

[77] J. R. Aronson, P. C. von Thuna, and J. F. Butler, Tunable Diode Laser High Resolution Spectroscopic Measurements of the ν_2 Vibration of Carbon Dioxide. Arthur D. Little, Inc., Cambridge, Massachusetts, August 1974.

[78] P. A. Wolff, *Phys. Rev. Lett.* **16**, 225 (1966).

[79] Y. Yafet, *Phys. Rev.* **152**, 858 (1966).

latter is the spin–flip transition. Experimental verification[80–85] immediately
followed the predictions from theory, but radiation from an additional
transition $\Delta l_c = 1$, $\Delta m_{sc} = 0$ was also present in the observed spectra.
Spontaneous Raman scattering of infrared radiation emitted by the $P(20)$
line of a CO_2 laser was observed from type III–V and IV–VI semiconductors
which have isotropic and anisotropic band structure, respectively. Referring
to (5.2.28) we find these inelastic scattering processes cause a change in
electronic energy

$$\Delta E_{\Delta lc} = \Delta l_c\, \hbar\omega_c + \Delta m_{sc}\, g\mu_B B, \qquad (5.2.32)$$

where definitions of the symbols were previously stated following (5.2.29).
For n-InSb the g-factor varies from 48 at $B = 0$ T to 36 at $B = 10$ T. A
positive change in electronic energy is characteristic of each of the three
transitions, thus Stokes radiation of frequencies

$$v_{s,\Delta l=0} = v_0 - (g\mu_B B/hc), \qquad (5.2.33)$$

and

$$v_{s,\Delta l \neq 0} = v_0 - \Delta l_c\, (\omega_c/2\pi c) \qquad (5.2.34)$$

were observed, where v_s and v_0 are consecutively the frequencies, in wavenumber units of reciprocal centimeters, of the Raman scattered and the
incident radiation. The spin–flip process in n-InSb, with $n_e \simeq 3 \times 10^{16}$
electrons/ cm^3, had a measured Raman cross section $\sigma \simeq 10^{-23}$ cm^2 sr^{-1}
which was an order of magnitude greater than that measured for the $\Delta l \neq 0$
transitions. The spin–flip Raman emission also had the narrowest linewidth
$\Gamma \simeq 2$ cm^{-1}, full width at one-half maximum intensity. Thus the spin–flip
transition in cleaved and polished, tiny parallelpiped n-InSb crystals was
chosen as a possible process for amplified stimulated Raman scattering.[86]

The Stokes–Raman scattering for the spin–flip process has optical gain
per unit length[87] (in reciprocal centimeters):

$$g_s = \{16\pi^2 c^2 \sigma n_e [f(E_F, B)]/\hbar\omega_s^3 n_0 n_s(\bar{n} + 1)\Gamma\} \qquad (5.2.35)$$

where $f(E_F, B)$ is a factor describing the statistics of the electronic distribution, E_F the Fermi energy, B the magnetic induction, $f(E_F, B) \leqslant 1$ for

[80] A. Mooradian and G. B. Wright, *Phys. Rev. Lett.* **16**, 999 (1966).

[81] R. E. Slusher, C. K N. Pastel, and P. A. Fleury, *Phys. Rev. Lett.* **18**, 77, 530 (1967).

[82] C. K. N. Patel, *Proc. Mod. Opt. Sym., April 1967*, pp. 19–52. Polytechnic Press, Brooklyn,
New York, 1967.

[83] A. Mooradian and A. L. McWhorter, *Phys. Rev. Lett.* **19**, 849 (1967).

[84] C. K. N. Patel and R. E. Slusher, *Phys. Rev.* **167**, 413 (1968).

[85] C. K. N. Patel and R. E. Slusher, *Phys. Rev.* **177**, 1200 (1969).

[86] C. K. N. Patel and E. D. Shaw, *Phys. Rev. Lett.* **24**, 451 (1970).

[87] W. D. Johnson, Jr. and I. P. Kaminow, *Phys. Rev.* **168**, 1045 (1968); **178**, E1528 (1969).

magnetic Raman scattering, $\omega_s = 2\pi\nu_s c$, $n_{0,s} \simeq 4$ are the indices of refraction of n-InSb for the incident and the Raman scattered frequencies, $\bar{n} = [\exp(g\mu_B B/kT) - 1]^{-1}$ the Boltzmann factor, and I the intensity (watts · reciprocal meters squared) of the incident radiation inside the crystal. The gain $(g_s/I) = 1.7 \times 10^{-5}$ cm^{-1}/W · cm^2 for $B = 5$ T and $\nu_0 = 944.14$ cm^{-1}, was calculated to be larger than the combined losses (~ 6 dB) from free-carrier absorption and reflection loss when noncollinear geometry was employed.[86] In this geometry $\mathbf{k}_0$, $\mathbf{k}_s$, and $\mathbf{B}$ are mutually perpendicular with each vector aligned normal to two parallel faces of the crystalline parallelipiped; $\mathbf{k}_{0,s}$ are the propagation vectors for the incident and the Raman scattered waves. Additionally, the incident wave from the pump laser is linearly polarized with the electric field vector $\mathbf{F}_0$ parallel to $\mathbf{B}$ in order to reduce absorption losses. Coherent, stimulated Raman scattering was obtained from the spin–flip transition of conduction electrons in n-InSb by Patel and Shaw at Bell Telephone Laboratories in 1970. The configuration and operating parameters of the experimental apparatus were succinctly described in their original article (p. 452).[86]

A repetitively Q-switched CO_2 laser with a diffraction grating inside the laser cavity was used to provide pump power of ~ 3 kW on a single transition at 10.5915 μ. Apertures in the pump laser cavity assured its operation in the lowest order transverse mode. The laser pulses were 250 nsec wide with a repetition rate of 120 Hz. The pump laser radiation was focused with a 30-cm fl lens into a $n_c = 3 \times 10^{16}$ cm^{-3} InSb sample ($T \approx 25$–$30°$K) in a superconducting solenoid. The sample geometry was $\bar{k}_0$, $\bar{k}_s \perp B_z$ where $\bar{k}_0$, $\bar{k}_s$ are the pump and Raman scattered light wave vectors, and $\bar{k}_0 \perp \bar{k}_s$. No reflective coatings were present on the sample. The Raman scattered radiation resonated in a cavity whose length was $l \approx 2.5$ mm and whose mirror reflectivity was 36% as determined by the refractive index of InSb. The InSb sample was ~ 5 mm long in the $\bar{k}_0$ direction. This is not the ideal geometry for stimulated Raman scattering since the pumped region is a small fraction of the Raman cavity length while the free carrier absorption occurs over the entire length l. A more desirable geometry would be one where $\bar{k}_0$ and $\bar{k}_s$ were collinear. However, the present geometry is convenient for the detection of Raman scattered light below the stimulated threshold. The pump radiation was linearly polarized along $\bar{B}_z$. The Raman scattered radiation was analyzed with a (3/4)-m spectrometer. A Ge:Cu detector was used. An ether absorption cell was used for varying the pump laser intensity.

Further experimental[88–90] and theoretical[91] investigations of pulsed CO_2 laser operated spin–flip Raman (SFR) lasers of n-InSb yielded the following eight important conclusions.

[88] C. K. N. Patel, *Proc. Symp. Submillimeter Waves*, pp. 135–155. Polytechnic Press, Brooklyn, New York, 1970.

[89] C. K. N. Patel, E. D. Shaw, and R. J. Kerl, *Phys. Rev. Lett.* **25**, 8 (1970).

[90] E. D. Shaw and C. K. N. Patel, *Appl. Phys. Lett.* **18**, 215 (1971).

[91] C. K. N. Patel and E. D. Shaw, *Phys. Rev.* **3**, B1279 (1971).

(1)   The upper limit of the spectral output width of the SFR laser was measured as 0.03 cm^{-1} by use of instruments that were not capable of greater spectral resolution. The Townes formula[92] establishes the spectral output width as

$$\Delta v_s \ (\text{cm}^{-1}) \approx 8\pi hc^2 v_s \Gamma^2 / P, \qquad (5.2.36)$$

where P is the output power of the laser under cw operation; thus $\Delta v_s \approx 5.65 \times 10^{-9}$ cm^{-1} (~ 1.7 Hz) at $v_s = 945$ cm^{-1} and $P = 10$ W. Indeed, this was an extremely narrow spectral width to be anticipated from future SFR infrared lasers.

(2)   The output from the SFR laser did not discontinuously shift from one cavity mode to another as B was continuously changed from about 5 to 10 T. Only a 10–20% modulation of output intensity was observed with a spectral periodicity corresponding to the calculated mode spacing.

(3)   A saturation of Raman output power was observed when the CO_2 laser was operated in the Q-switched or mode-locked phase. The saturation effect is due to the relatively long relaxation time for conduction electrons to decay from the $m_{sc} = -\frac{1}{2}$ state back to the $m_{sc} = +\frac{1}{2}$ state.

(4)   There was a threshold value B_{min} above which stimulated Raman scattering occurred. The minimum magnetic induction was required to shift the $l_c = 0$, $m_{sc} = -\frac{1}{2}$ energy level above the Fermi energy E_F so that all the conduction electrons were in the lower energy $l_c = 0$, $m_{sc} = +\frac{1}{2}$ state and thus available for the spin–flip transition. Therefore, when $g\mu_B B \leqslant 2kT$, the B_{min} threshold for stimulated emission is determined by the quantum limit $\gamma(E_F, B_{min}) \approx 1$ for the $m_{sc} = -\frac{1}{2}$ state. The threshold B_{min} decreased monotonically for n-InSb samples with decreasing carrier concentration n_e.

(5)   There was also an upper limit B_{max} above which g_s the Raman gain per unit length fell below the absorption and reflection losses and thus the stimulated Raman scattering ceased. The upper limit B_{max} was determined by the cross section σ, which decreases as B increases, and by free carrier absorption at both v_0 and v_s frequencies which increases as B increases.

(6)   Stimulated Raman scattering was theoretically feasible in n-InSb with $1.5 \times 10^{15} \leqslant n_e \leqslant 9 \times 10^{16}$ electrons/cm^3.

(7)   Continuous frequency tunability linear with respect to changing B was observed throughout the spectral range $g\mu_B(B_{max} - B_{min})/hc$.

(8)   Stimulated *anti*-Stokes SFR laser action, a complex process to manage theoretically, was observed by use of the following geometry: $\mathbf{k_0}$, $\mathbf{k_s}$, $\mathbf{k_{as}}$, and $\mathbf{k_e}$ all collinear and perpendicular to $\mathbf{B}$. Here $\mathbf{k_0}$, $\mathbf{k_s}$, $\mathbf{k_{as}}$, and $\mathbf{k_e}$ are wavevectors for the incident wave, the Stokes and *anti*-Stokes waves,

[92] C. H. Townes, "Advances in Quantum Electronics" (J. R. Singer, ed.). Columbia Univ. Press, New York, 1961.

and the physical entity responsible for the scattering, respectively. The collinear geometry provided for conservation of linear momentum

$$\mathbf{k}_0 = \mathbf{k}_s + \mathbf{k}_e, \qquad \mathbf{k}_0 = \mathbf{k}_{as} - \mathbf{k}_e, \qquad (5.2.37)$$

and

$$2\mathbf{k}_0 = \mathbf{k}_s + \mathbf{k}_{as}. \qquad (5.2.38)$$

Anti-Stokes scattering is a process whereby energy is added to the scattered wave thus increasing its frequency; this may be visualized as $\Delta m_{sc} = -1$ spin–flip transition giving

$$\nu_s = \nu_0 + (g\mu_B B/hc), \qquad (5.2.39)$$

thus combining the anti-Stokes and Stokes radiation effectively doubles the region of tunability.

The mathematical expression for the cross section σ for Raman scattering from conduction-band electrons[78, 79] contains an enhancement factor σ_{eh} of the form

$$\sigma_{eh} = [E_g^2/(E_g^2 - E_0^2)]^2, \qquad (5.2.40)$$

where E_g is the effective band–gap energy and E_0 the energy of the incident photons. The band-gap energy for InSb is $E_g = 1890$ cm^{-1} at $B = 0$. The energy of photons from the $P(20)$ line of a CO_2 laser is ~ 945 cm^{-1} and from the $P_{9-8}(12)$ line of a CO laser is 1888.3 cm^{-1}; thus considering all other terms in the cross section as constants, a relative enhancement of $\sim 10^5$ is achieved by use of the CO laser rather than the CO_2 laser. Both cw and pulsed CO laser-operated SFR lasers have been reported in the literature.[93–96] The efficiency of conversion of CO laser radiation to SFR laser radiation is nominally 40%. In this physical situation, a fraction of the large density of I-Stokes radiation interacts with another electron in the $m_{sc} = -\frac{1}{2}$ state causing a spin–flip to the $m_{sc} = +\frac{1}{2}$ state and thus providing II-Stokes radiation of frequency

$$\nu_{IIs} = \nu_0 - (2g\mu_B B/hc) \qquad (5.2.41)$$

from the same crystal. The II-Stokes frequencies are tunable over twice the spectral range of that for the I-Stokes frequencies. The measured value of the spectral output width of the cw SFR laser[97] operating near $\nu_s = 1887$

[93] A. Mooradian, S. R. J. Brueck, and F. A. Blum, *Appl. Phys. Lett.* **17**, 481 (1970).
[94] S. R. J. Brueck and A. Mooradian, *Appl. Phys. Lett.* **18**, 229 (1971).
[95] C. K. N. Patel, *Appl. Phys. Lett.* **18**, 274 (1971).
[96] C. K. N. Patel, *Appl. Phys. Lett.* **19**, 400 (1971).
[97] C. K. N. Patel, *Phys. Rev. Lett.* **28**, 649 (1972).

cm^{-1} was less than 3.33×10^{-6} cm^{-1} (1 kHz), which was the minimum resolution of the spectrum analyzer used as the measuring device. By reducing the free carrier concentration to 10^{15} electrons/cm^3 in n-InSb, the threshold magnetic flux density B_{min} was reduced to 0.04 T. Superconducting magnets, therefore, are not required for operation of the SFR laser if the semiconductor is properly grown and annealed and if the CO laser is the initial radiation source. C.K.N. Patel[86] again concisely expressed the pertinent operating parameters of this splendid instrument when he reported the discovery of the tunable cw SFR laser operating at magnetic flux density as low as 0.04 T; we quote (p. 452):

A cw CO laser with an intracavity grating was used to obtain a pump laser power of 3 W at a frequency of 1888.32 cm^{-1} [on the $P_{9\ 8}(12)$ transition of CO] which is very close to the zero magnetic field band gap of 1890 cm^{-1} for InSb. The InSb sample was $2- \times 2- \times$ 4-mm parallelepiped with the two $2- \times$ 2-mm surfaces optically polished. The electron concentration used was 1×10^{15} cm^{-3} (nominal) and the electron mobility was 3.5×10^5 cm^2/V sec (77° K). It should be noted that this mobility is more than three times that for a 1.3×10^{16}-cm^{-3} InSb sample. The InSb sample was mounted on a cold finger ($T \approx 18°$ K) in a superconducting solenoid (even though the superconducting solenoid is not essential for the very low magnetic fields used in most of the work reported here) such that k_0 was normal to the magnetic field, and the long dimension of the sample was along k_0. Thus the SFR radiation, if any, will be resonated collinearly with the pump radiation and normal to the magnetic field. The pump radiation was focused into the InSb sample with a 20-cm focal lens. The collinearly emerging pump laser as well as SFR scattered radiation was analyzed with a (3/4)-m grating spectrometer and detected with a Au:Ge (77° K) detector.

Figure 8 shows spectrometer analysis of the output for (a) $B = 2.53$ kG and (b) $B = 5.3$ kG. The output is plotted on a logarithmic scale and is normalized to the pump radiation. In Fig. 8(a) we see the cw operation of the anti-Stokes SFR laser together with cw operating I- and II-Stokes SFR lasers. Increasing the magnetic field to 5.3 kG in Fig. 8(b) shows the anti-Stokes, I-Stokes, and II-Stokes SFR laser lines moving away from the pump laser line. The maximum anti-Stokes SFR laser output occurred at $B \approx 3.14$ kG and was about 30 mW, i.e., 1% conversion. The maximum Stokes SFR laser power occurred at a somewhat higher magnetic field of $B \approx 16.9$ kG and ~ 0.75 W. Considering that the SFR laser power leaves equally from both ends of the InSb sample, $\sim 50\%$ conversion was achieved even at a low magnetic field of 16.9 kG. At $B \approx 12$ kG, the conversion was $\sim 30\%$. At $B \approx 500$ G, the anti-Stokes SFR laser output was ~ 15 mW, I-Stokes laser output ~ 200 mW, and II-Stokes laser output was ~ 15 mW.

The polarizations of the SFR lasers were checked with a wire-grid polarizer. These polarization measurements are as follows: (i) CO pump laser polarization-parallel to B; (ii) I-Stokes SFR laser polarization-normal to B; (iii) II-Stokes SFR laser polarization-parallel to $\mathbf{B}$; and (iv) anti-Stokes SFR laser polarization-normal to $\mathbf{B}$. These polarizations are in accordance with expectations from theory.

Spectra from a cw SFR laser pumped by a CO laser are shown in Fig. 8. The frequency tuning characteristics of the I-Stokes, II-Stokes, and anti-Stokes radiation from the cw SFR laser are presented graphically in Fig. 9. Because of its several unique and splendid characteristics, the SFR laser

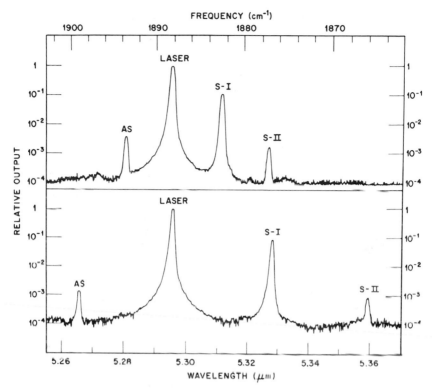

FIG. 8.   Spectra from a cw SFR laser operating in the collinear configuration at (top) $B = 2.53$ kG (0.253 T) and (bottom) $B = 5.3$ kG (0.53 T). The spectral lines are from left to right the anti-Strokes (AS), the $P_{9-8}(12)$ line from the CO laser, and the I-Strokes (S-I) and II-Strokes (S-II) emissions. Note, the scale of relative output intensity is logarithmic with normalization to the peak intensity of the $P_{9-8}(12)$ line. The peak intensities span four orders of magnitude. Reproduced by permission of C. K. N. Patel, *Appl. Phys. Lett.* **19**, 400 (1971).

was used recently for several spectroscopic applications[98] [102] and it is certainly destined for greater use in the future.

5.2.2.5. Dye Lasers.   Although it is a relatively new instrument the dye laser has achieved ubiquity among spectroscopic laboratories. It is the most practical of the tunable lasers because of its relatively simple construction and its broad spectral range of tunability; and because dye lasers can be

[98] L. B. Kreuzer and C. K. N. Patel, *Science* **173**, 45 (1971).

[99] L. B. Kreuzer, N. D. Kenyon, and C. K. N. Patel, *Science* **177**, 347 (1972).

[100] C. K. N. Patel, "Coherence and Quantum Optics" (L. Mandel and E. Wolf, eds.), pp. 567–593. Plenum Press, New York, 1973.

[101] C. K. N. Patel, E. G. Burkhardt, and C. A. Lambert, *Science* **184**, 1173 (1974).

[102] C. K. N. Patel, *Appl. Phys. Lett.* **25**, 112 (1974).

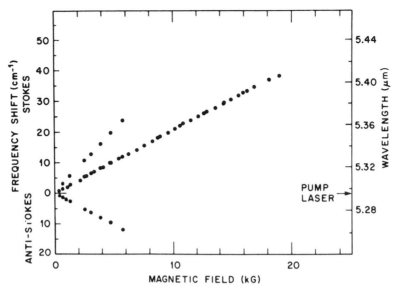

FIG. 9. Wavelength (microns) tuning characteristics of the cw SFR laser at low magnetic induction (10^4 G $= 1$ T). Tunability extended to $B = 10$ T for the I-Stokes emission. The experimental points, lying on the three different imaginary straight lines of successively decreasing slope, are for II-Stokes, I-Stokes, and anti-Stokes emission, respectively. Reproduced by permission of C. K. N. Patel, *Appl. Phys. Lett.* **19**, 400 (1971).

mode locked to yield radiant pulses of picosecond duration, pumped naturally with N_2 lasers to yield radiant pulses of nanosecond duration, pumped with flash lamps to yield radiant pulses of microsecond duration, or pumped with ion lasers to yield cw laser action. This subsection contains a brief discussion of the history of dye lasers, of the construction of N_2 laser-pumped dye lasers, and of the basic physics of these pulsed dye lasers. The reader seeking knowledge of topics such as flashlamp pumped dye lasers, cw dye lasers, mode locking, chemical structure and properties of laser dyes, and additional applications of dye lasers is referred to the book by Schafer and his colleagues[103] and to the references therein. A list of over 100 lasing dyes, their wavelengths, pumping conditions, and original references is tabulated in the Appendixes of the reports by Querry *et al.*[104, 105]

[103] F. P. Schafer, Dye lasers, "Topics in Applied Physics," Vol. 1. Springer-Verlag, Berlin and New York, 1973.

[104] M. R. Querry, R. C. Waring, and G. Mansell, Development and Construction of a New Spectrophotometer, 1st Annu. Rep., NASA-CR-130136, 30 July 1971. Available from Nat. Tech. Inform. Serv., Springfield, Virginia 22151.

[105] M. R. Querry, R. C. Waring, and W. E. Holland, Reflectance of Aqueous Solutions, Final Rep., NASA-CR-130136, 21 July 1972. Available from Nat. Tech. Inform. Serv., Springfield, Virginia 22151.

5.2.2.5.1. HISTORICAL REVIEW. Two independent discoveries of stimulated emission from organic dyes were reported in 1966.[106, 107] One of the discoveries resulted from the investigative use of a giant pulse ruby laser to produce enhanced Raman emission from chloroaluminum phthalocyanine dissolved in ethyl alcohol. Although the enhanced Raman emission was not observed, the spectra did contain a diffuse band centered at 755.5 nm in the near infrared which is the spectral region of intense emission from one fluorescence band of phthalocyanine. This diffuse band was judged to be incipient laser action. A cell containing the phthalocyanine solution was placed in a Fabry–Perot optical cavity. The giant pulse from the ruby laser was incident on the dye cell from a direction normal to axis of the cavity. A powerful secondary laser pulse of wavelength 755.5 nm due to stimulated emission from phthalocyanine was readily observed parallel to the axis of the optical cavity. Within a short time stimulated emission was observed from red-absorbing, *near-infrared-emitting*, organic dyes.[108–112] Stimulated emission of *visible* light from several different organic dyes [109, 113, 114] was reported in 1967. The stimulated emission of visible light was observed when the dyes were externally excited by second-harmonic light generated by passing a ruby or neodymium laser beam through a phase-matched ammonium-dihydrogen-phosphate (ADP) crystal. By late 1967 stimulated emission had been observed from 16 different dyes. The wavelengths of laser light from these dyes ranged discontinuously from 438.5 nm in the blue for acridone dissolved in ethyl alcohol[109] to 816 nm in the near infrared for 3,3′-diethylthiatricarbocyanine iodide[109] dissolved in dimethyl sulfoxide.

During the investigations of the ruby and neodymium laser-pumped organic dye lasers, it became obvious[106, 115, 116] that stimulated emission could be produced from organic dyes by flashlamps having rise times less than a few tenths of a microsecond and output power of ~100 kW. The fast rise time avoided an accumulative population of dye molecules in the lowest-energy metastable triplet electronic state. The high power produced the inverted population of dye molecules in the first excited singlet electronic

[106] P. P. Sorokin and J. R. Lankard, *IBM J. Res. Develop.* **10**, 162 (1966).

[107] F. P. Schafer, W. Schmidt, and J. Volze, *Appl. Phys. Lett.* **9**, 306 (1966).

[108] P. P. Sorokin, W. H. Culver, E. C. Hammond, and J. R. Lankard, *IBM J. Res. Develop.* **10**, 401 (1966).

[109] P. P. Sorokin, J. R. Lankard, E. C. Hammond, and V. L. Moruzzi, *IBM J. Res. Develop.* **11**, 130 (1967).

[110] M. L. Spaeth and D. P. Brotfeld, *Appl. Phys. Lett.* **9**, 179 (1966).

[111] M. Bass and T. F. Deutsch, *Appl. Phys. Lett.* **11**, 89 (1967).

[112] B. I. Stepanov, A. N. Rubinov, and V. A. Mostovinkov, *JETP Lett.* **5**, 117 (1967).

[113] F. P. Schafer, W. Schmidt, and K. Marth, *Phys. Lett.* **24A**, 280 (1967).

[114] B. B. McFarland, *Appl. Phys. Lett.* **10**, 208 (1967).

[115] P. P. Sorokin and J. R. Lankard, *IBM J. Res. Develop.* **11**, 148 (1967).

[116] W. Schmidt and F. P. Schafer, *Z. Naturforsch*, **22a**, 1563 (1967).

state, which was necessary for initiating the stimulated emission process. Flashlamp systems were constructed to the specified requirements and thus stimulated emission was readily observed from flashlamp-excited dye lasers in several independent laboratories.[115-124]

The first conveniently wavelength tunable organic dye laser was reported in 1967.[125] The dye rhodamine 6G dissolved in ethyl alcohol ($\sim 10^{-4}$ M concentration), when placed in an Fabry–Perot cavity composed of two dielectric coated first-surface mirrors produced a continuous 6.0-nm-wide band of stimulated emission when excited by the second-harmonic output of a gaint pulse ruby laser. It was reasoned that spectral narrowing of the output pulse from the dye laser could be accomplished by replacing one of the dielectric mirrors with a plane, reflective, diffraction grating. Output pulses of 0.06-nm spectral width were observed from the dye laser by use of a grating with 2160 lines/mm, blazed at 500 nm in first order, and reflectance of 80% for wavelengths greater than 552 nm. Furthermore, by rotating the grating, held in a Littrow mount, the 0.06-nm-wide output pulse was continuously wavelength-tuned from 552 to 595 nm. The energy per pulse for the spectrally narrowed output was 60% of that for the broad-band pulses. Other techniques for wavelength tuning the organic dye laser are:

(1) increasing the dye concentration in the optical cavity shifts dye laser output to longer wavelengths,[107, 109]

(2) making relatively large alterations in the length of the optical resonant cavity changes the macroscopic self-absorption characteristics of the dye solution and thereby shifts the spectral output,[110, 126]

(3) inserting a Fabry–Perot etalon[127] or interferometer[128] in the dye laser cavity, in addition to the diffraction grating further narrows the spectral

[117] P. P. Sorokin, J. R. Lankard, V. L. Moruzzi, and E. C. Hammond, *J. Chem. Phys.* **48**, 4726 (1968).

[118] B. B. Snaveley, O. G. Peterson, and R. F. Reithel, *Appl. Phys. Lett.* **11**, 275 (1967).

[119] M. Bass, T. F. Deutsch, and M. J. Weber, *Appl. Phys. Lett.* **13**, 120 (1968).

[120] D. W. Gregg and S. J. Thomas, *IEEE J. Quantum. Electron.* (Notes and Lines) **QE5**, 302 (1969).

[121] H. Furumoto and H. Ceccon, *J. Appl. Phys.* **40**, 4204 (1969).

[122] M. R. Kagan, G. I. Farmer, and B. G. Huth, *Laser Focus*, 26 (1968).

[123] D. W. Gregg *et al.*, Lawrence Radiat. Lab., Univ. of California, Livermore Rep. UCLR–72044, October 15, 1969.

[124] J. B. Marling, D. W. Gregg, and S. J. Thomas, *IEEE J. Quantum Electron.* **QE-6**, 370 (1960).

[125] B. H. Soffer and B. B. McFarland, *Appl. Phys. Lett.* **10**, 266 (1967).

[126] G. I. Farmer, B. G. Huth, L. M. Taylor, and M. R. Kagan, *Appl. Opt.* **8**, 363 (1969).

[127] D. J. Bradley, G. M. Gale, M. Moore, and P. D. Smith, *Phys. Lett.* **26**, A378 (1968).

[128] A. M. Bonch-Bruyevich, N. N. Kostin, and V. A. Khodovoi, *Opt. Spectrosc.* **24**, 547 (1968).

width of the output and allows wavelength tuning within the free spectral range of the etalon or inteferometer,

(4) replacing one of the dielectric mirrors with a rotating Littrow prism,[129]

(5) changing the temperature[130] of a dye solution from 351 to 156 K continuously decreased the dye laser wavelength by 20 nm,

(6) inserting a Q-switch in the dye laser cavity,[131] then opening the Q-switch when the gain per unit length is a maximum for the desired wavelength, and

(7) changing the gas pressure inside an additional Fabry–Perot interferometer which is inserted in the dye laser cavity.[132, 133]

Use of a pulsed N_2 laser as a more convenient means for exciting stimulated emission from organic dyes was reported in 1969.[134] The developers of this system obtained continuously tunable stimulated emission in the spectral range 350–650 nm by use of only nine dyes. The spectral output width ranged from 0.2 to 1.0 nm. The pulse repetition rate was 1–25 Hz. The energy per pulse was greatly stabilized and the repetition rate was increased to 100 Hz by adding a system to circulate cooled dye solutions through the laser cavity.[104] The N_2 laser-pumped dye laser was immediately incorporated as the basic component of a spectrophotometer system for investigating the optical properties of aqueous solutions.[104, 105, 135, 136] The N_2 laser-pumped dye laser is the type most widely used by spectroscopists. Dye lasers are currently available for purchase from several different manufacturers.[21] The N_2 laser-pumped dye lasers are wavelength tunable from 360 to 720 nm and by frequency doubling from 230 to 360 nm. Flashlamp-pumped dye lasers are tunable from ∼0.4 to 1.2 μm.

5.2.2.5.2. Dye-Laser Construction. The configuration of an N_2 laser-pumped dye laser[104] is represented by Fig. 10. Its operation is described as follows: Pulses of ultraviolet radiation of wavelength 337.1 nm and rectangular in cross section originated from the N_2 laser at repetition rates of 1 to 500 Hz. The duration of the pulse was about 10 nsec, and the peak power per pulse was 100 kw to 10 MW dependent on the type of N_2 laser. The pulses of uv radiation were focused to a line along the inner surface of a rectangular,

[129] S. Murakawa, G. Yamaguchi, and C. Yamanaka, *Jap. J. Appl. Phys.* **7**, 681 (1968).

[130] G. T. Schoppert, K. W. Billman, and D. C. Barnham, *Appl. Phys. Lett.* **13**, 124 (1968).

[131] M. Bass and J. I. Steinfeld, *IEEE J. Quantum. Electron.* **QE-4**, 53 (1968).

[132] R. Wallenstein and T. W. Hansch, *Appl. Opt.* **13**, 1625 (1974).

[133] R. Flach, I. S. Shahin, and W. M. Yen, *Appl. Opt.* **13**, 2095 (1974).

[134] J. A. Meyer, C. L. Johnson, E. Kierstead, R. D. Sharma, and I. Itzkan, *Appl. Phys. Lett.* **16**, 3 (1970).

[135] M. R. Querry, G. Mansell, and R. C. Waring, *J. Opt. Soc. Amer.* **61**, 680A (1971).

[136] M. R. Querry, R. C. Waring, W. E. Holland, and G. R. Mansell, *Proc. Int. Symp. Remote Sensing Environ., 7th, Univ. of Michigan, May* 1971 (J. Cook, ed). **2**, 1053–1069.

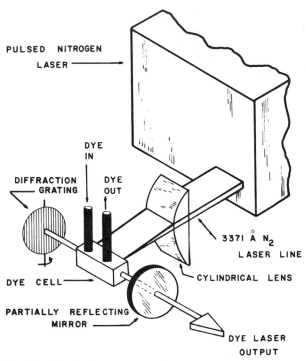

PULSED NITROGEN
LASER

DYE
IN

DIFFRACTION
GRATING

DYE
OUT

3371 Å N$_2$
LASER LINE

DYE CELL

CYLINDRICAL LENS

PARTIALLY REFLECTING
MIRROR

DYE LASER
OUTPUT

FIG. 10.   A drawing of the basic components of an organic dye laser as optically pumped by a pulsed N$_2$ laser; see text for discussion. Adapted from M. R. Querry, R. C. Waring, and G. Mansell, Development and Construction of a New Spectrophotometer, 1st Annu. Rep., NASA–CR–130136, 30 July 1971, and from J. A. Meyer, C. L. Johnson, E. Kierstead, R. D. Sharma, and I. Itzkan, *Appl. Phys. Lett.* **16**, 3 (1970).

flow-through, quartz dye cell. The dye cell was in an optical cavity having a reflectance grating at one end and a plane, first-surface mirror at the other end. The grating had 600 lines/mm, was blazed at 54°06', and was operated in 5th order. The laser cavity was about 12 cm long. The dye cell was 6 cm long and was optically pumped transversely along its entire length by the uv radiation. The dye cell was vertically askew in the laser cavity to prevent lasing action between its end faces. The cooled dye solution was recirculated continuously at ∼ 4 liter/min through the cavity by a variable speed mechanical micropump operating from an external reservoir. Wavelength tuning was accomplished by rotating the grating in its Littrow mount which was driven by a stepper motor assembly. The cavity radiation illuminated a small spot about 2 mm in diameter on the face of the grating. A mechanical counter (not shown in Fig. 10) coupled to the grating drive assembly rotated about

4×10^4 times faster than the grating and thus provided a wavelength reference

$$\sin^{-1}(mN\lambda/2) = a_1 c + a_2, \qquad (5.2.42)$$

where m is the diffraction order, N the number of grating (lines per millimeter), λ the wavelength, c the reading from the counter, and a_1 and a_2 constants characteristic of the drive assembly. An iris diaphram acting as the limiting aperture was placed in the dye laser beam. The output pulses were plane polarized and were of spectral width ~ 0.2 nm FWHM in the 570–600-nm wavelength region. The pulses were detected with an ITT FW114A photodiode of S-20 response. A boxcar integrator sampled the current from the photodiode during a 10–15 nsec interval. By integrating current, which was instantaneously linearly proportional to the power of the dye laser pulse, over the time duration of the pulse, the boxcar integrator provided a measure of the energy per dye laser pulse. The spectra were recorded on strip charts. Lengthening the dye laser cavity and then inserting an expanding–collimating telescope, which was composed of a spatial filter (40 × Leitz microscope objective focused on a 10-μm diameter pin hole centered in a thin opaque disk) and an Oriel acromat lens, No. A-18-161-66, spread the inner cavity radiation over the entire face of the diffraction grating, thus providing spectral output widths[137] of $\leqslant 0.02$ nm.

Depicted in Fig. 11 is the configuration of components of a pressure-tuned multielement dye laser[132] which provided spectral output widths of 25 MHz (3×10^{-5} nm) at 600 nm. The narrow spectral output width was continuously tunable over 150 GHz (0.18 nm). A dye cell 10 cm in length was transversely pumped by a N_2 laser operating at 100-kW peak power and 50-Hz repetition rate. The grating had 600 lines/mm and was blazed at 58°. A 25 × beam expanding–collimating telescope and the grating provided initial spectral narrowing. A tilted Fabry–Perot interferometer, invar spaced to $d = 2$–6 mm, provided intermediate spectral narrowing. Final spectral narrowing to 25 MHz was provided by an external invar-spaced confocal interferometer of 2-GHz free spectral range and finesse of 200. The grating, Fabry–Perot interferometer, and external confocal resonator were enclosed in a common vacuum chamber as shown in Fig. 11. The vacuum chamber was evacuated to a few Torr, and N_2 gas from an external tank was slowly admitted to the vacuum chamber through a needle valve. Allowing the N_2 pressure to go from a few Torr to 760 Torr tuned the output of the laser through a 146.2-GHz spectral range. The frequency stability was to better

---

[137] See M. R. Querry, R. C. Waring, and W. E. Holland, Reflectance of Aqueous Solutions. Final Rep., NASA–CR–130136, 21 July 1972, and interim progress reports associated with that contract. Also see T. W. Hansch, *Appl. Opt.* **11**, 895 (1972).

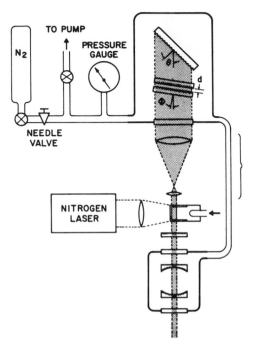

FIG. 11.   A drawing of the basic components of a pressure-tuned multielement dye laser. Reproduced by permission from R. Wallenstein and T. W. Hansch, *Appl. Opt.* **13**, 1625 (1974).

than ± 200 MHz over several hours of continual operation at a single frequency setting. Propane gas whose index of refraction is different from that of N_2 provided a tuning range of 142.5 GHz for the same range of pressure. Because the tuning range was so broad compared to the spectral output width of the laser, 7 hr were required to cover the 146.2-GHz tuning range at a scan rate of 6 MHz/sec. Spectral resolving power $\lambda/\Delta\lambda$ for this dye laser is about 2×10^7.

5.2.2.5.3. PHYSICS OF THE DYE LASERS. In terms of electronic potential energy, there are four energy states of large organic dye molecules that are significant to the lasing process. The four energy states are the ground singlet, first excited singlet, lowest-energy triplet, and first excited triplet which are separated by potential energies characteristic of electronic transitions. Each electronic state is further composed of a broad band of vibrational–rotational states so closely spaced that they approach a continuum. The first excited singlet state is metastable with a fluorescence lifetime of about 10 nsec.[138]

[138] M. Bass, T. F. Deutsch, and M. J. Weber, NASA Rep. CR–1374, Contract NASA 12–635, 11 July 1969.

The lowest triplet state is also metastable, with a phosphorescence lifetime greater than 0.1 msec.[138] In solution the organic dye molecules can also perform nonradiative transitions from the first excited singlet to the ground singlet state or nonradiative intersystem crossing from the first excited singlet to the lower triplet state. The nonradiative transitions are associated with phonon emission.[139]

Photons from a laser or flashlamp are absorbed by a dye molecule thus causing a rapid transition from the ground singlet state to some vibrational level in the first or higher excited singlet state. Radiative and nonradiative transitions then occur with the molecule being left in some lower vibrational level of the first excited singlet state. At this point the molecule can experience either a natural fluorescent or stimulated radiative transition to some vibrational level in the ground singlet state, or an undesirable nonradiative transition to some vibrational level of the lower triplet state. Transitions from the triplet state to the ground singlet state are spin forbidden; the triplet state, therefore, acts as a trap that holds molecules from further participation in the lasing process. To further quench the lasing process, absorption from the lower triplet to an excited triplet can occur and may be of an energy equal to that of photons emitted by singlet–singlet transitions. The accumulation of molecules in the triplet state is extremely detrimental to the operation of dye lasers.

The transition from the first excited singlet to the ground singlet, which is the stimulated transition important to the lasing process, is represented in Fig. 12 as a plot of potential energy versus some generalized coordinate of the molecule. The singlet–singlet transition can be interpreted in terms of the Frank–Condon principle[140] as follows. Organic dyes that have nearly identical absorption and emission spectra, which is the general case, can be described as having vibrational potential energy functions whose classical turning points are nearly vertically collinear as shown in Fig. 12. Self-absorption would be expected to have a significant role in quenching the laser action at some frequencies in materials possessing such properties because the absorbed and emitted photons have nearly the same energy. If the classical turning points are displaced due to nonradiative transitions, then there is a Frank–Condon shift and the absorption and emission spectra will peak at different frequencies, thus the laser output will not display pronounced self-absorption effects. Stimulated emission from chloroaluminum phthalocyanine, which has a large amount of self-absorption, and from 3,3'-diethylthiatricarbocyanine which exhibits a limited amount of self-absorption, was observed by using giant pulse ruby lasers or specially

[139] D. E. McCumber, *Phys. Rev.* **134**, A299 (1964).
[140] E. U. Condon and H. Odishaw, "Handbook of Physics." McGraw-Hill, New York, 1958.

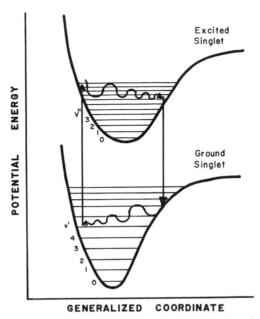

GENERALIZED COORDINATE

FIG. 12. Hypothetical potential energy graphs and related vibrational energy levels for two typical singlet electronic states of an organic dye molecule. Rotational energy levels were deleted in order to simplify the diagram.

designed flashlamps[109] as the optical pumping source for the dye laser.

Analytical studies of the population dynamics of the electronic states, the photon densities, and gain per unit length were reported in 1967–1969.[108, 117, 119, 138, 141] A set of coupled rate equations,[142] which treat the populations of electronic states and photon densities as rapidly varying functions of time, were modified to include the effects of the triplet states as follows:

$$dN_1/dt = N_0 P(t) - (N_1/\tau), \tag{5.2.43}$$

$$dN_T/dt = k_{ST} N_1 - (N_T/\tau_T), \tag{5.2.44}$$

$$dN_0/dt = -P(t)N_0 + [(1/\tau) - k_{ST}]N_1 + (N_T/\tau_T), \tag{5.2.45}$$

where N_1, N_T and N_0 were the number of dye molecules per cubic centimeter in the first excited singlet, the triplet, and the ground electronic states, respectively; $P(t)$ the number of excitation photons per unit time, τ the exponential fluorescence (singlet–singlet transitions) lifetime, τ_T the phosphorescence (triplet–singlet transitions) lifetime, and k_{ST} was the lifetime for nonradiative

[141] B. B. Snaveley and O. G. Peterson, *IEEE J. Quantum. Electron.* **QE-4**, 540 (1968).
[142] D. Roess, *J. Appl. Phys.* **37**, 2004 (1966).

intersystem crossing (singlet–triplet plus phonons). The gain $G(\lambda)$ per unit length was[141]

$$G(\lambda) = A(\lambda)N_1 - N\varepsilon(\lambda)_{ss}(L_2/L_1) - N_T[\varepsilon(\lambda)_{TT} - \varepsilon(\lambda)_{ss}] + (1/2L_1)\ln(r_1r_2), \tag{5.2.46}$$

where

$$A(\lambda) = [\lambda^4 E(\lambda)n(\lambda)/8\pi\tau c] + \varepsilon(\lambda)_{ss}, \tag{5.2.47}$$

and λ was the wavelength, $E(\lambda)$ the lineshape for spontaneous emission, $n(\lambda)$ the index of refraction of the dye solution, $\varepsilon(\lambda)_{ss}$ and $\varepsilon(\lambda)_{TT}$ extinction coefficients for singlet–singlet and triplet–triplet absorption, respectively, L_1 the length of optically active region of the dye solution, L_2 the total length of the dye cell, r_1 and r_2 reflectance values for the end mirrors, and N the total number of dye molecules per cubic centimeter. Analysis of (5.2.46) at the threshold of laser oscillation provides

$$N_T = \left(\frac{a_1 - a_4}{a_5 - a_2}\right)\frac{NL_2}{L_1} + \frac{a_3}{a_5 - a_2}, \tag{5.2.48}$$

where

$$a_1 = \varepsilon(\lambda)_{ss}/A(\lambda), \tag{5.2.49}$$

$$a_2 = [\varepsilon(\lambda)_{TT} - \varepsilon(\lambda)_{ss}]/A(\lambda), \tag{5.2.50}$$

$$a_3 = \ln(r_1r_2)/2L_1A(\lambda), \tag{5.2.51}$$

$$a_4 = \varepsilon'(\lambda)_{ss}/A'(\lambda), \tag{5.2.52}$$

$$a_5 = [\varepsilon'(\lambda)_{TT} - \varepsilon'(\lambda)_{ss}]/A'(\lambda). \tag{5.2.53}$$

In (5.2.52) and (5.2.53) the primes denote differentiation with respect to the wavelength λ. A relation between N_1 and N_T was obtained from the rate equations: assuming (1) that $P(t)$ increased linearly with time, (2) that N_1 was proportional to the intensity of the pump radiation, (3) that $t \ll \tau_T$, and (4) that the triplet state was quenched gave

$$N(t)_T/N(t)_1 = (k_{sT}/2)\tau_T. \tag{5.2.54}$$

Because space limitations prevent further analysis of the dye laser in this section, the reader is referred to a full treatment of the subject by Schafer,[103] Sorokin et al.,[108, 117] Bass et al.,[119, 138] and Snaveley and Peterson.[142] We return to consider some of these equations further when discussing intracavity enhanced absorption in Section 5.2.3.7.

5.2.2.6. Four-Wave Parametric Conversion. Four-wave parametric conversion[143] is a resonantly enhanced third-order nonlinear optical process whereby coherent electromagnetic waves of frequency ν_L and ν_p from two separate pulsed dye lasers, and a third wave of frequncy ν_s originating from

[143] P. P. Sorokin, J. J. Wynne, and J. R. Lankard, Appl. Phys. Lett. 22, 342 (1973).

stimulated Stokes–Raman scattering of the wave of frequency v_L, beat together in an alkali metal vapor thus forming an electrical polarization $\mathbf{P}^{(3)}$ which coherently radiates at infrared frequencies

$$v_r = v_L - v_p - v_s. \tag{5.2.55}$$

The frequency v_r, in principle, is continuously tunable throughout the range $5000 \text{ cm}^{-1} \geqslant v_r \geqslant 20 \text{ cm}^{-1}$ by use of proper combinations of Na, K, Rb, or Cs vapors which assure phase matching:

$$\mathbf{k}_r = \mathbf{k}_L - \mathbf{k}_s - \mathbf{k}_p, \tag{5.2.56}$$

where $\mathbf{k}$ is the wave propagation vector.

The pertinent physical phenomena associated with this process were illustrated [143] on the simplified electronic energy level diagram for potassium which appears here in Fig. 13. The laser system for this application bore great similarity to the one diagrammed in Fig. 14 in Section 5.2.2.7; the quarter-wave plate, however, was not present in the system for four-wave parametric conversion. The two dye lasers, pumped by the same 100-kW peak power N_2 laser, were each equipped with an intracavity telescopic expander–collimator[137] which reduced their spectral output widths Γ_L and Γ_p to $\sim 0.1 \text{ cm}^{-1}$. The two synchronous dye laser beams, of orthogonal linear polarization and frequencies v_L and v_p (v_1 and v_2 in Fig. 14), were combined collinearly in a glan prism and then focused by a lens, of $+50 \text{ cm}$ focal length, into a heat-pipe oven[144, 145] containing the alkali metal vapor. The region containing the vapor was about 30 cm long. Radiation emanating from the heat-pipe oven was passed through a monochromator selective of v_r and observed with a Ge:Cu detector. The dye laser wave of frequency v_L was tuned near resonance with the $5p_{1/2}$ and $5p_{3/2}$ electronic states of potassium having energies $E(5p_{1/2}) = 24701.4 \text{ cm}^{-1}$ and $E(5p_{3/2}) = 24720.2 \text{ cm}^{-1}$. Other electronic energies for potassium are $E(5s) = 21026.8 \text{ cm}^{-1}$, $E(4p_{3/2}) = 13042.9 \text{ cm}^{-1}$, and $E(4p_{1/2}) = 12985.2 \text{ cm}^{-1}$.

The wave of frequency v_L had three primary functions: first, it provided for stimulated electronic Stokes–Raman scattering at frequencies $v_s = v_L - E(5s)$. Second, it partially provided for the radiant polarization through the dominant term in the third-order nonlinear susceptibility (esu)[146, 147]

$$\chi^{(3)} \simeq \frac{iN}{h^3} \frac{\langle 4s|\mu_e|4p\rangle \langle 4p|\mu_e|5s\rangle \langle 5s|\mu_e|5p\rangle \langle 5p|\mu_e|4s\rangle}{[E(4p) - v_p][E(5p) - v_L]\Gamma_L}, \tag{5.2.57}$$

[144] C. R. Vidal and J. Cooper, *J. Appl. Phys.* **40**, 3370 (1969).

[145] C. R. Vidal and M. M. Hessel, *J. Appl. Phys.* **43**, 2776 (1972).

[146] A theoretical introduction to nonlinear stimulated Raman scattering is given by N. Bloembergen, "Nonlinear Optics," pp. 37–44. Benjamin, New York, 1965.

[147] J. J. Wynne, P. P. Sorokin, and J. R. Lankard, *Proc. Laser Spectrosc. Conf.*, *Vail, Colorado.*, 1973.

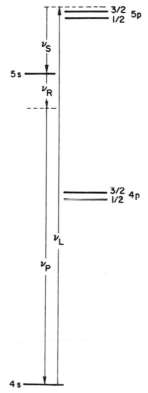

FIG. 13. Energy-level diagram for the four-wave parametric mixing process in potassium vapor. Waves of frequency ν_L and ν_p were from two separate dye lasers. Waves of frequency ν_r and ν_s were the desired infrared waves and the Stokes–Raman scattering of waves of frequency ν_L, respectively. Reproduced by permission of P. P. Sorokin, J. J. Wynne, and J. R. Lankard, *Appl. Phys. Lett.* **22**, 342 (1973).

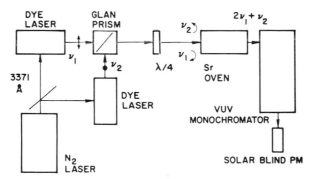

FIG. 14. Block diagram of the laser system for sum–frequency mixing in atomic vapors. The block diagram also applies to the four-wave parametric converter when the quarter-wave plate is removed and alkali metal vapors are present in the heatpipe oven. Reproduced by permission of R. T. Hodgson, P. P. Sorokin, and J. J. Wynne, *Phys. Rev. Lett.* **32**, 343 (1974).

where N was the density of alkali atoms (cubic centimeters) and the angle brackets indicated expectation values between appropriate electronic states for μ_e the electric dipole moment. Third, v_L partially provided for the third-order mixing of v_L, v_p, and v_s. The collinearity of $\mathbf{k}_L$ and $\mathbf{k}_p$, and therefore the similar collinearity $\mathbf{k}_s$ and $\mathbf{k}_r$, yielded the largest volume for interaction in the vapor, and together with conservation of energy (5.2.55), also reduced (5.2.56) to a scalar equation

$$\Delta n_L\, v_L = \Delta n_r\, v_r + \Delta n_s\, v_s + \Delta n_p\, v_p, \qquad (5.2.58)$$

where Δn was the amount by which the index of refraction exceeded unity. Values for Δn were calculated with the Sellmeier equation[148]

$$(2 + \Delta n)\, \Delta n = \sum_k \left[\rho_k/(v_k{}^2 - v^2) \right], \qquad (5.2.59)$$

with oscillator strengths ρ_k provided by an NBS publication.[149] The index k ranged over all the electronic states of the alkali atom. In pure potassium and pure rubidium vapor, phase matching limited the range of tunability to $2500\ \mathrm{cm}^{-1} \leqslant v_r \leqslant 5000\ \mathrm{cm}^{-1}$ and $1851\ \mathrm{cm}^{-1} \leqslant v_r \leqslant 3448\ \mathrm{cm}^{-1}$, respectively. Vapor mixtures of sodium and potassium yielded phase matching over broader regions of v_L and v_p, and provided tunability in the range $400\ \mathrm{cm}^{-1} \leqslant v_r \leqslant 5000\ \mathrm{cm}^{-1}$. The spectral output width Γ_r was $\sim 0.1\ \mathrm{cm}^{-1}$, the same as that for the dye lasers. The output power was 0.1 W and 0.1 mW at v_r of $5000\ \mathrm{cm}^{-1}$ and $400\ \mathrm{cm}^{-1}$, respectively.

What can be expected from the four-wave parametric process in the future? Theoretically, vapor mixtures provide phase matching that sets the lower limit for frequency tunability at about $20\ \mathrm{cm}^{-1}$. The radiant power of the wave of frequency v_r, however, is proportional to $v_r{}^2$ and the power therefore decreases at lower frequencies. The nonlinear polarization $\mathbf{P}^{(3)}$ is directly proportional to the electric fields of the three waves of frequencies v_L, v_p, and v_s, and thus the power generated at frequency v_r is proportional to the product of the individual power of these waves. Pulsed N_2 lasers of 10-MW peak power are presently available from commercial sources; these lasers could theoretically provide a 10^6 increase in power at frequencies v_r, provided of course that saturation effects do not limit the process. Further-more, the pressure-tuned dye laser[132, 133] has a spectral output width $\Gamma_L \simeq 10^{-5}\ \mathrm{cm}^{-1}$, which according to (5.2.57) would provide a 10^4 further enhancement in $\chi^{(3)}$. The wave of frequency v_r would have $\Gamma_r \simeq 10^{-5}\ \mathrm{cm}^{-1}$; a resolving power of $\sim 5 \times 10^8$ at $v_r = 5000\ \mathrm{cm}^{-1}$. It appears, therefore,

[148] See M. Born and E. Wolf, "Principles of Optics," p. 97. Pergamon, Oxford, 1965.

[149] "Atomic Transition Probabilities," Vol. 2, NSRDS–NBS Publication No. 22, U.S. Govt. Printing Office, D.C., 1969.

that the future will yield significant advances in four-wave parametric technology.

5.2.2.7. Sum-Frequency Mixing in Atomic Vapors. There are several recent papers[150–153] devoted to investigations of sum-frequency mixing in atomic vapors. We will discuss the results from only one of those investigations.[150] The reader may consult Hodgson *et al.*,[150] Miles and Harris,[151] Young *et al.*,[152] and Harris and Miles,[153] and the references therein, for further detail.

Sum-frequency mixing is an enhanced, double-resonance, third-order, nonlinear, optical process whereby coherent electromagnetic waves of fixed frequency v_1 and tunable frequency v_2 from two separate pulsed dye lasers beat together in an atomic vapor thus creating an electric polarization $P^{(3)}(v_{uv})_i = \chi_{ijkl}^{(3)} E(v_2)_j E(v_1)_k E(v_1)_l$, which coherently radiates at vacuum ultraviolet (vuv) frequencies

$$v_{uv} = 2v_1 + v_2. \tag{5.2.60}$$

In strontium vapor tunability was obtained throughout spectral ranges of ~ 3500 and ~ 1200 cm^{-1} centered at 189.5 and 179.8 nm, respectively.

The laser system for sum-frequency mixing is diagrammed in Fig. 14. The N_2 laser emitted pulses of radiation at wavelength 337.1 nm. The pulses were of 7.5-nsec duration and possessed ~ 1-MW peak power at a repetition rate of 15 pps. The two dye lasers, pumped by the same N_2 laser, were equipped with intracavity beam-expander collimators[137] which reduced the spectral output widths to $\Gamma_1 \simeq 0.1$ cm^{-1} and $\Gamma_2 \simeq 1$ cm^{-1}. The two synchronous dye laser beams of orthogonal linear polarization were combined collinearly in a glan prism and then passed through a component acting as a quarter-wave plate for the wave of frequency v_1. The quarter-wave plate fully converted the waves of frequency v_1 from linear to circular polarization. The linearly polarized wave of frequency v_2 was partially converted to waves that were circularly polarized in an opposite sense to that of the waves of frequency v_1. The collinear laser beams next passed through a vapor-cell oven made of nickel and containing strontium at pressures of 10–30 Torr and temperatures of 800–900°C. About 450 Torr of xenon was also present in the vapor cell thus facilitating phase matching. The vapor cell was equipped with Pyrex and lithium–fluoride input and output windows, respectively. Radiation from the vapor cell was passed through a vuv monochromator and detected with a solar blind pm-tube

[150] R. T. Hodgson, P. P. Sorokin, and J. J. Wynne, *Phys. Rev. Lett.* **32**, 343 (1974).

[151] R. B. Miles and S. E. Harris, *IEEE J. Quantum. Electron.* **9**, 470 (1973).

[152] J. F. Young, G. C. Bjorklund, A. H. Kung, R. B. Miles, and S. E. Harris, *Phys. Rev. Lett.* **27**, 1551 (1971).

[153] S. E. Harris and R. B. Miles, *Appl. Phys. Lett.* **19**, 385 (1971).

which gave no response at optical frequencies less than $\sim 22{,}200$ cm^{-1} (~ 450 nm).

The physical phenomena associated with sum-frequency mixing were illustrated on the Grotrian diagram for strontium which appears in Fig. 15. The ground state electronic configuration of strontium is KLMN$_{sp}$(5s^2). Very strong third harmonic generation (THG), i.e., $\nu_{uv} = 3\nu_1$, was observed when only linearly polarized waves of frequencies $\nu_1 = 18{,}587$, 18,487, 17,841, or 17,602 cm^{-1} from a sodium–fluorescein dye laser were passed through the strontium vapor. These frequencies correspond to four exact half-frequencies of the double-resonance even-parity $^1S_0$, $^1D_2$, $^3P_2$, and $^3P_0$ states of strontium with a final electronic configuration KLMN$_{sp}$(5p^2). The odd-parity states

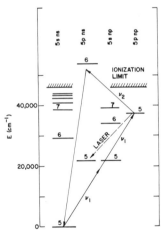

FIG. 15.   Partial energy-level diagram of strontium atoms showing enhanced sum–frequency mixing $2\nu_1 + \nu_2$ by use of the double-resonant 5p^2 intermediate electronic state. The dashed line designated "laser" represents superradiant transitions between the even parity 5p^2 and odd parity 5p5s electronic states. Reproduced by permission of R. T. Hodgson, P. P. Sorokin, and J. J. Wynne, *Phys. Rev. Lett.* **32**, 343 (1974).

$^3S_1{}^\circ$, $^1P_1{}^\circ$, $^3D_1{}^\circ$, and $^3D_3{}^\circ$ were not active. Thus in the case of THG, the doubly resonant states at energy $2\nu_1$ beat together with the remaining unabsorbed waves of frequency ν_1 to create a third-order polarization which radiates at frequency $3\nu_1$. The dashed line in Fig. 15 represents superradiant laser actions due to electronic transitions 5p^2 to 5p5s; transitions $^3P_2-^3P_2{}^\circ$, $^1D_2-^1P_1{}^\circ$, and $^1S_0-^1P_1{}^\circ$ in order, provided frequencies 20,785, 15,267, and 15,299 cm^{-1}.

The quarter-wave plate quenched the THG because in isotropic media circularly polarized waves cannot be frequency tripled since it violates conservation of angular momentum. Sum-frequency mixing was then observed by adding circularly polarized waves of tunable frequency v_2. The nonlinear susceptibility[150, 154] in this case was

$$\chi_{ijkl}^{(3)} \propto \sum_{g,m,n,o} \frac{N_g \langle g|\mu(3)_i|0\rangle \langle 0|\mu(2)_j|n\rangle \langle n|\mu(1)_k|m\rangle \langle m|\mu(1)_l|g\rangle}{(v_{og} - v_3)(v_{ng} - 2v_1)(v_{mg} - v_1)} + \cdots,$$

(5.2.61)

where the summation of g, m, n, and o was over all the electronic states of strontium, N_g the density of occupied states in the quantum state g. The matrix elements were for the i, j, k, and l components of the electric dipole vector $\mu(\alpha)$ which were parallel to the electric field of the wave of frequency v_α. The term v_{ng} represented the energy difference between electronic states n and g. The continuation of the sum $(+ \cdots)$ represents similar summation terms with the frequencies permuted. When $v_{ng} = 2v_1$, then $\chi_{ijkl}^{(3)}$ was resonantly enhanced, assuming the double quantum transition n to g had nonvanishing dipole matrix elements. Furthermore, double resonance provided the same enhancement of $\chi^{(3)}$ as did single resonance ($v_{mg} = v_1$), but absorption is relatively weak for the double-resonance process. Additional enhancements of $\chi^{(3)}$ were observed when v_3 was resonant ($v_3 = v_{og}$) with discrete excited states such as 6s5p which were above the ionization limit. The maximum power was the order of 10^9 photons/pulse at $v_1 = 17,370$ cm^{-1}, $v_2 = 21,090$ cm^{-1}, and $v_{uv} = 5583$ cm^{-1}. The spectral output width Γ_{uv} was $\simeq \Gamma_1 = 0.1$ cm^{-1}. Theoretically, third-order sum-frequency mixing could provide tunable coherent vuv radiation from 43,480 to 125,000 cm^{-1} (230–80 nm) with $\Gamma_{uv} < 0.1$ cm^{-1}.

5.2.2.8. Frequency Mixing in Solids. Frequency mixing in solids is generally a second-order nonlinear optical process whereby two coherent electromagnetic waves of either equal or unequal frequencies v_1 and v_2 are superimposed in a crystal thus creating an induced polarization which coherently radiates sum or difference frequencies $v_3 = v_1 \pm v_2$. Phase matching, $k_3 = k_1 \pm k_3$, is critical to this process. The two best known examples of this phenomenon are the LiNbO$_3$ parametric oscillator, and the frequency doubling, i.e., second harmonic generation, of fundamental laser frequencies in crystalline $(NH_3)H_2PO_4$, ammonium dihydrogen phosphate (ADP), or in lithium niobate LiNbO$_3$. Second-order nonlinear processes, however, do not occur in optically isotropic crystals, e.g., those having the cubic (sphalerite) structure; thus the lowest-order nonlinear effects observable from isotropic crystals would be the third-order processes. This is directly analogous to the absence of second-order and presence of third-order process in the atomic vapors which, of course, are optically isotropic. Here the

[154] J. A. Armstrong, N. Bloembergen, J. Ducuing, and P. S. Pershan, *Phys. Rev.* **127**, 1918 (1962). Also see Bloembergen, "Nonlinear Optics," pp. 37–44. Benjamin, New York, 1965.

discussion is selectively restricted to a few examples of optically aniso-tropic crystals that are known to provide frequency tunable coherent radia-tion through second-order nonlinear effects. The tunability of v_3 is achieved in these devices by simultaneously varying v_1 and/or v_2, and the orientation of the optical axes of the crystal relative to $\mathbf{k}_1$ and $\mathbf{k}_2$, which are usually collinear or nearly collinear, so that the phase-matching condition is satisfied. Tunability can also be achieved by continuously varying the temperature of the crystal, thus continuously varying its index of refraction, and thereby continuously altering the phase-matching condition.

Parametric oscillators[155] gave frequency tunability from 2740 to 18,180 cm^{-1} (3.65–0.55 μm). The parametric oscillators were optically ex-cited by radiation from a flashlamp-pumped Q-switched Nd:YAG laser which could be tuned to 13 discrete frequencies in the range 14,730 to 21,140 cm^{-1}. Intracavity Brewster angle prisms and a LiIO$_3$ crystal provided respective means for tuning and doubling the fundamental frequencies of the Nd:YAG laser. The nonlinear element in the parametric oscillator was an a-axis crystal of LiNbO$_3$ contained in an oven. The crystal, grown from a congruent melt,[156] was from 3 to 5 cm long. The crystal was placed inside a 9.8-cm-long optical cavity consisting of two nearly confocal mirrors. The dielectric coatings on the mirrors transmitted 70–95% of the radiation from the Nd:YAG laser but transmitted less than 10% of the resonant frequency from the LiNbO$_3$ crystal. Four pairs of mirrors with different dielectric coat-ings were required to span the entire range of frequency tunability. The para-metric oscillator was tuned by varying its temperature from 110 to 450°C. The range of frequency tunability of the oscillator was about 3000 cm^{-1} for each discrete optical excitation frequency from the Nd:YAG laser. The spectral output width was a "small fraction of a reciprocal centimeter." Para-metric oscillators are now available for purchase from some manufacturers of laser products.

Other examples of frequency mixing are as follows: coherent waves from a ruby laser and a ruby laser-pumped DTTC iodide dye laser were difference-frequency mixed in proustite[157] thereby providing tunability throughout the frequency range 1770 cm$^{-1} \leqslant v \leqslant 3125$ cm^{-1}. In a crystal of tellurium[158] sum-frequency mixing of waves from a CO$_2$ laser operating at $v_1 = 943.39$ cm^{-1} and from a CO$_2$ laser-pumped n-InSb SFR laser operating at $v_2 = v_1 - g\mu_B B/hc$ gave continuous frequency tunability from 1725 to 1835 cm^{-1} when the magnetic induction was changed from 9.5 to 2.5 T, respectively. Collinear difference-frequency mixing in AgGaSe$_2$ gave tun-

[155] R. W. Wallace, *Appl. Phys. Lett.* **17**, 497 (1970).

[156] R. L. Byer, J. F. Young, and R. S. Feigelson, *J. Appl. Phys.* **41**, 2320 (1970).

[157] C. D. Decker and F. K. Tittle, *Appl. Phys. Lett.* **22**, 411 (1973).

[158] C. R. Pidgeon, B. Lax, R. L. Aggarwal, and C. E. Chase, *Appl. Phys. Lett.* **19**, 333 (1971).

ability from 665 to 1430 cm^{-1}; the initial waves were of frequency $v_1 =$ 7575.7 cm^{-1} and 5880 cm^{-1} $\leqslant v_2 \leqslant$ 6670 cm^{-1} and were from a Nd:YAG laser and a Nd:YAG laser-pumped LiNbO$_3$ parametric oscillator, respectively.[159] In GaAs crystals[160] difference mixing of discrete frequencies from P and R branch lines of the 961.5 and 1052.6 cm^{-1} bands of two CO$_2$ lasers yielded step-tunable far-infrared radiation from 5 to 145 cm^{-1}. Dual frequencies from a single double-grating ruby laser-pumped DTTC iodide dye laser[161] produced tunability in the far infrared from 20 to 190 cm^{-1} when they were difference mixed in a LiNbO$_3$ crystal, and difference-frequency mixing in ZnGeP$_2$ of waves from two step-tunable CO$_2$ lasers[162] furnished far-infrared frequencies from 70 to 110 cm^{-1}. The spectral output widths of waves from the mixing processes corresponded to that of the input lasers. The signal-to-noise ratios of the power output from these devices was large enough to render them useful for spectroscopic applications.

5.2.3. Spectroscopic Applications

Currently radiant emissions from tunable lasers and optically nonlinear devices span major portions of the electromagnetic spectrum from 5 cm^{-1} $\leqslant$ $v \leqslant$ 56,200 cm^{-1}. During the forthcoming decade we can expect lasers to provide frequency tunability throughout the vacuum ultraviolet, ultraviolet, visible, near- and far-infrared, and the microwave regions of the electromagnetic spectrum. Furthermore, improved stability of these devices together with their very narrow spectral output width will provide for new strides forward in all branches of qualitative and quantitative high-resolution spectroscopy. It is not surprising, therefore, that the diversity of possible spectroscopic applications for frequency-tunable lasers corresponds directly to a composite of the interests of individual spectroscopists. In this section we present a critique of a few of the past and present spectroscopic applications of frequency tunable lasers. Applications yet to come are left in the creative realm of the reader's adroit imaginative consciousness.

5.2.3.1. Reflection Spectroscopy. Coherent and noncoherent electromagnetic waves of moderate power incident on an interface between two optically dissimilar materials are partially reflected, partially transmitted, and refracted in accordance with the Fresnel equations and Snell's law. The portion of the waves either reflected or transmitted at the interface is dependent on the angle of incidence, and on n_r and n_i the real and imaginary parts of the

[159] R. L. Byer, M. M. Choy, R. L. Herbst, D. S. Chemla, and R. S. Feigelson, *Appl. Phys. Lett.* **24**, 65 (1974).

[160] B. Lax, R. L. Aggarwal, and G. Favrot, *Appl. Phys. Lett.* **23**, 679 (1973).

[161] K. H. Yang, J. R. Morris, P. L. Richards, and Y. R. Shen, *Appl. Phys. Lett.* **23**, 669 (1973).

[162] G. D. Boyd, T. J. Bridges, and C. K. N. Patel, *Appl. Phys. Lett.* **21**, 553 (1972).

complex refractive indices of both material media.[1] In spectral regions where condensed materials exhibit either very small or very large characteristic values of n_i it is more convenient to investigate their optical properties by use of refraction or reflection techniques, respectively. The normal course of events in the latter case is to measure the absolute reflectance,[163] relative reflectance,[164] ratio reflectance,[165] or attenuated total reflectance[166] spectra of the interface. Dispersion relations or algorithms consistent with the Fresnel equations are then applied to the reflectance spectra to obtain values for n_r and also n_i.[†] Theoretical analyses of physical models based on the electronic, atomic, molecular, and/or crystalline structure then hopefully provide a deeper understanding of the classical and quantum-mechanical properties of the material media. Frequently, experiment confirms preexistant theoretical analysis.

Frequency-tunable organic dye lasers were used to investigate the absolute-reflectance spectra of aqueous solutions containing organic salts.[104, 105, 136, 167] The nearly collimated nature of the dye laser beam greatly simplifies the optical system required for making the beam incident on the sample at a known angle of incidence θ. A flat first-surface mirror intersected the laser beam at a point on the circumference of a circle of radius $\simeq 25$ cm which was scribed on a large slab of phenolic material. The beam was reflected to the center of the circle where it was incident on the horizontal surface of the aqueous sample. The surface of the sample was parallel to a diametrical line bisecting the circle. The laser beam was partially reflected from the sample to a photodiode detector located a distance $2r\theta$ along the circumference of the circle from the first-surface mirror. The sensitive surface of the detector was rotated to obtain maximum response as observed from a boxcar integrator. An initial series of continuous reflectance spectra was repeatedly recorded throughout the spectral region of tunability of a single dye. The sample was removed thus allowing the laser beam to pass from the flat mirror to the diametrical conjugate point on the circumference of the circle. The detector was moved to the conjugate point, rotated to maximize the signal, and then a second series of spectra were recorded. The ratio of the initial to second series of spectra provided the absolute reflectance. Series of spectra were obtained at several angles of incidence θ. A high-speed digital

[163] M. R. Querry, R. C. Waring, W. E. Holland, G. M. Hale, and W. Nijm, *J. Opt. Soc. Amer.* **62**, 849 (1972).

[164] G. M. Hale, W. E. Holland, and M. R. Querry, *Appl. Opt.* **12**, 48 (1973).

[165] M. R. Querry and W. E. Holland, *Appl. Opt.* **13**, 595 (1974).

[166] E. E. Remsberg, *Appl. Opt.* **12**, 1389 (1973).

[167] M. R. Querry, G. Mansell, and R. C. Waring, *J. Opt. Soc. Amer.* **61**, 680A (1971).

[†] The optical constants n_r and n_i are elsewhere denoted by n and k (Chapter 4.1) and n and κ (Chapter 4.4).

computer was programmed to search systematically $n_r n_i$-space to obtain points (n_r, n_i) consistent with the Fresnel equations and all sets of absolute reflectance spectra. Because n_i was very small ($\simeq 10^{-8}$), only values of n_r could be obtained from reflectance data having three significant digits. Future refinement of these techniques should yield data providing for a quantum-statistical analysis of the interaction of the atomic and molecular solutes with the complex liquid-water substance.

5.2.3.2. Absorption Spectroscopy. In spectral regions where condensed materials and gases exhibit moderate values of n_i it is more convenient to investigate their optical properties, and thus their spectroscopic properties by use of absorption techniques. Applying the Poynting flux theorem separately on surfaces of two imaginary parallel planes, perpendicular to the Poynting vector and situated in the same material medium, yields

$$I = I_0 e^{-\alpha L}, \tag{5.2.62}$$

where I_0 and I are the electromagnetic intensities on the initial and final planes, respectively, L the distance between the planes, and $\alpha = 4\pi n_i \bar{\nu}$ is the Beer–Lambert absorption coefficient of the material. When working with gases in the infrared it is customary to cast the αL product in the altered form $k w^0$, where

$$k = S\Gamma/\pi[(\nu - \nu_0)^2 + \Gamma^2], \tag{5.2.63}$$

$$w^0 = (P_a/760 \text{ mm Hg/atm})(273 \text{ K}/T)L, \tag{5.2.64}$$

and

$$S = \int_0^{-\infty} k(\nu) \, d\nu. \tag{5.2.65}$$

Here k is the absorption coefficient in units of reciprocal atmosphere · centimeters, ν the wavenumber (in reciprocal centimeters), Γ(in reciprocal centimeters) the half-width of an individual rotational line of a vibrational–rotational band, S the line strength (in reciprocal centimeters per atmosphere · centimeter), P_a the partial pressure (in millimeters of Hg) of the absorbing gas, T the absolute temperature, and L the path length (in centimeters).

Until just recently, with a few exceptions, tunable lasers have not provided the stable output required for precise quantitative measurements of α, k, or S. This is evident from the absence of a numerical scale on the ordinate of graphical representations of absorption or transmission spectra. The recent advent of the dual-channel boxcar integrator and similar modular electronic systems provide an excellent means for circumventing the problems associated with fluctuating power output from lasers; we will see more quantitative investigations reported in the future.

The very narrow spectral output width associated with many tunable lasers, however, yields highly resolved spectra and is thereby an excellent means for confirming the precise central frequencies and half-widths of spectral lines. One such example,[77] the spectra for the Q-branch of the v_2-band of CO_2, is shown in Fig. 16. The spectra were obtained with two $Pb_{1-x}Sn_xSe$ diode lasers, with $x \simeq 0.06$. Nearly continuous frequency tunability was achieved by independent current tuning of the two lasers which had different mode characteristics; thus the superconducting magnet was not required for the experiment. The laser emission was collimated by use of a 19-mm focal-length KRS-5 lens. The absorption path was 1 m long. The absorption cell

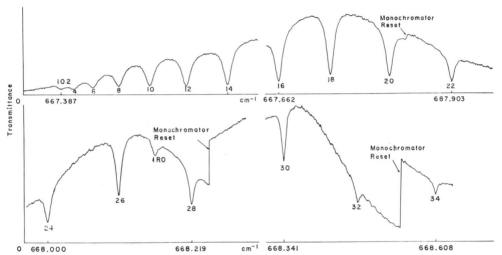

FIG. 16. Spectra of the Q-branch of the v_2-band of CO_2 obtained with a $Pb_{1-x}Sn_xSe$ diode laser ($x \simeq 0.06$). The spectral output width of the laser was less than 10^{-4} cm^{-1}. The absorption path was 1 m long. The gas mixture contained 320 ± 3 ppm of CO_2 in N_2. The total gas pressure was 30 Torr. The temperature was 199 °K. Reproduced by courtesy of Arthur D. Little, Inc. Cambridge, Massachusetts from a report prepared under NOAA contract No. 3–35386.

contained a Matheson Gas Products primary standard mixture of 320 ± 3 ppm of CO_2 in N_2 at a total absolute pressure of 30 Torr. The temperature was 199 K. The notation 1QJ designated the Q-branch of the ground-state v_2 vibrational transition having rotational quantum number J; only J was denoted on rotational lines $4 \leqslant J \leqslant 34$. The spectra were observed by use of a Perkin–Elmer 98 G monochromator as a bandpass filter; the grating had 75 lines/mm and the spectral slitwidth was 0.34 cm^{-1}. In Fig. 16 "monochromator reset" denoted shifting the 0.34-cm^{-1}-wide bandpass to an adjacent position of the spectrum. Average values of measured linewidths exceeded slightly those predicted by theory.[168]

[168] G. Yamamoto, M. Tanaka, and T. Aoki, *J. Quant. Spectrosc. Radiat. Transfer* **9**, 371 (1969).

Continuously tunable systems such as (a) the Zeeman-tuned He–Ne and He–Xe lasers, (b) the spin–flip Raman laser, (c) the p–n junction diode laser, (d) the dye laser, and (e) the four-wave parametric converter have been used for absorption spectroscopy. The Zeeman-tuned He–Ne laser,[14] for example, was used for obtaining absorption profile measurements in methane–other-gas mixtures. Absolute pressure broadening data or rotational lines provided the information needed to calculate collisional diameters. In methane–neon the collisional diameter was the sum of the radii of the two molecules. In collisions between methane and more polarizable atoms or molecules such as argon, nitrogen, hydrogen, chloride, and methane itself the collisional diameter was greater than the sum of the individual radii. For collisions between methane and helium the collisional diameter was less than the sum of their radii. Also, as another example, the absorption spectrum of CH_3F was observed with a tunable He–Xe laser.[17] Diode p–n junction lasers were used to obtain high-resolution absorption spectra of NO, NO_2, CH_4, CO, CO_2, NH_3, SO_2, SF_6 as reported in Hinkley,[72] Nill et al.,[73, 75] Hinkley et al.,[74] Antcliffe et al.,[76] and Aronson et al.[77] The high stability of a spin–flip Raman laser facilitated spectroscopic measurements of NO and H_2O vapor in the stratosphere. The SFR laser system was lifted by a balloon to an altitude of 28 km where it obtained data in situ for about 7.5 hr.[169]

5.2.3.3. Saturation Spectroscopy. The breadth of spectral lines is traceable to four primary physical phenomena:

(1) the natural radiative linewidth which, through the uncertainty principle, is related to the lifetime of the initial state,

(2) shifting of the natural radiative frequencies to lower and higher frequencies during collisions with other molecules, i.e., pressure broadening,

(3) Doppler shifts of the natural frequencies due to the Maxwellian distribution of thermal velocities of the atoms or molecules, and

(4) power broadening, which in the case of high-power lasers, is associated with a saturation of the absorption process.

When investigating absorption spectra of gases, the pressure and power broadening of spectral lines can be reduced greatly by lowering the total pressure in the absorption cell and by reducing the intensity of the radiant source, respectively. Doppler broadening, however, cannot be reduced significantly, because that requires temperatures below which most gases condense to the liquid or solid state. The full-width Doppler broadening is given by

$$\Delta\nu_D = (6.08 \times 10^{-7})(T/M)^{1/2}\nu_0, \qquad (5.2.66)$$

where T is the absolute temperature, M the mass in atomic mass units, and ν_0 the natural frequency for the radiant transition. For example, H_2O vapor

[169] C. K. N. Patel, E. G. Burkhardt, and C. A. Lambert, Science 184, 1173 (1974).

at $T = 300$ K has $\Delta v_D = (2.48 \times 10^{-6})v_0$, which at infrared frequencies gives Δv_D from 100 to 150 MHz. The natural radiant linewidth Γ_n for many transitions in H_2O vapor, however, is from 10^2 to 10^3 kHz. Thus for H_2O vapor, as for other molecules, the natural linewidth is concealed two to three orders of magnitude below the Doppler width, and therefore is impossible to observe directly by use of the techniques of ordinary absorption spectroscopy. Many frequency-tunable lasers fortunately possess spectral output widths Γ_l that are much less than Δv_D, and in some cases less than Γ_n. The natural width and profile of the spectral line can be observed directly by use of those lasers having very narrow Γ_l, and by use of the technique known as *saturation spectroscopy*.

The spin–flip Raman (SFR) laser with a spectral output width $\Gamma_l < 1$ kHz, therefore, was ideally suited for saturation spectroscopy,[102] as are many other types of laser systems. The arrangement of a SFR laser and other necessary equipment for saturation spectroscopy is represented in Fig. 17. The cw pump laser was a grating-tuned liquid-nitrogen-cooled CO laser operating on the $P_{9\,8}(11)$ transition with output power $\simeq 2$ W. The pump-laser beam was focused by a 10-cm-focal-length lens into the InSb semiconductor crystal. The SFR laser was an n-InSb crystal, 9 mm long, with $\sim 5 \times 10^{14}$ free carriers/cm^3, cooled to 4.2 K, and producing output power of $\simeq 1$ W. A 4-in. Varian electromagnet provided gross tuning of the SFR laser to a frequency near $v_0 = 1889.58$ cm^{-1}, the natural frequency for the $v_2(5_{3,2} \to 6_{4,3})$ transition in H_2O vapor. Fine tuning over a frequency interval of ± 450 MHz centered at 1889.58 cm^{-1} (5.66×10^{13} Hz) was accomplished with an additional trimmer-coil solenoid having only a few turns per unit length. The long-term (10–20 min) instability of the SFR laser linewidth Γ_l was 50–100 kHz. The laser beams were chopped and then traveled to a polarizer which rejected the residual pump-laser beam and passed the beam from the SFR laser. Next, the remaining SFR laser beam was attenuated to

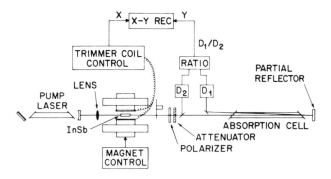

FIG. 17. Diagram of a SFR laser system used for saturation spectroscopy of H_2O vapor. Reproduced by permission of C. K. N. Patel, *Appl. Phys. Lett.* **25**, 112 (1974).

about 10 mW, to reduce power broadening in the absorbing gas, and was then divided by a beamsplitter. Part of the beam was reflected to detector D_2. The remainder, which became the saturation beam, traversed a 40-cm-long absorption cell containing H_2O vapor at 30 m Torr and was then incident on a partially reflecting plate. A small fraction of the beam, which became the probe beam, was reflected back through the absorption cell and was then made to be incident on detector D_1. The ratio of the signal from D_1 to that from D_2 was proportional to the transmittance of the probe beam through the cell. The SFR laser beams were about 2 cm in diameter inside the absorption cell. The ratio signal D_1/D_2 and a signal proportional to the current through the trimmer-coil solenoid were monitored on the Y and X terminals of an XY recorder, respectively.

The purpose of the saturation beam was to interact with those H_2O molecules in the $v_2(5_{3,2})$ state. Only a small subclass of the H_2O molecules had Maxwellian thermal velocities providing for the correct Doppler shift of their natural frequency v_0 to interact with the $\Gamma_l < 1$ kHz band of frequencies in the saturation beam. This interaction caused a depleted population of molecules within a small subclass of the Maxwell velocity distribution. The probe beam of weaker intensity, traversing the absorption cell in the direction opposite to that of the saturation beam, interacted similarly with the subclass of H_2O molecules having oppositely directed velocities As the SFR laser was tuned to frequencies approaching v_0, the saturation beam and the probe beam began to interact with molecules in overlapping velocity subclasses; thus the probe beam was absorbed less and a dip, corresponding to the increasing transmittance as shown in Fig. 18, was observed in the absorption spectrum.

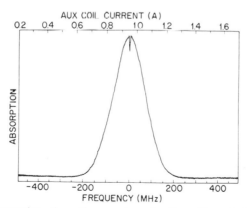

FIG. 18. Laser saturation–absorption spectrum for the $v_2(5_{3,2} \rightarrow 6_{4,3})$ transition of H_2O vapor with a pressure of 30 mTorr and $l = 40$ cm. The primary magnetic induction was 0.117 T. Absorption is plotted as a function of the trimmer-coil current (see Fig. 17). Frequencies are plotted on the lower scale; zero frequency corresponds to the central position of the spectral line. Reproduced by permission of C. K. N. Patel, *Appl. Phys. Lett.* **25**, 112 (1974).

As the SFR laser was tuned through ν_0 the shape of the dip directly corresponded to the natural line profile for the $\nu_2(5_{3,2} \rightarrow 6_{4,3})$ transition in H_2O. The full width of the natural line was $\Gamma_n \simeq 200$ kHz, while the width of the Doppler-broadened line was measured as $\simeq 165$ MHz FWHM.

The pressure-tuned pulsed dye laser has a very narrow spectral output width which also makes it ideally suited for saturation spectroscopy. The fine-structure components of the Balmer H_α line in hydrogen were resolved by use of that laser system[170] and the techniques of saturation spectroscopy. We discuss this application further in the next section.

5.2.3.4. Measurement of the Rydberg Constant. The precision and accuracy of the Rydberg constant is directly related, through Bohr's formula corrected for finite nuclear mass, Lamb shifts, and Dirac fine structure,[171] to precise and accurate measurements of the central frequencies or wavelengths of only the radiatively broadened part of individual atomic spectral lines. A more precise value for the Rydberg constant,

$$R_\infty = 109,737.3143(10) \text{ cm}^{-1},$$

was recently measured by applying the technique of saturation spectroscopy to the $2P_{3/2}-3D_{5/2}$ component of the H_α and D_α Balmer lines of atomic hydrogen and deuterium.[172]

The diagram of the laser system used for measuring the Rydberg constant is reproduced in Fig. 19. A pressure-tuned N_2 laser-pumped pulsed dye laser (see Fig. 11, Section 5.2.2.5) provided a primary output beam of spectral width (FWHM) $\Gamma \simeq 30$ MHz when operated at 71 pps. A signal ξ proportional to the total radiant energy per laser pulse was obtained by observing a small fraction of the dye laser output with a photodiode–integrator system. The frequency-tuning (ramp) rate of the dye laser was also observed in electronic form. The primary dye laser beam was divided in two parts by a partially reflecting plate. The transmitted part of the primary beam went to a wavelength comparator which is described in the next paragraph. The reflected part of the primary beam went to the saturation experiment where it was further divided in three parts by transmission through and reflection at the two faces of an optical flat. The transmitted beam, intersected by a chopper on alternate pulses, became the saturation beam. The two reflected beams became the probe and reference-probe beams. The reference-probe beam did not intersect the saturation beam. The two probe beams were ob-

[170] T. W. Hansch, I. S. Shahin, and A. L. Schawlow, *Nature (London) Phys. Sci.* **235**, 63 (1972).

[171] G. W. Series, *Contemp. Phys.* **14**, 49 (1974).

[172] T. W. Hansch, M. H. Nayfeh, S. A. Lee, S. M. Curry, and I. S. Shahin, *Phys. Rev. Lett.* **32**, 1336 (1974).

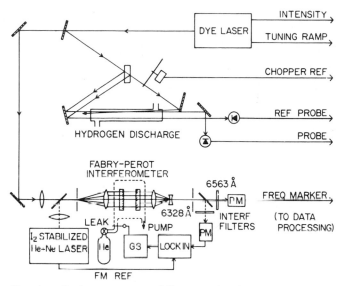

FIG. 19.   Drawing of a laser system used for measuring the Rydberg constant in atomic hydrogen. A pulsed pressure-tuned dye laser was used to obtain Doppler-free saturation spectra of the hydrogen. A highly stabilized He–Ne laser and a pressure-tuned Fabry–Perot interferometer were used as a comparator to measure the frequencies of waves from the dye laser. Reproduced by permission of T. W. Hansch, M. H. Nayfeh, S. A. Lee, S. M. Curry, and I. S. Shanin, *Phys. Rev. Lett.* **32**, 1336 (1974).

served with separate photodiode detectors connected to a dual channel integrator system. The signals ξ_p and ξ_r from the probe and reference probe, respectively, as well as all other data from the experiment, were collected for each dye laser pulse and were stored on a magnetic disk for later processing. The frequency-tuning range of the dye laser was equally divided into an 800-element array. At each spectral position the saturation signal S, which occurred on alternate pulses was given by

$$S = (\xi_p - \xi_r)[1 - \langle(\xi_p^0 - \xi_r^0)/\xi^0\rangle\xi], \qquad (5.2.67)$$

where the superscript zero denoted the absence of the saturation beam, and the square brackets denoted the quotient averaged over two appropriate neighboring pulses. Six spectra were obtained in this manner, were individually smoothed by convolution with a Gaussian or Lorentzian profile, and were then averaged. The absorption cell was an 8-mm-diameter Pyrex Wood-type cold Al cathode dc discharge tube. Electronically generated hydrogen or deuterium continuously flowed through the cell. The folded absorption path was 1 m long. The pressure was 0.1–1 Torr.

The wavelength comparator provided for precise wavelength measurement of the output from the dye laser. The comparator, which is also diagrammed in Fig. 19, consisted of an iodine-stabilized He–Ne laser[173] and a Fabry–Perot interferometer. The wavelength of the output from the He–Ne laser was known to 1.4 parts in 10^9. The He–Ne and dye laser beams were focused through a common pinhole to a quartz collimating lens and were then incident normally on a pressure-tuned plane Fabry–Perot interferometer. The plates of the interferometer were flat to $\lambda/200$ and were separated by a 31.76730356(15)-mm-thick Invar spacer. Light emerging from the interferometer was focused by a Galilean telescope ($20\times$) on a second pinhole to eliminate ghost fringes. The laser beams were then split, separated by optical filters, and observed with separate photomultiplier tubes. The signal from the pm tube integrator system monitoring the dye laser was sent to data storage. The output from the pm tube monitoring the He–Ne laser became a servo signal for stabilizing the pressure-tuned interferometer.

The Rydberg constant was determined by fitting the measured frequencies of the H_α and D_α lines to the corrected Bohr formula. Systematic corrections and residual systematic errors were conservatively estimated as follows (parts in 10^9): pressure shifts, 2; intensity shifts, 2; Stark shifts, 3; unresolved hyperfine splitting, 4; wavelength standard, 4; refractive index of tuning gas, 2; and phase dispersion of etalon coatings, 4. Greater precision in the value for R_∞ is expected from future investigations of metastable H atoms by use of two-photon spectroscopy, which also provides the Doppler-free profile of individual spectral lines.

5.2.3.5. Two-Photon Absorption Spectroscopy. Two-photon absorption in atomic vapors is phenomenologically a second-order quantum-mechanical and a third-order nonlinear optical process. Doppler-free two-photon absorption from a single loser beam was recently observed[174] by transversely interacting a frequency-tunable laser beam with an atomic beam. Two-photon absorption by atoms from two spectrally narrow frequency-tunable contrapropagating laser beams was theoretically proposed[175] four years prior to its recent observation in alkali metal vapors.[176–179] The technique of absorption from contrapropagating laser

[173] W. G. Schweitzer, Jr., E. G. Kessler, Jr., R. D. Deslattes, H. P. Layer, and J. R. Whetstone, *Appl. Opt.* **12**, 2827 (1973).

[174] D. Pritchard, J. Apt. and T. W. Ducas, *Phys. Rev. Lett.* **32**, 641 (1974).

[175] L. S. Vasilenko, V. P. Chebotaev, and A. V. Shishaev, *Pis'ma Zh. Eksp. Teor. Fiz.* **12**, 161 (1970) [*English transl.: JETP Lett.* **12**, 113 (1970)].

[176] F. Biraben, B. Cagnac, and G. Grynberg, *Phys. Rev. Lett.* **32**, 643 (1974).

[177] M. D. Levenson and N. Bloembergen, *Phys. Rev. Lett.* **32**, 645 (1974).

[178] N. Bloembergen, M. D. Levenson, and M. M. Salour, *Phys. Rev. Lett.* **32**, 867 (1974).

[179] J. E. Bjorkholm and P. F. Liao, *Phys. Rev. Lett.* **33**, 128 (1974).

beams was similar to saturation spectroscopy, except (1) here the two laser beams were of nearly equal intensity, and (2) the absorption was detected by observing perpendicular to the laser beams, the fluorescent radiation from the decay of the excited electronic state.

The contrapropagating scheme was superior to the single-laser-beam absorption technique because the two beams can interact with nearly all the ground-state atoms of the Maxwell velocity distribution as follows: Consider an atom that moved with thermal velocity v, relative to the laboratory coordinate frame, through a standing wave of frequency v. The atom, therefore, interacted in its rest frame with oppositely traveling photons of energy $hv(1 - v_x/c)$ and $h(1 + v_x/c)$. If the combined energies of the two oppositely traveling photons corresponded to E, the energy difference of the ground state and an excited state of similar parity (2nd-order process), then two-photon absorption occurred independent of v_x:

$$E = hv(1 - v_x/c) + hv(1 + v_x/c) = 2hv. \qquad (5.2.68)$$

There was a subclass of atoms within the Maxwell velocity distribution that could interact with two photons traveling in the same direction. This effect, however, was relatively small because $\Gamma_1 \ll \Delta v_D$, and the effect was eliminated entirely for 3S–5S transitions in Na by circularly polarizing[177] the two laser beams in the same sense (opposite helicity with respect to the lab frame) and thereby not satisfying conservation of angular momentum for the same-beam two-photon transition.

In this manner Doppler-free Zeeman spectra for the 3S → 5S transition was observed in Na.[177, 178] The source of the two contrapropagating beams was a pulsed N_2 laser-pumped dye laser. The dye laser was gross frequency tuned with an intracavity expander–collimator telescope and diffraction grating, and was fine tuned by rotating a Fabry–Perot etalon that had 5.00-cm plate spacing. The linearly polarized beam from the dye laser passed through a quarter-wave plate, thus converting it to circular polarization, and was then focused into the center of the sodium vapor cell by a 15-cm-focal-length lens. Pressures in the cell were $\simeq 5$ mTorr. The cell was in an oven. The beam emerging from the cell was recollimated by a similar lens, passed through a quarter-wave plate, and was incident on a flat first-surface mirror which reflected the beam back through the optical system. Proper time delays prevented reflection back to the dye laser. The spontaneous emission at 330 nm from the 4P–3S transition, which followed the two-photon excitation to 5S, was monitored with a pm tube, signal A. Filters eliminated extraneous radiation. A second pm tube, signal B, monitored the intensity of a small fraction of the primary dye laser beam. Signals from both pm tubes were detected with a dual-channel boxcar integrator. The ratio A/B^2 was plotted by a chart recorder. The hyperfine splitting $\Delta v_{I \cdot s}$ and the hyperfine interaction constant A_{5S} of ^{23}Na were measured to be 808 ± 5 and 78 ± 5 MHz, respectively.

If the frequencies of the waves were different for each of the two contra-propagating laser beams, but satisfied conservation of energy and angular momentum for the two-photon absorption, then slight residual Doppler broadening was observed in the spectra.[177, 178] A compensating occurence in this case, however, was a resonant enhancement of the two-photon absorption process. The cross section (in squared centimeters) for two-photon absorption[179] was

$$\sigma(v_1) = 1.15 \times 10^{-34} \left| \sum_n \langle f|Z|n\rangle \langle n|Z|g\rangle [(E_n - hv_1)^{-1} \right.$$
$$\left. + (E_n - hv_2)^{-1}] \right|^2 v_1 \rho(v_1 + v_2) I_2, \qquad (5.2.69)$$

where v_1 and v_2 were the frequencies of the laser beams, I_2 the intensity (in watts per square centimeter) of the radiation field for waves of frequency v_2, $\rho(v_1 + v_2)$ the normalized lineshape function for the transition, the matrix elements were in units of the Bohr radius, f and g designated final and ground states for the two-photon absorption, and the summation was over all intermediate states of energy E_n expressed in rydbergs. A 10^7 enhancement was obtained for the two-photon transitions 4S–4D in Na when hv_2 was fixed at slightly less than E_{3p}, and when v_1 was continuously tuned through the range $2\Delta v_1$ so that $(E_{4D} - h\,\Delta v_1) \leqslant h(v_1 + v_2) \leqslant (E_{4D} + h\,\Delta v_1)$.

Resonant enhancement of two-photon absorption from contrapropagating laser beams of different frequencies was discussed in Section 5.2.2.7. Looking to the future, the contrapropagating laser beams, combined with the physical concepts presented in Section 5.2.2.7, should yield a broad band of resonantly enhanced stimulated vuv radiation from alkali metal vapors. The bandwidth would correspond to the Doppler width

$$\Delta v_0 = (6.08 \times 10^{-7})(T/M)^{1/2}(2v_1 + v_2), \qquad (5.2.70)$$

where we combined Eq. (5.2.60) and (5.2.66). Furthermore, the intensity profile would be the same as the profile for the Maxwell velocity distribution.

Continuous wave dye lasers provide a much narrower spectral output width than that of the pulsed laser. Thus from investigators using cw dye lasers we expect to see extremely high-resolution two-photon spectroscopy and further precision in the measurement of the fundamental spectroscopic constants.

5.2.3.6. Fluorescence and Luminescence Spectroscopy. Frequency-tunable lasers with their narrow spectral output width and relatively great radiant intensity are ideally suited for selective one- or two-photon excitation of electronic, vibrational, or rotational energy states of atoms, molecules, or condensed matter. As the atom, molecule, or solid decays from the excited eigenenergy state it emits a cascade of fluorescent or luminescent radiation. The spectral distribution of the radiation is characteristic of the microscopic physical nature of the entity. For example, the observation of luminescence spectra from selectively excited electronic energy states of ions in solids will surely yield new knowledge of the spin–orbit and ligand–field interactions. Of particular interest are the one-, two-, and three-phonon assisted optical transitions in crystalline solids. In addition to an elucidation of the quantum-mechanical properties, investigations of luminescence spectra may lead to the discovery of phonon-broadened optical transitions providing for continuously tunable solid-state lasers. The theory of optical interactions in solids[180] and a review of the optical properties of ions in solids[181] are presented elsewhere and therefore will not be repeated here. In the United States, Europe, and Japan, as well as other parts of the world, many laboratories dedicated to investigations of luminescence spectra are currently being equipped with frequency-tunable lasers; much is to be expected during the next decade.

In the previous section we were introduced to the importance of fluorescent spectra for observing the Zeeman effect in the two-photon absorption for the 3S–5S transition in atomic sodium. Another example, a segment of the fluorescent rotational spectrum of diatomic iodine is shown in Fig. 20. This spectrum was obtained by monitoring the fluorescent radiation emitted perpendicular to a frequency-tunable beam from a pressure-tuned N_2 laser-pumped dye laser. The intensity plotted on the ordinate was that of the total fluorescence which was monitored during the frequency scan of the dye laser. The pressure plotted on the abcissa was that of propane which was the tuning gas in the dye laser (see Fig. 11, Section 5.2.2.5). The spectral output width of the dye laser was 1.3 GHz.

Another application of tunable-laser fluorescence spectroscopy is the monitoring of chemical reactions in crossed molecular beams.[182] For investigations of this type, tunable lasers provide a detection efficiency that is comparable to mass spectrometry. Unlike the mass spectrometer, however, laser-induced fluorescence is a highly specific detection mechanism but is not as universal, e.g., the OH radical could be detected at 300 nm (frequency-doubled dye laser), whereas with mass spectrometry this radical is extremely difficult to quantify due to the large background signal. In one sample of

[180] B. DiBartolo, "Optical Interactions In Solids." Wiley, New York, 1968.
[181] B. DiBartolo, "Optical Properties of Ions In Solids." Plenum Press, New York, 1975.
[182] R. N. Zare and P. J. Dagdigian, *Science* **185**, 739 (1974). This is a review article containing more than fifty references.

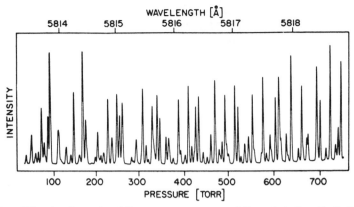

FIG. 20.   Vibrational-rotational fluorescence spectrum of diatomic iodine. Excitation of the iodine was provided by a pressure-tuned pulsed dye laser having spectral output width of $\simeq 1.3$ GHz. Reproduced by permission of R. Wallenstein and T. W. Hansch, *Appl. Opt.* **13**, 1625 (1974).

BaO it was possible to detect a particular vibrational–rotational level[183] in a small sample containing 5×10^4 molecules cm^3.

5.2.3.7. Intracavity Enhanced Absorption Spectroscopy. Placement of an absorbing material in the cavity of a dye laser altered the laser gain in such a manner that the laser emission was quenched or reduced in intensity. This technique was recently used to observe enhanced optical absorption by chlorophyll[184] and other samples.[185–190] With narrow-band dye lasers, e.g., grating tuned, the maximum enhanced absorption was about $10 \times$ greater than absorption measured with a conventional dual-beam spectrophotometer. If the dye laser is operated in the broad-band mode, e.g., flat mirrors at the ends of the cavity, then the absorption was enhanced as much as 10^3 times. This did not indicate the absorption coefficient of the sample

[183] A Schultz, H. W. Cruse, and R. N. Zare, *J. Chem. Phys.* **57**, 1354 (1972).

[184] M. R. Querry, W. E. Holland, R. C. Waring, M. D. Herrman and G. M. Hale, High Sensitivity Laser Absorption Spectroscopy: A Feasibility Study, Rep. OWRR A–058–MO, August 1973, Available as PB–225431/6 from Nat. Tech. Inform. Serv., Springfield, Virginia 22151.

[185] T. W. Hansch, A. L. Schawlow, and P. E. Toschek, *IEEE J. Quantum. Electron.* **QE-8**, 802 (1972).

[186] N. C. Peterson, M. J. Kurylo, W. Braun, A. M. Bass, and R. E. Keller, *J. Opt. Soc. Amer.* **61**, 746 (1971).

[187] R. A. Keller, E. F. Zalewski, and N. C. Peterson, *J. Opt. Soc. Amer.* **62**, 319 (1972).

[188] R. A. Keller, J. D. Simmons, and D. A. Jennings, *J. Opt. Soc. Amer.* **63**, 1552 (1973).

[189] R. J. Thrash, H. Von Weyssenhoff, and J. S. Shirk, *J. Chem. Phys.* **55**, 4659 (1971).

[190] G. H. Atkinson, A. H. Laufer, M. J. Karylo, *J. Chem. Phys.* **59**, 350 (1973).

had increased when it was in the laser cavity. It did indicate greatly quenched lasing action in the spectral region of the absorption band of the sample but with normal laser action in adjacent spectral regions, thus an "enhanced" absorption.

An example of enhanced absorption is shown in Fig. 21. A 2-cm long cell containing 0.95 μgm/liter and 0.4 μgm/liter of chlorophyll a and b, respectively, dissolved in 80% acetone was placed in the optical cavity of a grating-tuned N_2 laser-pumped dye laser. The dye was cresyl violet. The output mirror of the dye laser had a reflectance of 4%. The laser operated at 665.2 nm with a spectral output width of $\simeq 0.5$ nm. The laser voltage plotted on the abscissa was that for the voltage supply to the N_2 laser and was a measure of the relative optical pumping power delivered to the dye laser. The symbol I denoted a signal proportional to the energy per pulse from the dye laser,

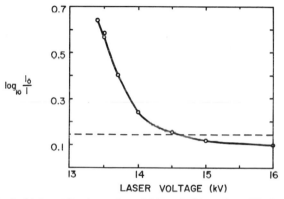

FIG. 21.   Enhanced intracavity absorption of chlorophyll in acetone. The laser was a grating-tuned dye laser. The symbol I denotes radiant energy per laser pulse with the chlorophyll solution in the laser cavity. The subscript zero denotes the intracavity absorption cell filled only with acetone. The dashed line represents the optical density of the same chlorophyll solution as measured with a Cary 14 spectrophotometer. The laser voltage is that for the N_2 laser; see text for additional details.

as observed exterior to the laser with a pm-tube and a boxcar integrator; the subscript zero denoted absence of the absorption cell from the optical cavity. At lower N_2 laser voltages the dye laser operated at threshold gain when the absorption cell was present. As the voltage was increased the dye laser gain increased, and the laser action became more intense. At higher N_2 laser voltages the dye laser action was intense enough to optically saturate the chlorophyll sample, thus reducing the ratio I_0/I to values below those obtained for the same sample with a Cary 14 dual-beam spectro-photometer as indicated by the dashed line. The data appear as circles. The solid line demonstrates the trend of the data. The dye laser action was quenched for N_2 laser voltages less than $\simeq 13.5$ kV.

An analytical description of enhanced absorption was obtained[185–188] by including a term for the attenuation by the intracavity absorber in the rate equations (5.2.43)–(5.2.45). The resultant rate equations were then solved numerically to obtain values for minimum detectable absorption as a function of the dye laser parameters. The analysis was made for both the broad-band and narrow-band dye lasers. The narrow-band laser was analytically capable of a maximum of $10 \times$ enhancement in the absorption, which is consistent with the data shown in Fig. 21.

5.2.3.8. Optoacoustic Detection. A frequency-tunable modulated beam of radiation passing through an absorbing gas composed of diatomic or polyatomic molecules will be absorbed by the frequency-selective vibrational–rotational transitions of the molecules. After the absorption process the molecules transfer the excitation energy to their neighbors by collisional deexcitation. A large fraction of the excitation energy is thereby converted to translational kinetic energy. The change in kinetic energy of the molecules macroscopically corresponds to a change in temperature and pressure. The pressure changes are detectable with a microphone provided the radiation is modulated at acoustic frequencies. The process of absorption of optical radiation by a gas and subsequent detection by acoustical means is known as the *optoacoustic effect*. The effect was discovered in 1881 by Röntgen,[191] Tyndall,[192] and Bell.[193] Ironically, 90 years later Kreuzer,[194] a scientist at the Bell Telephone Laboratories, published an analysis of the optoacoustic effect as applied to detecting ultralow gas concentrations by use of frequency-tunable infrared lasers with narrow spectral output widths. All prior analyses of this effect[195, 196] were made for thermal sources of broad-band infrared radiation. In this section we restrict our discussion to the analysis made by Kreuzer.

At a temperature of 2000°C the ideal blackbody radiates 4×10^{-3} $W \cdot cm^{-2} \cdot sr^{-1}$ in a spectral band 1 cm^{-1} wide centered at a frequency of 5000 cm^{-1}. Pulsed frequency-tunable lasers, however, easily radiate 10^6 $W \cdot cm^{-2} \cdot sr^{-1}$ into spectral bands $\leqslant 10^{-2}$ cm^{-1} wide, and thus are better radiant sources for optoacoustic applications than chopped beams from broad-band blackbody sources. The analysis of the laser-related application of the optoacoustic effect was made in three parts: (1) absorption of radiation

[191] W. C. Röntgen, *Phil. Mag.* **11**, 308 (1881).
[192] J. Tyndall, *Proc. Roy. Soc. London* **31**, 307 (1881).
[193] A. G. Bell, *Phil. Mag.* **11**, 510 (1881).
[194] L. B. Kreuzer, *J. Appl. Phys.* **42**, 2934 (1971).
[195] M. E. Delany, *Sci. Progr.* **47**, 459 (1959).
[196] R. Kaiser, *Can. J. Phys.* **37**, 1499 (1959).

by the gas, (2) signal generation, and (3) signal detection and noise analysis. In part (1) an equation was derived for the population density of molecules in the first excited vibrational level. The equation was a function of induced, radiative, and collisional lifetimes, a function of the individual probabilities that the upper rotational levels of the ground and excited vibrational states were occupied, and a function of the amplitude and frequency of the modulation. In part (2) the population densities were used to derive equations for the time-dependent fundamental and second-harmonic pressure variations in the radiation-absorbing gas. In part (3) the ratio of a pressure-induced signal S_p to a noise signal S_B due to Brownian motion of the diaphram of the microphone was found to be

$$\frac{S_p}{S_B} = \frac{1}{6}\,\xi\,\frac{N}{\tau_c}\,\frac{\alpha\delta}{\beta}\,\frac{\tau_1}{(1 + \omega^2\tau_T^2)^{1/2}}\left(\frac{\tau_A A}{8\sigma k T\,\Delta f}\right)^{1/2},\qquad(5.2.71)$$

where N was the total number of absorber molecules per cubic centimeter, ξ the energy difference between ground and excited vibrational states, τ_c the collisional lifetime of the excited state, and τ_T the thermal relaxation time for heat exchange between the absorbing gas and the walls of the absorption cell. The symbol β denoted the sum $\alpha + \gamma$, where α was the probability that a molecule in the ground state was also in a rotational state for an allowable transition, and γ was the probability that a molecule was in the excited vibrational state and also in a rotational state corresponding to the same allowed transition. The angular frequency and amplitude of the modulation were denoted by ω and δ, respectively, with $0 \leqslant \delta \leqslant 1$. The Boltzman constant and the absolute temperature were k and T, respectively. The remaining symbols were defined as follows: A the area of the diaphram of the microphone, σ the mass density of the diaphram of the microphone, τ the mechanical damping time for the diaphram, and Δf the bandwidth of detectable acoustic frequencies. A comparable expression was derived for the ratio of S_p to a signal S_J due to Johnson noise in the electronic instrumentation. Assuming the Johnson noise was reducible to a negligible level, (5.2.71) was an expression for the ultimate sensitivity of optoacoustic detection. A numerical example: $\alpha \simeq \gamma$, $\alpha/\beta \simeq \frac{1}{2}$, $\tau_c \simeq 10^{-6}$ sec, $\tau_A \simeq 10^{-4}$ sec, $\xi = hvc$ with $v = 3333$ cm^{-1}, $\delta = 0.1$, $\omega\tau_T \ll 1$, $A = 1$ cm^2, $\sigma = 5 \times 10^{-3}$ gm/cm^2, $\Delta f = 1$ Hz, $\omega = 2\pi \times 10^3$ Hz, and $T = 300$ K gave

$$S_p/S_B = (2 \times 10^{-7}\text{ cm}^3)N.\qquad(5.2.72)$$

This implies for a unity signal-to-noise ratio, the minimum density of detectable radiation-absorbing molecules would be $N = 5 \times 10^6$ cm^{-3}. At standard temperature and pressure the density of gas molecules was $\rho \simeq 3 \times 10^{19}$ cm^{-3}, thus the ultimate detection limit $N\rho^{-1}$ was about two radiation-absorbing molecules per 10^{13} nonabsorbing molecules. The power needed from the laser for optimum detection was calculated to be 10^5 W/cm^2.

Of course the Johnson noise could not be totally eliminated from the electronic system. Initial experiments with a Zeeman-tuned He–Ne laser, radiating 15 mW at 2950 cm^{-1} (3.39 μm) and modulated at 400 Hz, gave a detection limit of one part methane per 10^6 parts of air. In later experiments[99] the more powerful CO and CO$_2$ lasers were used to successfully detect five parts ethylene in 10^9 parts of air. Absorption of radiation by the windows and walls of the gas cell introduced an extraneous signal that was approximately 125 × greater than that due to absorption by the ethylene; thus measurements were required with an accuracy greater than one part in a hundred. Optoacoustic detection, in combination with spin–flip Raman lasers, was also used for measuring the pressure broadening of water vapor lines[100] and for spectroscopic measurements of nitric oxide and water vapor concentrations in the stratosphere.[101]

5.2.3.9. Stark Spectroscopy.   When an atom or molecule is placed in an external electric field, the electric monopole, dipole, and quadrupole moments interact with the external electric potential, electric field, and gradient of the electric field, respectively. The Stark effect, which is the electrical analog of the magnetic Zeeman effect, originates from the dipole–electric-field interaction. The Hamiltonian for this interaction is $H = -\mathbf{p} \cdot \mathbf{E}$, where $\mathbf{p}$ is the vector dipole moment of the atom or molecule and $\mathbf{E}$ the electric field. If the $\mathbf{p} \cdot \mathbf{E}$ interaction is weak, which includes most electric fields easily generated in a gas sample, then the calculation of the corresponding shifting and splitting of the atomic or molecular eigenstates proceeds from perturbation theory. The selection rules for radiative absorption or emission then follow from time-dependent perturbation theory.

Lasers have not been frequency tuned by use of the Stark effect, but the eigenstates of the gas molecules, observed in absorption with Zeeman-tuned He–Ne and He–Xe lasers,[16, 197, 198] have been frequency tuned by use of the Stark effect. Zeeman-tuning the laser and Stark-tuning the sample brings more eigenstates of the molecule into resonance with radiation from the laser. Additionally, the relative orientation of the Stark field and the direction of polarization for the laser beam further restrict the set of selection rules for allowed transitions between eigenstates. In this manner particular spectral lines can be absolutely identified in the midst of a complex absorption band.

A drawing of a tunable laser system used to obtain direct absorption spectra or Stark-modulated absorption spectra of gases is reproduced in

[197] K. Sakurai, K. Uehara, M. Takami, and K. Shimoda, J. Phys. Soc. Japan 23, 103 (1967).
[198] K. Uehara, T. Shimizu, K. Shimoda, IEEE J. Quantum. Electron. QE-4, 728 (1968).

Fig. 22. The laser was a Zeeman-tuned He–Ne or He–Xe system operating at central frequencies 2950 and 2849 cm^{-1}, respectively. After being chopped and linearly polarized by a quarter-wave plate the laser beam passed through either (1) an ordinary gas absorption cell or (2) a Stark-modulated gas absorption cell. A cooled gold-doped germanium detector was used at the exit window of each cell. Considering case (1), the ordinary absorption cell, the signal from the detector was monitored with a lock-in amplifier which was connected to the Y-terminal of an XY-recorder. A signal proportional to the current in the solenoid of the Zeeman-tuned laser provided a

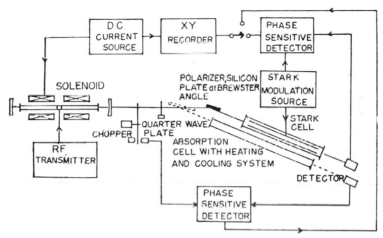

FIG. 22. Diagram of a Zeeman-tuned He–Ne and He–Xe laser system used for Stark modulation spectroscopy of polyatomic molecules. Reproduced by permission of K. Sakurai and K. Shimoda, *Jap. J. Appl. Phys.* **5**, 938 (1966).

wavelength reference to the X-terminal of the recorder. An absorption spectrum for H_2CO, in the frequency range ± 4 GHz about the central laser frequency, was obtained and appears here in Fig. 23a. The pressure in the absorption cell was 20 Torr. The cell was 60 cm long. Note the fluctuations in the spectra from two different frequency scans; this was due to variations in radiant power from the laser.

Returning now to case (2), the Stark-modulated absorption cell, the signal from the detector was again monitored with the lock-in amplifier. In this case, however, the amplifier was triggered in synchronization with Stark modulation at 400 Hz. A static electric potential of 200 V and a modulated potential of 150 V were applied to obtain the spectra shown in Fig. 23b. The separation between the Stark electrodes was about 0.5 mm. Note the $2\times$ difference in frequency scales on the abscissa of Fig. 23a and 23b, and

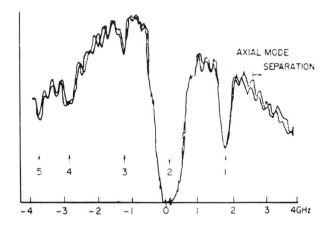

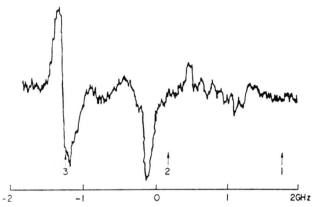

FIG. 23. The upper graph is the ordinary infrared absorption spectrum of H_2CO at $\rho = 20$ Torr as obtained with a Zeeman-tuned He–Ne and He–Xe laser. The lower graph is the three Stark-modulated absorption spectrum of H_2CO at lower pressure $\rho = 1.7$ Torr, with Stark voltages of 150 V (ac) and 200 V (dc). Note the difference in scales on the abcissa. Reproduced by permission of K. Sakurai and K. Shimoda, *Jap. J. Appl. Phys.* **5**, 938 (1966).

also the differences in pressure. The Stark-modulated spectra were clearly selective to particular transitions, e.g., only the shoulder line on the strong absorption line marked 2 in Fig. 23a appears in Fig. 23b. Other examples of Stark spectroscopy with a grating-tuned molecular laser are found in the work of Johns and Mckeller[199] and Johns *et al.*[200]

[199] J. W. C. Johns and A. R. W. McKeller, *J. Mol. Spectrosc.* **48**, 354 (1973).
[200] J. W. C. Johns, A. R. W. McKeller, and A. Trombetti, *J. Mol. Spectrosc.* **55**, 131 (1975).

AUTHOR INDEX

Numbers in parentheses are reference numbers and indicate that an author's work is referred to although his name is not cited in the text. Numbers in italics show the page on which the complete reference is listed.

Ablett, S., 203
Abrahamson, E. W., 149
Adam, G., 200
Ade, P. A. R., 100
Afsar, M. N., 199, 200
Aggarwal, R. L., 51, 322, 323
A'Hearn, M. F., 76
Ahern, F. J., 76, 77
Ahmed, S. A., 275
Ahrenkiel, R. K., 47
Akabane, K., 132
Alexander, F. B., 290
Alguard, M. J., 248, 254
Allen, H. C., 2
Altshuler, H. M., 159
Amat, G., 2, 34
Andersen, N., 219, 221, 225, 229
Andersen, T., 222, 236, 237, 238, 241, 242 (62), 243 (62), 244
Anderson, J. C., 149
Anderson, J. E., 186
Anderson, P. W., 39
Anderson, R. E., 127
Anderson, W. A., 82
Andrä, H. J., 215, 239, 245, 247, 248, 249, 253, 254 (115)
Angress, J. F., 75
Antcliffe, G. A., 291, 297 (63), 298, 299, 327
Antony, A. A., 196, 198 (169)
Aoki, T., 326
Applequist, T., 256, 257
Apt, J., 332
Arai, T., 149
Armstrong, J. A., 321
Arnold, J. W., 202

Arnoult, R., 188
Aronson, J. R., 299, 327
Aspinall, D., 87
Atkinson, G. H., 336
Atkinson, R. d'E., 213
Auton, J. P., 60, 84 (46)
Auty, R. P., 139

B

Baker, D. J., 61, 78 (55), 100 (55)
Balashov, V. V., 271
Baldecchi, N. G., 58, 60
Bardsley, J. N., 268
Barlow, H M , 163
Barnes, R B , 6, 17
Barnett, E. F., 167
Barnham, D. C., 309
Baron, P. A., 123
Barr, E. S., 43
Barrette, L., 233
Barrie, I. T., 148
Barrow, G. M., 90
Bashkin, S., 213, 214, 215, 218, 219, 224, 243 (15, 16), 245, 248, 252, 253, 256, 257, 258
Bass, A. M., 336, 338 (186)
Bass, M., 307, 308, 309, 312, 313 (138), 314 (119, 138), 315
Bate, R. T., 299, 327 (76)
Bauer, E., 187
Beauchemin, G., 252
Beckenridge, R. G., 199
Bednar, J. A., 262, 263, 267, 272
Beers, A. C., 51
Bekefi, G., 97
Bell, A. G., 114, 338
Bell, E. E., 23, 70, 73

Bell, J. W., 66
Bell, W. E., 275
Bellamy, L. J., 43
Bells, R. J., 61
Benedict, W. S., 30, 37, 39
Benesch, W., 24, 58
Bennett, R. G., 142, 144, 148 (10), 152, 153, 159
Bentley, F. F., 44, 88
Benz, C. A., 189
Bergmann, F., 185
Bergmann, K., 180, 196, 198 (169)
Bergström, I., 225, 226, 240
Berkner, K., 214, 232
Berne, B. J., 88
Berry, H. G., 215, 222, 225, 228, 229, 230, 231, 236 (30), 238, 239, 243 (15), 248, 252, 254, 255, 256
Bertie, J. E., 45
Bethe, H. A., 246, 247 (98)
Betz, H., 219
Bevacqua, S. F., 290
Bhalla, C. P., 272
Bhardwaj, S. N., 231
Bickel, W. S., 214, 215, 217, 225, 226, 229, 240, 245 (9), 247, 248 (100), 252, 253
Biggerstaff, J. A., 247, 248 (101)
Biggs, A. J., 148
Billman, K. W., 309
Biraben, F., 332
Birch, J. R., 100
Bird, R., 42
Birnbaum, G., 42, 59
Bishop, W. S., 201
Bjorkholm, J. E., 332, 334 (179)
Bjorklund, G. C., 319
Blakemore, J. S., 293
Blanken, R. A., 84, 98, 100
Blass, W. E., 36
Bloeurbergen, N., 316, 321, 332, 333 (177, 178), 334 (177, 178)
Bloom, S. L., 275
Bloor, D., 52, 94
Blum, A. L., 299, 303, 327 (73, 75)
Blumlein, A. D., 146
Boggs, F. W., 162
Bohr, N., 235
Boleu, R., 219, 221, 225, 229
Bolton, H. C., 157, 158 (63)
Bonch-Bruyevich, A. M., 308
Born, M., 61, 273, 318, 324 (1)
Bos, F. F., 202

Bose, E., 5
Bosomworth, D. R., 42, 71
Bottreau, A. M., 180
Boullet, C., 188
Boyce, J. C., 225
Boyd, G. D., 323
Boyle, W. S., 52
Bozman, W. R., 243
Brackett, C. C., 94
Brackett, E. B., 118
Brackman, R. T., 262, 264
Bradley, D. J., 308
Braithwaite, W. J., 263, 266, 269 (155)
Brand, J. H., 244
Braun, W., 336, 338 (186)
Breit, G., 260, 262
Bridges, J. M., 244
Bridges, T. J., 280, 323
Bridwell, L. B., 219
Briotta, D. A., Jr., 87
Britt, C. O., 123
Broadhurst, M. G., 138, 139, 163 (5)
Brockman, R. J., 39, 40, 41
Brodsky, S. J., 256, 257
Bromander, I., 215, 224, 229, 232, 238, 240, 243 (15)
Brossier, P., 84, 98, 100
Brot, C., 188, 191
Brotfeld, D. P., 307, 308 (110)
Brown, A. S., 149
Brown, M., 263, 266
Brown, N. L., 200
Brzozowski, J., 234, 244 (56)
Brueck, S. R. J., 303
Bruma, M., 188
Brunet, H., 275, 280 (15)
Buchanan, T. J., 151, 152 (43, 46), 172, 188, 198
Buchet, J. P., 225, 229, 232, 233
Buchet-Poulizac, M. C., 229, 232, 233 (53)
Buchta, R., 225, 226, 229, 234, 238, 240, 244 (56)
Budo, A., 195, 196 (166)
Buhl, D., 132
Bukow, H. H., 254
Bur, A. J., 138, 139, 163 (5)
Burch, D. E., 28, 39
Burenin, A. V., 104, 108 (15), 114 (15), 128 (15), 131
Burkhardt, E. G., 305, 327, 340 (101)
Burns, D. J., 248, 254
Burns, G., 290
Burroughs, W. J., 100

Burrus, C. A., 110, 113
Bush, E. T., 114
Butler, J. F., 291, 299, 326 (77), 327 (77)
Butterworth, J., 77
Button, K. J., 96, 97
Byer, R. L., 322, 323

C

Cagnac, B., 332
Calawa, A. R., 290, 291, 299, 327 (73, 74, 75)
Calderwood, J. H., 142, 148 (10), 153, 159
Callaway, J., 293, 298
Calvert, R., 146
Carlson, R. O., 290
Carré, M., 238, 252
Carrington, A., 103, 121, 131
Carriveau, G. W., 219, 244
Case, T. W., 10
Caughey, W. S., 94
Ceccon, H., 308
Chamberlain, J. E., 70, 73, 74, 78 (78), 80, 81, 93, 94, 95, 100, 191, 192, 199, 200
Chambers, W. G., 75
Chantry, G. W., 53, 80, 88 (15), 93, 94, 95, 99, 192
Charette, J., 43
Chase, C. E., 322
Chasmar, R. P., 2, 4 (9), 10 (9), 12 (9), 25 (9)
Chebotaev, V. P., 332
Chemla, D. S., 323
Cheng, K., 270, 271
Chin-Fatt, C., 99
Chojnacki, D. A., 238
Choplin, A., 118, 120 (44)
Choy, M. M., 323
Church, D. A., 239, 252, 256
Clark, F. O., 132
Clark, H. A. M., 146
Clemett, C., 201
Clendenin, J. E., 262, 264
Clifton, B. J., 57, 113
Coblentz, W. W., 10
Cocke, C. L., 219, 231, 244, 262, 263, 266 (23), 267, 270, 272
Cohen, M. H., 186, 240
Cole, K. S., 175, 182
Cole, R. H., 16, 139, 141, 146, 154, 167, 173, 175, 182, 183, 186, 188, 198, 200, 201, 202
Collie, C. H., 155, 162, 177, 188, 203
Colpa, J. P., 42

Colthup, N. B., 43
Compton, K. T., 225
Condon, E. U., 313
Conley, R. T., 2
Connes, J., 61, 64, 69 (49), 71 (49), 73, 76 (49), 78, 100
Connes, P., 85, 100
Cook, J. S., 199
Cook, R. L., 102, 115 (7), 126, 131 (38)
Cooper, J., 52, 316
Cooper, W. S. III, 214, 232 (10), 245 (10)
Corfield, M. G., 160
Corliss, C. H., 243
Cosens, B. L., 256, 257
Costley, A. E., 100
Couder, Y., 54, 56, 98
Cowley, A. M., 110, 113 (21)
Crawford, F. H., 41
Cross, F., 54
Cross, P. C., 2, 43
Crouch, G. E., 157, 160
Cruse, H. W., 336
Cullen, A. L., 163
Culver, W. H., 307, 314 (108), 315 (108)
Cupp, R. E., 127
Curbelo, R., 69, 77 (70)
Curnutte, B., 39, 40, 41, 45, 46 (116), 219, 244, 262, 263, 266 (23), 267, 270 (23, 158), 272
Curry, S. M., 330, 331
Curtis, L. J., 225, 234, 236, 238, 239, 244 (56), 255, 256
Czerny, M., 23

D

Dafta, D., 90
Dagdigian, P. J., 335
Dakin, T. W., 162
Dale, E. B., 38
Dalgarno, A., 240, 263, 264, 266
Dall'Oglio, G., 52
Daly, L. H., 43
Dane Rigden, J., 287
Daniel, V. V., 92
Dannhauser, W., 188, 198
Dansas, P., 202
Dasgupta, S., 202
Datz, S., 219
Daumézon, P., 202
Davidson, D. W., 183, 188, 202
Davies, G. J., 199, 200

Davies, M., 187, 191, 192, 197 (137), 198, 201
Davis, S. P., 22
Dayhoff, E. S., 256, 257
Daykin, P. N., 84
Dean, P., 45
Dean, T. J., 52
Debye, P., 173, 194
Decius, J. C., 43
Decker, C. D., 322
Decker, J. A., Jr., 85
Delany, M. E., 338
del Bene, J., 198
de Loor, G. P., 204
De Lucia, F. C., 54, 56, 104
Dempster, A. J., 213
Denis, A., 225, 232, 252
Denney, D. J., 186, 202
Derr, V. E., 55, 84 (30)
Desesquelles, J., 225, 228, 232, 233 (53), 248
Deslattes, R. D., 332
Despain, A. M., 66
Deutsch, T. F., 307, 308, 312, 313 (138), 314 (119, 138), 315 (119, 138)
Dev, S. B., 163
de Wijn, H. W., 123
DiBartolo, B., 335
Dick, R., 101
Dickson, F. N., 88
Dill, F. H., 290
Dimmock, J. O., 291, 298
Ditchburn, R. W., 16, 20, 273
Dmitriev, I. S., 219, 269, 271
Dobberstein, P., 254
DoCao, G., 232, 233 (53)
Dollish, F. R., 44
Donnally, B., 263, 266
Dowling, J. M., 66, 76, 77, 80
Downing, H. D., 46, 47, 48
Doyle, H. T., 264
Drake, C. W., 248, 254
Drake, G. W. F., 229, 260, 262, 263, 264, 266, 267
Drouin, R., 225, 233
Druetta, M., 252
Dryden, J. S., 146, 199, 201 (183), 202
Ducas, T. W., 332
Ducuing, J., 321
Dufay, M., 219, 220, 225, 228, 232
Dunike, W. P., 290
Dunn, M. H., 273
Dunsmuir, R., 162

E

Eastman, D. P., 26
Eastman, L. F., 55
Ebers, E. S., 5
Ebert, H., 23
Eck, T. G., 248, 256
Edlén, B., 231
Ellis, D. G., 255
Elsley, R. K., 55
Elton, R., 266
Emery, R. J., 100, 101
Engman, B., 223, 229 (31)
Eppers, W. C., Jr., 280, 282 (32), 283, 288 (32)
Erickson, G. W., 257, 258, 260
Ernst, R. R., 82
Evans, H. M., 80
Evans, J. C., 244
Eyring, H., 187

F

Fahlen, T. S., 280
Fahrenfort, J., 48
Fairchild, C. E., 254
Falkenhagen, H., 193
Fan, C. Y., 257, 258, 259, 263, 266
Fano, U., 254
Faries, D. W., 57
Farmer, G. I., 308
Farrands, J. L., 202
Fastie, W. G., 9, 23, 28 (37), 58
Fastrup, B., 236
Fateley, W. G., 44
Fatuzzo, E., 159, 160, 178
Faust, W. L., 280
Favrot, G., 323
Feigelson, R. S., 322, 323
Feinberg, G., 260, 262, 267
Feldman, P. D., 270, 271
Fellgett, P., 53, 86
Fellner-Feldegg, H., 167
Fenner, G. E., 290
Fetterman, H. R., 57, 113
Filler, A. S., 64
Finch, A., 88
Findlay, J. W., 112, 113 (27)
Fink, D., 214, 245 (9)
Fink, U., 37, 219
Finn, M., 291
Fischer, E., 195, 196 (166)

Fish, K., 196, 198 (169)
Fite, W. L., 262, 264
Flach, R., 309, 318 (133)
Fleming, J., 29
Fleming, J. W., 80, 93, 94, 95, 131
Fleury, P. A., 300
Fong, F. K., 196, 198 (169)
Forest, E., 202
Fork, R. L., 275
Forman, M. L., 70, 78
Foskett, C. T., 69, 77 (70)
Fowles, G. R., 273
Fox, J., 54
Fraley, P. E., 6
France, W. L., 39
Franck, F., 45
Franks, F., 203
French, M. J., 60
Frenkel, L., 127
Frieden, B. R., 66
Fröhlich, H., 173, 184, 189
Froome, K. D., 54
Fulks, R. G., 144
Fuoss, R. M., 184
Furumoto, H., 308

G

Gabriel, A. H., 266, 267
Gaillard, F., 219
Gaillard, M. L., 229, 238, 252
Gale, G. M., 308
Gallagner, J. J., 127
Garcia, J. D., 232
Garcia-Munoz, M., 257, 258 (131), 259
Garg, S. K., 157, 188
Garret, C. G. B., 280
Garstang, R. H., 266
Gates, P. N., 88
Gaupp, A., 215, 239, 249, 254, 255
Gebbie, H. A., 6, 42, 52, 61, 73, 74, 78 (78), 80,
   81, 95, 191, 192
Gehrz, R. D., 27
Geltman, S., 268
Genzel, L., 69, 84
Gerritsen, H. J., 275, 279 (14), 327 (14)
Gibbs, J. E., 73, 74, 78 (78), 81
Gierer, A., 194
Giese, K., 203
Gillot, D., 202
Ginzton, E. L., 154

Girard, A., 85, 86
Glarum, S. H., 154, 177, 178, 185, 186
Glass, E., 54
Glasstone, S., 187
Glennon, B. M., 263, 266
Golant, M. B., 128
Golay, M. J. E., 8, 51, 85, 86
Goody, R. M., 38
Gordon, R. G., 88, 89 (124), 127
Gordy, W., 54, 56, 102, 103, 104, 111 (1), 112,
   113 (13), 115 (7), 118, 128
Gore, R. C., 17
Gottlieb, C. A., 132
Gottlieb, M., 47
Gould, H., 263, 266, 267
Goulon, J., 93, 94, 95
Goy, P., 54, 56, 98
Grainger, J. F., 86, 87
Grant, E. H., 151, 152 (46), 203, 204
Green, S., 132
Gregg, D. W., 308
Griem, H. R., 99, 267
Griffin, P. M., 229, 247, 248 (101), 249, 250, 270,
   271
Griffiths, P. R., 69, 77 (70)
Gross, F., 108
Gross, P. M., 146, 167
Grubb, E. L., 196, 198 (169)
Grynberg, G., 332
Gryvnak, D. A., 28, 39
Gush, H. P., 42, 70

H

Hadeishi, T., 252, 256
Haggis, G. H., 172, 188, 198
Hague, B., 146
Hale, G. M., 324, 336
Hall, H. P., 144
Hall, R. N., 290
Hall, R. T., 77, 80 (86)
Hallin, R., 229, 232
Ham, N. S., 20
Hammond, E. C., 307, 308 (109), 314 (108, 109,
   117), 315 (108, 117)
Hamon, B. V., 163, 207, 211
Hancock, W. H., 248
Hanna, F. F., 196, 198 (172)
Hansch, T. W., 309, 311 (132), 312, 316 (137),
   318 (132), 319 (137), 330, 331, 336, 338 (185)
Hard, T. M., 60

Hardy, J. D., 23
Harman, T. C., 291, 299, 327 (73, 75)
Harp, G. D., 88
Harrick, N. J., 48
Harries, J. E., 100
Harris, D. O., 123
Harris, J. L., 66
Harris, S. E., 319
Harrison, G. R., 22
Harshbarger, W. R., 114
Hartshorn, L., 148
Harwit, M., 84, 85, 87
Haselton, H. H., 249, 250, 270, 271
Hassion, F. X., 175, 188
Hasted, J. B., 155, 162, 172, 177, 183, 188, 198, 199, 200, 203, 208
Hastie, R. J., 100
Hastings, E., 177
Hathaway, C. E., 28
Havriliak, S., 200, 201
Hawkins, R. E., 202
Hedrick, L. C., 123
Heller, M. E., 275, 279 (14), 327 (14)
Helminger, P., 54, 104
Hemmerich, K., 79
Hennelly, E. J., 157
Henry, L., 288
Herbst, R. L., 323
Herman, R. C., 3, 39
Heroux, L., 214, 240
Herrman, M. D., 336
Hersh, J. F., 146
Hershberger, W. D., 114
Herzberg, G., 1, 3, 28, 34 (6), 41 (5), 283, 284, 286
Hese, A., 240
Hessel, M. M., 316
Heston, W. M., 157
Higasi, K., 185, 196
Hilke, J., 223, 229 (31)
Hill, J. C., 94
Hill, J. J., 146
Hill, N. E., 187, 189, 190, 191, 195, 197 (137)
Hinkley, E. D., 299, 327
Hodgson, R. T., 317, 319, 320
Hoeft, J., 127
Hoffman, J. D., 202
Högberg, G., 222, 236
Holah, G. D., 60, 84 (46)
Holland, W. E., 306, 309 (105), 311, 316 (137), 319 (137), 324 (105, 136), 336

Hollis, J. M., 132
Holøien, E., 268
Holnyak, N., Jr., 290
Honijk, D. D., 74
Hooge, F. N., 42
Hoover, G. M., 28
Hopkins, F. F., 263, 266, 269 (154)
Hornbeck, G., 3
Horner, F., 162
Horzelski, J., 160
Houck, J. R., 84
Houldin, J. E., 148
Howard, J. N., 28, 39
Hrubesh, L. W., 127, 132
Huber, M. C. E., 244
Hufnagel, F., 151
Hughes, R. H., 103, 116
Hummer, D. G., 262, 264
Humphreys, C. J., 2, 16, 30 (46)
Hunt, R. H., 37
Huth, B. G., 308
Hvelplund, P., 219, 236
Hyde, P. J., 164

I

Ibett, R. N., 87
Ichijo, B., 149
Ichimura, H., 203
Iguchi, T., 132
Ijuuin, Y., 132
Ingram, D. J. E., 102
Itzkan, I., 309, 310

J

Jackson, W., 162
Jacquinot, P., 52
Jahoda, F. C., 98
James, J. F., 85
Jamieson, J. A., 2
Jamison, K. A., 263, 266, 269 (154)
Jansson, P. A., 37
Jefferts, K. B., 113, 132
Jenkins, F. A., 273
Jennings, D. A., 336, 338 (188)
Jensen, K., 219, 221, 225, 229
Jessen, K. A., 222, 236
Johns, J. W. C., 342
Johnson, B. C., 57
Johnson, C. E., 263, 266

Johnson, C. F., 141
Johnson, C. L., 309, 310
Johnson, D. R., 118, 131, 132 (63)
Johnson, W. D., Jr., 300
Johnson, W. R., 263, 267, 270, 271
Jones, D. E. H., 60
Jones, F. E., 2, 4 (9), 10 (9), 12 (9), 25 (9)
Jones, G. E., 115, 126, 131 (38)
Jones, G. O., 52
Jonscher, A. K., 193
Jordon, C., 217, 266, 267
Junker, B. R., 268

K

Kadara, P. K., 202
Kagan, M. R., 308
Kahler, H., 10
Kaifu, N., 132
Kaiser, R., 338
Kaminow, I. P., 300
Kaplan, J. N., 214, 232 (10)
Karylo, M. J., 336
Kasai, P. H., 118
Kasuya, T., 275, 280, 327 (17)
Kauffman, R. L., 263, 266, 269 (154), 272
Kay, T., 214, 219 (11)
Kay, R. L., 139, 163 (7)
Keller, R. F., 336, 338 (186, 187, 188)
Kelley, P. L., 299, 327 (74)
Kenney, C. N., 102
Kenyon, N. D., 305, 340 (99)
Kerl, R. J., 301
Kerr, A. R., 110, 113 (24)
Kessler, E. G., Jr., 332
Ketelaar, J. A. A., 42
Kewley, R., 118
Keyes, R. J., 290
Khodovoi, V. A., 308
Kierstead, E., 309, 310
Kilp, H., 157
Kimmit, M. F., 53, 88 (16), 99
Kinch, M. A., 52
King, L. W., 87
King, R. B., 237, 243 (67)
King, W. C., 54, 112
Kingsley, J. D., 290
Kinnaird, R. F., 17
Kirke, H. L., 146
Kirkwood, J. G., 172, 184
Kiss, Z. J., 42

Kittel, C., 293
Klages, G., 151, 196, 198 (172)
Klemperer, W., 132
Klug, D. D., 178
Kneubühl, F., 13, 57, 59
Knight, C. W., 264
Knoll, D. B., 141
Knystautas, E. J., 225, 233
Kocher, C. A., 262, 264
Kocher, R. C., 275
Kohl, J. L., 238
Kolbe, W., 256
Komm, D. S., 84, 100
Kormangos, K., 236
Kostin, N. N., 308
Kovner, M. A., 43
Krainov, E. P., 43
Kranbuehl, D. E., 178
Kreuzer, L. B., 114, 305, 338, 340 (99)
Kroon, D. J., 55
Krupnov, A. F., 104, 108, 114, 128, 131
Kruse, P. W., 2
Kubo, R., 89, 90 (126), 178
Kugel, H. W., 257, 258
Kung, A. H., 319
Kuroiwa, D., 207, 208
Kurylo, M. J., 336, 338 (186)
Kutner, M. L., 132

L

Labbé, H. J., 45
Lacques, H. L., 157
Laegsgard, E., 219, 236, 242 (62), 243 (62)
Lahmann, W., 79
Laidler, K. J., 187
Lamb, J., 162
Lamb, W. E., Jr., 256, 257
Lambert, C. A., 305, 327, 340 (101)
Landau, L., 204, 206
Lane, J. A., 156, 157 (57)
Langenberg, D. N., 256, 257
Lankard, J. R., 307, 308 (109, 115), 314 (108, 109, 117), 315 (108, 117), 316 (143), 317
Larkin, I., 191
Lasher, G. J., 290
Lassier, B., 191, 192
Laubert, R., 249, 250
Laufer, A. H., 336
Laurent, J. M., 180
Laurie, V., 118, 120 (44)

Lawrence, G. M., 240, 243
Lawrence, G. P., 257, 258
Lax, B., 57, 96, 97, 290, 322, 323
Layer, H. P., 332
Leavitt, J. A., 219, 221, 224
Lebrun, A., 188
Leck, G. W., 114
Lecomte, J., 1
Lee, E., 1
Lee, S. A., 330, 331
Legay, F., 288
Legay-Sommaire, N., 288
Lengyel, B. A., 273, 277, 291
Lennard, W. N., 231, 244
Lennard-Jones, J., 198
Leskovar, B., 104, 105, 106
Leslie, W. H. P., 146
Levenson, M. D., 332, 333 (177, 178), 334 (177, 178)
Leventhal, M., 256, 257, 258
Levin, P. M., 149
Levine, H. B., 42
Levitt, M., 270, 271
Levy, G., 101
Lewitt, R. S., 290
Liao, P. F., 332, 334 (179)
Lichtenberg, A. J., 99
Lichtenstein, M., 127
Lide, D. R., Jr., 103, 127
Lifshitz, E. M., 204, 206
Lightman, A., 66
Lin, C., 263, 267, 270, 271
Lin, C. C., 118
Lindhard, J., 235, 236
Lindskog, J., 232
Lipworth, E., 256, 257
Little, A. D., 292, 295, 296, 326
Little, V. I., 160
Liu, C. H., 239, 252, 253, 256
Locke, J. L., 41
Loeb, H. W., 166
Loewenstein, E. V., 66
Looyenga, H., 206
Lord, R. C., 60
Lovas, F. J., 127, 131, 132
Lovell, W. S., 157
Low, F. J., 52
Lüders, G., 246, 247 (98)
Lundin, L., 223, 225, 226, 229 (31), 240
Lutz, H. O., 219, 244
Lynch, A. C., 139, 146

M

Macdonald, J. R., 219, 262, 263, 266 (23), 270 (23)
MacDonald, R. S., 17
Macek, J., 249, 252, 254
MacLaughlin, D. E., 77
Maczak, J., 176
Madsen, O. H., 237, 238
Maillard, J. P., 100
Maitland, A., 273
Mallikartjun, S., 195
Malmbug, P. R., 214, 219
Mandel, M., 74
Mankin, W. G., 84
Mansell, G., 306, 309 (104), 310, 324 (104, 136)
Marelius, A., 232
Marling, J. B., 308
Marrus, R., 260, 262, 263, 264, 265, 266 (138), 267 (159)
Marth, K., 307
Martin, D. H., 52, 53, 79, 88 (17), 94
Martin, G. A., 240
Martinez-Garcia, M., 237, 243 (67)
Martinson, I., 215, 223, 224, 225, 226, 229 (31), 234, 240, 244 (56), 245 (15)
Marzat, C., 180
Mason, P. R., 159, 160, 178, 183
Masterson, K. D., 239
Mathews, C. W., 280, 288, 289
Matossi, F., 1, 4, 15 (2), 20
Matthews, D. L., 263, 266, 269 (155)
Mawer, P. A., 52
McGlauchlin, E. D., 2
McCubbin, T. K., 6
McCumber, D. E., 313
McCutchen, C. W., 66
McFarland, B. B., 307, 308
McFarlane, R. A., 280
McFee, J, H., 280, 282 (30)
McFee, R. H., 2
McKeller, A. R. W., 342
McLaughlin, R. D., 252
McQuistan, R. B., 2
McPetrie, J, S., 213
McWhorter, A. L., 300
Meakins, R. J., 195, 198, 199, 201 (183), 202, 210
Meinel, A, B., 214
Melchiorri, B., 52, 58, 60
Melchiorri, F., 52

Melngailis, I., 290, 291
Meredith, R. E., 206
Mertz, L., 61, 69 (51), 71 (51), 78 (51), 80 (51), 86
Meyer, J. A., 309, 310
Michel, M. C., 252, 256
Mickey, D. L., 243
Miles, R. B., 319
Miller, A. P., 146
Miller, R. C., 196
Milward, R. C., 80
Minglegrin, U., 127
Mitsuishi, A., 26
Miyamoto, S., 195, 196 (166)
Moak, C. D., 219, 247, 248 (101)
Moeller, G., 287
Mohr, P. J., 263, 266, 267 (159)
Möller, K. D., 2, 53, 88 (18)
Montgomery, J. A., Jr., 132
Mooradian, A., 300, 303
Moore, C. E., 276, 277
Moore, C. F., 263, 266, 269
Moore, G. E., 39
Moore, L., 183
Moore, M., 308
Moos, H. W., 263
Mopsik, F. I., 138, 153, 154, 183, 186
Moreau, J., 180
Moriamez, C., 188
Moriamez, M., 188
Morimoto, M., 132
Morino, Y., 103
Morris, D., 146
Morris, J. R., 51, 323
Morozzi, V. L., 307, 308 (109), 314 (109, 117), 315 (117)
Moss, B. C., 77
Mossberg, T., 231
Mostovinkov, V. A., 307
Mowat, J. R., 249, 250, 263, 266, 270, 271
Mungall, A. G., 146
Murakawa, S., 309
Murnick, D. E., 257, 258
Myers, R. J., 118

N

Nagane, K., 132
Narasimham, M. A., 256, 257
Natale, V., 52
Nathan, M. I., 290
Nayfeh, M. H., 330, 331

Nee, S. F., 84, 100
Nee, T. W., 178
Nernst, W., 5
Ney, E. P., 27
Nichols, E. F., 25
Nicolson, A. M., 166.
Nielsen, A. H., 36
Nielsen, A. K., 242
Nielsen, H. H., 2, 5, 23, 24, 34
Nielsen, J. R., 38
Nielsen, V., 236
Nijm, W., 324
Nikolaev, V. S., 269, 271
Nill, K. W., 299, 327
Noble, R. H., 23
Nordén, H., 222, 236 (30)
North, A. M., 163
Northcliffe, L. C., 219, 235
Novick, R., 257, 262, 264, 270, 271
Nozal, V., 73

O

Odishaw, H., 313
O'Dwyer, J. J., 177
Oepts, D., 63, 66, 67, 89 (67), 91, 92
Ohlman, R. C., 94
Oka, T., 133
Okabayashi, H., 160
Öline, A., 229
Oliner, A. A., 159
Olsen, J. Ø., 219, 236, 242 (62), 243 (62)
Onsager, L., 171
Osgood, R. M., Jr., 280, 283
Oster, G., 172
Oxholm, M. L., 41
Ozawa, Y., 207, 208

P

Palmer, P., 132
Papoular, R., 51
Pardoe, G. W. F., 191, 192
Paren, J. G., 207, 208
Parker, C. D., 113
Parker, S. G., 291, 298, 299, 327 (76)
Parker, T. J., 75
Parker, W. H., 256, 257
Parkinson, W. H., 244
Passchier, W. F., 74

Patel, C. K. N., 275, 280, 281 (24, 31), 282 (27, 29, 30), 286, 287 (24, 31), 288 (29), 300, 301 (86), 303, 304, 305, 306, 323, 327, 328 (102), 329, 340 (99, 100, 101)
Paul, J. W. M., 100
Pauly, H., 176
Pearl, A. S., 263, 266
Pearson, R., Jr., 118, 120 (44)
Pechacek, R. E., 100
Pedersen, E. H., 219, 236, 242 (62), 243 (62)
Pegg, D. J., 229, 234, 263, 266, 268, 269, 270, 271, 272 (168)
Penfield, H., 113
Penner, S. S., 38
Penzias, A. A., 113, 132
Perrin, F., 195
Perry. C. H., 52
Pershan, P. S., 321
Peterson, N. C., 336, 338 (186, 187)
Peterson, R. S., 249, 250, 263, 266
Peterson, R. W., 98
Peterson, O. G., 308, 314, 315 (141)
Pfund, A. H., 9, 23, 28 (37)
Pfeiffer, H. G., 202
Phelan, R. J., Jr., 290, 291
Phillips, P. G., 87
Phillips, T. G., 113
Pidgeon, C. R., 322
Pihl, J., 232
Pimentel, G. C., 43
Pinnington, E. H., 229, 244, 248, 254, 255
Pinson, P., 86
Pitt, D. A., 196
Plass, G. N., 2
Plyler, E. K., 6, 37, 43
Polder, D., 204
Poley, J. Ph., 157, 189
Pople, J. A., 172, 198
Posel, K., 144
Potts, A. D., 202
Poulizac, M. C., 229
Poulson, O., 254
Powell, F. X., 131, 132
Powles, J. G., 157, 160, 162, 177, 190
Pradhan, M. M., 84
Pressley, R. J., 280
Price, A. H., 160, 187, 197 (137)
Prichet, C., 77
Prior, A. C., 99
Prior, M. H., 262, 263, 264, 266
Pritchard, D., 332

Provder, T., 196, 198 (169)
Puplett, E., 79
Purcell, W. P., 196, 198 (169)
Puthoff, H. E., 57
Putley, E. H., 51, 52, 95
Pyle, R. V., 214, 232 (10)

Q

Querry, M. R., 306, 309 (104, 105), 310, 316 (137), 319 (137), 324 (104, 105, 136), 336
Quickenden, P. A., 166, 203
Quist, T. M., 290

R

Radcliffe, K., 88
Radford, H. E., 118, 127
Radziemski, L. J., Jr., 222, 236
Kampolla, R. W., 196
Randall, H. M., 23
Randall, R., 219, 263, 266 (23), 267, 270 (23, 158)
Rank, D. H., 16, 26, 30 (46), 37
Rao, B. S., 26
Rao, D. A. A. S. N., 160, 199
Rao, K. N., 6, 16, 22, 24, 26, 27, 30 (46), 280, 288, 289
Rawlins, F. I. G., 1
Reddish, W., 163
Redheffer, R. M., 157
Rediker, R. H., 290, 291
Reid, D. S., 203
Reithel, R. F., 308
Renk, K. F., 84
Retherford, R. C., 256
Richard, P., 263, 266, 269, 270, 272
Richards, P. L., 57, 58, 61, 76 (40), 84, 94, 97 (103), 323
Richards, R. G., 2
Rieck, H., 290, 291 (54)
Rinehart, E. A., 127
Ring, J., 86
Ritson, D. M., 155, 162, 177, 188, 203
Rivail, J. L., 93, 94, 95
Roberti, D. M., 180, 196, 198 (169)
Roberts, D. D., 159
Roberts, J. R., 241, 244
Roberts, S, 144
Roberts, V., 19, 99
Robertson, C. W., 45, 46 (116)
Robin, M. B., 114

Robinson, D. W., 24, 54
Robinson, L. C., 2, 53, 57, 88 (19)
Robiscoe, R. T., 256, 257
Robson, J. W., 219, 221
Rocard, Y., 190
Roeder, S. B., 196, 198 (169)
Roesler, F. L., 221
Roess, D., 314
Rogal, B., 146
Rogers, K. F., Jr., 52
Roland, G., 77
Rollin, B. V., 52
Ronn, A. M., 118
Röntgen, W. C., 338
Ross, G. F., 166
Rothschild, W. G., 2, 53, 88 (18)
Rubens, H., 25, 85
Rubinov, A. N., 307
Rupert, C. S., 5

S

Saito, S., 103
Sakai, H., 61, 76 (52)
Sakurai, K., 275, 279 (16), 340 (16), 341, 342
Salour, M. M., 332, 333 (178), 334 (178)
Saltpeter, E. E., 246, 247 (98)
Samulon, H. A., 165
Sanderson, R. B., 70, 74
Sastry, K. V. L. N., 118
Sautter, C. A., 236
Savage, B. D., 240
Saxton, J. A., 156, 157 (57)
Scalo, J. M., 244
Schaefer, C., 1, 4, 15 (2), 20
Schafer, F. P., 306, 307, 308 (107, 116), 315
Schallamach, A., 202
Scharff, M., 235, 236
Schawlow, A. L., 102, 111 (3), 330, 336, 338 (185)
Schectman, R. M., 236, 238, 255, 256
Schering, H., 142
Scherr, R. E., 264
Schilling, R. F., 235
Schiøtt, H. E., 236
Schmidt, W., 307, 308 (107, 116)
Schmieder, R. W., 260, 262, 263, 264, 265, 266 (138), 267
Schwan, H. P., 176, 182
Schwartz, J., 118, 120 (44)
Schwarz, G., 176, 203

Schweitzer, W. G., Jr., 332
Schoffstall, D. R., 238
Schoppert, G. T., 309
Schultz, A., 336
Scofield, J. H., 272
Selby, M. J., 86
Sellin, I. A., 229, 247, 248, 249, 250, 257, 258 (131), 259, 263, 266, 268, 270, 271, 272 (44)
Senashenko, V. S., 271
Series, G. W., 330
Sesnic, S. S., 58, 99
Sessler, G. M., 114
Shahin, I. S., 309, 318 (133), 330, 331
Shannon, C., 165
Shapiro, J., 260, 262
Sharma, R. D., 309, 310
Sharpless, W. M., 112
Shaw, E. D., 300, 301 (86), 304 (86)
Shaw, J., 214
Shaw, J. H., 3
Shcahen, Th, P., 79
Shen, Y. R., 57, 323
Shimizu, H., 90
Shimizu, T., 340
Shimoda, K., 275, 279 (16), 340 (16), 341, 342
Shirk, J. S., 336, 338 (189)
Shishaev, A. V., 332
Shivanandan, K., 84
Shugart, H. A., 263, 266
Shyn, T. W., 256, 257
Sievers, A. J., 55
Sillars, R. W., 204, 207, 209, 210 (223)
Sills, R. M., 231, 244
Silvera, I. T., 59
Silverman, S., 4, 39
Simmons, J. D., 336, 338 (188)
Sinanoğlu, O., 240, 241
Sinclair, D. E., 144
Singleton, E. B., 39
Sinton, W. M., 6
Sixou, P., 202
Sjödin, R., 229, 232
Sloan, R., 3
Slusher, R. E., 300
Smith, D. R., 66
Smith, L. G., 6
Smith, M. W., 240, 263, 266
Smith, P. D., 308
Smith, P. L., 217, 219, 222, 243, 244
Smith, R. A., 2, 4, 8, 10 (9), 12, 25 (9)
Smith, W. V., 102, 111 (1), 273

Smith, W. W., 229, 263, 266, 272 (44)
Smyth, C. P., 157, 180, 188, 196, 198 (169), 201, 202
Snaveley, B. B., 308, 314, 315
Snyder, L. E., 132
Soffer, B. H., 308
Soltys, T. J., 290
SooHoo, J., 57
Sørensen, G., 222, 236, 237, 238, 241, 242, 243, 244
Sorenson, H. O., 110, 113 (21)
Sorokin, P. P., 273, 307, 308 (109, 115), 314 (108, 109, 117), 315, 316 (143), 317, 319, 320, 321 (150)
South, G. P., 203, 204
Spaeth, M. L., 307, 308 (110)
Spanbauer, R., 6
Stair, A. T., Jr., 61, 78 (55), 100 (55)
Stark, A., 244
Stebbings, R. F., 262, 264
Steel, W. H., 53, 70, 80
Stein, W. A., 27
Steinbach, W., 113
Steinfeld, J. I., 309
Steinmetz, W. E., 126
Stell, J, H., 86
Stepanov, B. I., 307
Sterling, S. A., 84, 97 (103)
Sternberg, R. S., 85
Stewart, G. W., 189
Stewart, J. E., 2, 48
Stiefvater, O. L., 118
Stone, N. W. B., 42
Stoner, J. O., Jr., 219, 221, 222, 236, 239
Strandberg, M. W. P., 102, 111 (2)
Stratton, J. A., 204
Strauss, A. J., 290, 291
Strecker, G., 236
Strecker, D. W., 27
Stroke, G. W., 20, 21
Strombotne, R. L., 256, 257
Strong, J., 5, 20, 24, 80
Subtil, J, L., 229, 238, 239, 248, 254, 255
Sucher, J., 260, 262, 267
Sugden, T. M., 102
Suggett, A., 166, 203
Sullivan, T. E., 127
Surber, W. H., 157, 160
Surh, M. T., 69
Sussman, S. S., 57
Sutherland, G. B. B. M., 1

Sverdlov, L. M., 43
Sweet, A. A., 109
Swenson, R. W., 200
Szymanski, H. A., 2

T

Tait, M. J., 203
Takagi, K., 132
Takami, M., 340
Takashima, S., 203
Talpey, T. E., 157
Tanaka, M., 326
Tannenwald, P. E., 57, 113
Tarrago, G., 2, 34
Taylor, A. M., 1
Taylor, B. N., 97, 256, 257
Taylor, L. M., 308
Taylor, T. A., 162
Teich, M. C., 87
Tekou, B., 271
Teplova, Y. A., 269, 271
Testerman, L., 244
Thaddeus, P., 132
Thoe, R. S., 270, 271
Thomas, S. J., 308
Thompson, A. M., 137, 146
Thompson, H. W., 157, 188
Thompson, S. W., 22
Thornton, V, 38
Thrash, R. J., 336,
Tiemann, E., 127, 132
Tien, P. K., 280
Tilford, S. G., 214, 219
Tillmann, K., 253, 254 (115)
Tinkham, M., 94, 97
Tittle, F. K., 322
Tobias, C. W., 206
Torrey, H. C., 110
Törring, T., 110, 120, 127
Toschek, P. E., 336, 338 (185)
Townes, C. H., 102, 111 (3), 302
Trambarulo, R. F., 102, 111 (1)
Triebwasser, S., 256, 257
Trivelpiece, A. W., 84, 99, 100
Trombetti, A., 342
Tsao, C. J., 39
Tubbs, L. D., 26, 30, 31, 32, 33, 35, 39
Tucker, K. D., 132
Tuma, D., 99
Turnbull, D., 186

Turner, A. F., 23
Turner, B. E., 132
Tuttle, W. N., 143
Twiss, R. Q., 61
Tyndall, J., 338

U

Uehara, H., 132
Uehara, K., 340
Ulich, B. L., 132
Ullman, R., 186
Ulrich, R., 60, 84
Unger, H. G., 273

V

Valkenburg, E. P., 55, 84 (30)
van Aalst, R. M., 90
Vanasse, G. A., 61, 70, 76 (52), 78 (55), 80, 100 (55)
van Beek, L. K. H., 204, 210, 211
Vanderlyn, P. B., 146
Van Dyck, R. S., 263, 266
van Gemert, M. J. C., 167
van Kranendonk, J., 42
van Nieuwland, J. M., 55
van Santen, J. H., 204
van Turnhout, J., 164, 167
Varghese, S. L., 272
Vasilenko, L. S., 332
Vaughan, W. E., 157, 178, 187, 196, 197 (137), 198 (169)
Veji, E., 219, 221, 225, 229
Véron, D., 98
Victor, G. A., 263, 264
Vidal, C. R., 316
Vidulich, G. A., 139, 163 (7)
Volze, J., 307, 308 (107)
von Hippel, A. R., 141, 159
von Thuna, P. C., 299, 326 (77), 327 (77)
Von Weyssenhoff, H., 336
Vrabec, D., 77, 80 (86)

W

Wagner, K. W., 204, 207, 209, 210 (223)
Walker, W. I. O., 204
Wallace, R. W., 322
Wallenstein, R., 309, 311 (132), 312, 318 (132), 336

Walsh, A., 19, 20
Wangsness, R. K., 214, 245 (9)
Ward, W. H., 148
Waring, R. C., 306, 309 (104, 105), 310, 311, 316 (137), 319 (137), 324 (104, 105, 136), 336
Watson, H. A., 110
Watson, J. K. G., 131,
Watts, D. C., 163
Weber, M. J., 308, 312, 313 (138), 314 (119, 138), 315 (119, 138)
Weinreb, S., 110, 113 (24)
Weir, K. G., 146
Weise, H. P., 240
Weiser, K., 290
Weiss, A. W., 240, 241
Welsh, H. K., 202
Welsh, H. L., 41
Wemelle, R., 188
Wenstrand, D. C., 39, 40, 41
West, J. E., 114
Westhaus, P., 240, 241
Westphal, W. B., 141, 160
Whaling, W., 217, 219, 222, 231, 237, 243 (67), 244
Whalley, E., 45
Wheeler, J., 188
Wheeler, R. G., 94
Whetstone, J. R., 332
Whiffen, D. H., 157, 188, 192
White, A. H., 201
White, H. E., 273, 278
White, J. U., 17
Whitmer, C. A., 110
Wiberly, S. E., 43
Wien, W., 213
Wiese, W. L., 240, 244, 263, 266
Wiggins, T. A., 26, 37
Willardson, R. K., 51
Williams, D., 3, 26, 28, 30 (76), 32, 39, 40, 41, 42, 43, 45, 46 (116), 47, 48, 274
Williams, G., 160, 163
Williams, T., 66
Williams, V. Z., 1, 17
Willis, J. B., 20
Wilson, E. B., 120
Wilson, E. B., Jr., 43, 103, 116, 118,
Wilson, G. J., 202
Wilson, R. W., 132
Winkler, E. D., 157
Winnewisser, M., 118
Wintle, H. J., 167

Wirgin, A., 80
Wirtz, K., 194
Witte, W., 79, 85
Wittmann, W., 239,253, 254 (115), 255
Wodarczyk, F. J., 120
Wolf, E., 61, 273, 318, 324 (1)
Wolff, P. A., 299, 303 (78)
Wollrab, J. E., 2, 102
Wood, J. L., 60
Wood, R. W., 16, 85
Woods, C. W., 263, 266, 269 (154), 272
Woods, D., 144
Woods, R. C., III, 118
Woodworth, J. R., 263
Works, C. N., 162
Wright, G. B., 298, 300
Wright, R. H., 84
Wrixon, G. T., 113
Wrobel, J. S., 291
Wyllie, G., 192
Wyman, J., 149, 179 (40)
Wynne, J. J., 315, 316 (143), 317, 319, 320, 321 (150)

Y

Yafet, Y., 299, 303 (79)
Yamada, Y., 26

Yamaguchi, G., 309
Yamamoto, G., 326
Yamanaka, C., 309
Yang, K. H., 51, 323
Yariv, A., 273, 295 (8)
Yasumi, M., 156, 158 (58)
Yates, D. J. C., 6
Yen, W. M., 309, 318 (133)
Yéou, T., 54
Yoshinaga, H., 26, 57, 78
Young, G. M., 166
Young, J. F., 319, 322
Youtz, P., 291

Z

Zafar, M. S., 199, 200
Zahl, H., 8
Zaidins, C., 219
Zajc, W. A., 244
Zalewski, E. F., 336, 338 (187)
Zare, R. N., 335, 336
Zentek, A., 196, 198 (172)
Ziman, J. M., 96
Zipoy, D., 76
Zuckerman, B., 132
Zwanzig, R. W., 178

SUBJECT INDEX FOR PART B

A

Absorption
  coefficient, 46
  enhanced, 336
  index, 46, 135
  laser studies of, 325
  two-photon, 332
Amplitude spectroscopy, 73
Apodization, 62, 64
Astronomical applications, 48, 100, 243
Asymmetric interferograms, 70
Atomic spectra in the infrared, 2
Attenuated total reflectance (ATR), 48
Azbel–Kaner resonance, 97

B

Backward wave oscillators (BWO), 107
Beam foil spectroscopy, 213, 218
  astronomical applications, 243
  high energy beams, 232
  sources, 216
Blazed gratings, 16, 58
Bolometers, 7, 8, 52, 113
Bridge measurements of impedance, 142
Burch's law, 28

C

Capacitance cells, 136
Carbon arc, 5
Carbon rod furnace, 6
Cavity resonators, 161
Cascading, 236
Christiansen filters, 25
Clausius–Mosotti relation, 171
Cole–Cole diagrams, 94, 175
Collisional broadening, 38
Condenser microphone detectors, 114
Correlation spectroscopy, 28
Crystal diode detectors, 111
Cyclotron resonance, 95
Czerny–Turner spectrometers, 23

D

Debye absorption, 92
Debye equations, 171, 174
Deconvolution, 37
Defect diffusion, 185
Depolarization spectra, 167
Detectivity, 12
Detectors
  Condenser microphone, 114
  crystal diode, 110, 111
  far infrared, 51
  infrared, 10, 11, 13
  microwave, 109
  photo, 6
  pneumatic, 8, 51
  quantum, 6
  thermal, 6
Dielectric constants, 92, 134
Dielectric relaxation spectra, 168, 175, 194
Diode injection lasers, 290
Dispersion curves, 15, 192
Doppler broadening, 219
Dye lasers, 305

E

Ebert–Fastie spectrometers, 23
Echellettes, 16, 58
Electrical properties of matter, 134
Electrode polarization, 140
Emission spectra in infrared, 3
Enforced dipole transitions, 41
Etalon, 304
Extinction coefficient, 135

F

Fabry–Perot interferometers, 55, 84, 127
Far-Infrared Region, 50
  astronomical applications, 100
  detectors, 51
  interferometry, *see* Fourier transform spectroscopy

Far-Infrared Region (*Continued*)
  spectra of gases, 88
  spectra of solids, 90
Fellgett advantage, 53
Filters, 25, 26, 29, 59, 84
Fluorescence, 334
Focal isolation, 85
Forbidden decays, 260
Fourier transform spectroscopy, 53, 60
  amplitude, 73
  noise, 75
Free wave rf techniques, 154
Frequency mixing, 319, 321
Fresnel zone plates, 84

G

Girard grille spectrometer, 85
Globar, 4
Golay cell, 8, 51
Gratings, 20
  blazed, 16, 58
  echellette, 16, 58
  holographic, 22
  lamellar, 80
  replica, 16, 21
  spectrometers, 20, 57
Gunn effect, 108

H

Hadamard spectrometer, 87
Harmonic generation, 128
Heterodyne spectroscopy, 87
Hindered rotation and translation, 44

I

Impedance measurements, 142, 147
Infrared Region, 1
  astronomical applications, 48
  atomic spectra, 2
  detectors, 10, 11, 13
  emission spectra, 3
  far-, *see* Far-infrared
  near-, *see* Optical region
  optical components, 13
  sources, 2
  spectra of gases, 28

  spectra of liquids and solids, 44
  windows, 14
Interferograms, 61, 68, 70
Interferometers
  Fabry–Perot, 55, 84, 127
  lamellar grating, 80
  Michelson, 62
  Mock, 86
  resolving power, 62
Interferometry, 60 69, *see also* Fourier trans-
  form spectroscopy

J

Jacquinot advantage, 52
Josephson junctions, 97

K

Klystrons, 105
Kramers–Kronig relations, 47, 179

L

Lamb shifts, 256
Lamellar grating, 80
Langevin function, 171
Lasers, *see also* Tunable lasers
  application to absorption spectra, 325
  diode injection, 290
  millimeter wave, 54
  molecular, 280
  spin–flip Raman, 299

M

Michelson interferometer, 62, 79
Microwave region, 102
  absorption cells, 116, 127
  cavity resonators, 161
  detectors, 109, 113
  harmonic generators, 128
  modulation techniques, 114
  radio astronomy, 132
  sources, 104, 105, 108
  spectrometers, 121
Millimeter, waves 54, 128, *see also* Far-
  infrared and microwave regions
Molecular correlation, 175

Molecular spectra, 1
Multiple relaxation, 180
Multiplex advantage, 53
Multiply excited states, 224

N

Nernst glower, 5

O

Onsager equation, 172
Optical constants, 43, 134, *see also* Dielectric
    constants
  metals, 94
  semiconductors, 94
  solids and liquids, 92
Optoacoustic effect, 338
Oscillator strengths, 239

P

Parametric conversion, 315
Parametric oscillators, 322
Phase modulation, 70
Photodetectors, 6, 52
Plasmas, 97
Pneumatic detectors, 8, 51
Point contact diodes, 111
Prism, 15
  spectrometers, 17

Q

Quantum beats, 244, 245, 249, 252
Quantum detectors, 6

R

Radiation chopping, 7
Radiofrequency region, 134
  astronomy, 132
  capacitance cells, 136
  time-domain techniques, 163
Reflection spectra, 47, 323
Refractive index, 46, 135
Relaxation
  activation energy, 187
  dielectric, 168, 175

flexible molecules, 196
solids, 199
spectra, 168, 198
times, 173, 181, 194
Replica gratings, 16, 21
Resolving instruments, *see* Prisms, gratings,
    and interferometers
Resolving power of interferometer, 62
Resonance methods in rf region, 147
Resonance processes, 189
Resonant cavity, 55, 127
Responsivity, 12
Reststrahlen, 25
Rotation spectra, 29
  hindered, 44
Rotation–vibration spectra, 30
Rydberg constant, 330
Rydberg states, 231

S

Signatures, 66
SISAM, 85
Sources, *see* appropriate spectral region
Spectrographs and spectrometers
  Czerny–Turner, 23
  Ebert–Fastie, 23
  Girard grille, 85
  grating, 20
  Hadamard, 87
  prism, 17
  submillimeter wave, 129
  Walsh double pass, 19
Standing wave techniques, 158
Stark beats, 245
Stark effect modulation, 115
Stark spectra, 340
Submillimeter waves, *see* Far infrared and
    microwave region
Superconducting bolometer, 8

T

Thermisters, 8
Thermal detectors, 6
Thermocouples, 7
Transitions
  enforced dipole, 41
  forbidden, 260
  probabilities, 234

Translational absorption, 42
  hindered, 44
Tunable laser spectroscopy, 273
  absorption, 325
  diode injection, 29
  dyes, 305
  millimeter region, 54
  molecular, 280
Two-photon absorption, 332

V

Vibration–rotation spectra, 30
Viscosity, 194

W

Water spectrum, 44
Waveguide techniques, 54, 116, 127, 128, 150
Wavelengths, 26

Z

Zeeman, effect
  modulation, 115
  quantum beats, 249
  tuned lasers, 275
Zero-field quantum beats, 252

SUBJECT INDEX FOR PART A

A

Absorption
  coefficients, 22, 110
  spectra, 22, 248
  theory, 87
Alternate decay modes, 118
Angular
  correlations, 134
  momentum of e-m fields, 40
  momentum of molecules, 106
Arcs, 267, 269
Aspherical gratings, 229
Atomic spectra, 6, 100ff, 148ff, 204ff, 253ff
  analysis of, 336
  nuclear effects in, 341
Astigmatism, 229
Auger transitions, 178

B

Bands, 21
Beta–gamma correlations, 178
Blazed gratings, 230, 288
Bremsstrahlung, 186

C

Classical theory of e-m fields, 32ff
Coherence, 50, 75
Collision broadening, 23
Collisions, ion–atom, 200
Cold-cathode discharge, 215
Comparators, 330
Concave grating mounts, 291
Continuum, 21, 206
Cornu prism, 285
Coulomb excitation, 145
Counters, 150
Crystal properties, 161

D

Decay rates, 100, 104, 115, 118
Detectors, 26
  gamma-ray, 121
  for optical region, 314
  photoconductive, 323
  photoelectric cells, 321
  solid state, 153
  thermal, 325
  ultraviolet, 241, 244
  x ray, 149
Diffraction gratings, 156, 160, 227, 287
Dipole, 115, see also Multipoles
Dispersive devices, see also Gratings, Prisms,
    and Interferometers
Doppler broadening, 23
Doppler shift, 144
Doppler-tuned spectrometer, 164

E

Eagle mounting, 293
Ebert-Fastie spectrometer, 235
Echelle, 299
Echelon, 301
Electrodeless discharge, 261
Electromagnetic fields, 32, 38, 61
Electromagnetic spectrum, 2
Emission spectra, 20, 87
Etalon, 304, 306, 309, 314

F

Fabry–Perot interferometer, 304
Far-ultraviolet region, 204
Flames, 260
Fluorescence, 145
Fourier representations, 55
Free e-m fields, 60
Frequency, 2, see also Wavenumber

G

Gamma-ray region, 16, 115
  detectors, 121
  spectrometers, 131
Gamma–gamma correlation, 138
Gauge transformations, 32
Geissler tubes, 265
Grating, 227, 287
  aspherical, 229
  concave, 226
  crystal, 156, 160
  fabrication, 297
  holographic, 229
  mounts, 291, 293
  phase, 231

H

Hamiltonian formulation, 60
Handedness, 45
Helicity, 45
History, 3
Holographic gratings, 229
Hollow cathode discharge, 218, 272
Hot filament discharge, 219
Hydrogen continuum, 206
Hyperfine structure, 342
Hypersatellite lines, 172

I

Induced emission, 87
Infrared region, 2, 8
  near, *see also* Optical region
Intensity, 20, 73
Interaction of light and matter, 79
Interferograms, 333
Interferometers, 304

K

King's furnace, 260

L

Lasers, 221
Lifetime measurements, 141
Light, 32
  intensity, 20, 73
  interaction with matter, 79

Line sources for uv, 215
Linear momentum of radiation, 39
Lines, 20, 24, 254
Littrow mounting, 286
Lorentz line shape, 24
Luminosity, 311

M

Mechanical properties of fields, 38
Methods of spectroscopy, 19
Microwave, 10
  discharge, 220
  excitation, 261
Molecular states, 106
Molecular x rays, 182
Monochromators, 234
Mössbauer effect, 25, 146
Mountings, *see also* Gratings
Multipass systems, 296
Multipoles, 93, 115
Muonic x rays, 189

N

Natural broadening, 23
Nuclear, *see also* Gamma rays
  effects in atomic spectra, 341
  reactions, 137
  transitions, 100

O

Optical region, 253
  detectors, 314
  grating spectrographs, 287
  prism spectrographs, 279
Octupoles, *see also* Multipoles

P

Parametric calculations, 347
Paschen–Runge mounting, 293
Perturbation calculations, 79
Perturbed correlations, 140
Phase gratings, 231
Photoconductive cells, 321
Photoelectric cells, 321
Photoelectron spectroscopy, 162, 250

Photographic emulsion, 316
Photon, 67
Photon sources, 205
Photon states, 73
Plane waves, 41
Plasmas, 191, 208
Plasmon satellites, 180
Polarization, 73
Polarized targets and beams, 139
Polarizers, 239
Potentials, 32
Prism, 3, 7, 279
    spectrographs, 279
    constant deviation, 282
    Cornu, 285

Q

Quadrupoles, *see also* Multipoles
Quantum conditions, 65
Quantum theory of e-m fields, 64
Quasimolecular x rays, 182

R

Radiation theory, 31
Radiative electron capture, 186
Radiative transitions, 31
Radiofrequency region, 12
Rare gas continuum, 206
Rayleigh criterion, 22, 276
Reflective coatings, 232
Resolving instruments, *see also* Gratings,
        Interferometers, and Prisms
Resolving power, 21, 276
Resonance fluorescence, 145
Resonant capture, 136
Rowland circle, 227, 292

S

Satellite lines, 166
Selection rules, 97
Sliding spark, 273
Solar x rays, 193
Solid state detectors, 153
Sources, 24, 204, 224, 245, 246, 259
Sparks, 210, 220, 269, 273
Spectral line, *see also* Line

Spectrographs and spectrometers
    Doppler-tuned, 164
    Ebert-Fastie, 235
    gamma ray, 131
    optical region, 274
    prism, 274
    ultraviolet, 234
    x ray, 149
Spontaneous emission, 87
Standard sources, 224
State vector, 73
Submillimeter waves, 10
Synchrotron radiation, 210

T

Theory of e-m radiation, 31
Thermal detectors, 325
Tokamaks, 191
Transition probabilities, 80, 100, 116, 141
Triple correlations, 139
Twentieth-century physics, 17

U

Ultraviolet region, 13, 204
    absorption spectra, 248
    continuum sources, 206
    detectors, 241
    emission spectra, 247
    gratings, 226
    near, *see also* Optical region
    sources, 204, 215, 259
    spectra of gases, 247
    spectra of solids, 251
Units, 2, 149

V

Vector potential, 73
Visible region, *see also* Optical region
Vodar monochromator, 237

W

Wadsworth mounting, 294
Wavelength, 2, 246, 326, 327
Wavenumber, 2
Widths of spectral lines, 23
Windows for ultraviolet, 238

X

X-ray region, 14, 148
  cross sections, 200
  detectors, 149
  focusing, 158
  muonic, 189
  plasmas, 191

  quasimolecular, 182
  solar, 193
  spectra, 166
  spectrometers, 156, 160

Z

Zeeman effect, 341

A 6
B 7
C 8
D 9
E 0
F 1
G 2
H 3
I 4
J 5